COMPSTAT

Proceedings in
Computational Statistics

7th Symposium held at Rome 1986

Edited by
F. De Antoni, N. Lauro
and A. Rizzi

With 59 Figures

Physica-Verlag
Heidelberg Wien

Dr. F. De Antoni, Dipartimento di Statistica, Probabilità e Statistiche applicate, Università degli Studi di Roma „La Sapienza“, Piazzale Aldo Moro, 5, I-00185 Roma/Italia

Professor N. Lauro, Dipartimento di Matematica e Statistica, Università degli Studi di Napoli, Via Partenope, 36, I-80121 Napoli/Italia

Professor A. Rizzi, Dipartimento di Statistica, Probabilità e Statistiche applicate, Università degli Studi di Roma „La Sapienza“, Piazzale Aldo Moro, 5, I-00185 Roma/Italia

ISBN-13: 978-3-7908-0355-6 e-ISBN-13: 978-3-642-46890-2
DOI: 10.1007/978-3-642-46890-2

CIP-Kurztitelaufnahme der Deutschen Bibliothek
COMPSTAT:
COMPSTAT ... : proceedings in computational statistics; . . . symposium. - Heidelberg; Wien: Physica-Verlag
Teilw. nur mit Verlagsort Wien
7. 1986. Held at Roma 1986. - 1986. -

Cover: Erich Kirchner, Heidelberg
Offsetprinting: Kiliandruck Grünstadt
Bookbinding: J. Schäffer OHG, Grünstadt
7120/7130-543210

Preface

As recalled by the President of the International Association for Statistical Computing, in its talk at the IASC meeting, in Amsterdam during the 45th session of the International Statistics Institute (ISI), it is by now one century since the introduction of automatic computing in Statistics. In fact at the first ISI session in 1885, Herman Hollerith showed his tabulating machine to representatives of statistical offices from across the world. That machine was largely used in the analysis of census data at the end of the last century giving birth to Computational Statistics (C. S.).

Since then great developments have affected Computational Statistics towards to both the introduction and diffusion of powerful computers.

Among the significant dates of C.S. history we would like to recall the first Compstat conference in Vienna in 1974 and the foundation, in 1977 of a section of ISI, the International Association for Statistical Computing, a section among whose activities there is the organization of Compstat.

By reading the proceedings of the past Compstat conferences it is possible to have an idea of the main changes which took place in the last two decades in Computational Statistics. Considered, at the beginning, a tool only for the elite, C. S. is becoming, nowadays, thanks to the friendness of the latest generations of computers and softwares, an indispensable tool for every statistician. For this reason the teaching of Computational Statistics has been included among Compstat'86 topics.

Furthermore we point out that Computational Statistics not only has contributed to the diffusion of Statistics in different fields of applied research, but also it has influenced, in a significant manner, methodological researches contributing to the development of new and interesting subjects such as: Non-parametric Procedures, Robust Methods, Exact Tests, Simulations, Exploratory Data Analysis (EDA), Cluster Analysis and other multivariate procedures.

Some of these themes and certain applications of computational statistics in traditional as well as in emerging sectors (econometrics, clinical trials) have been included in the scientific programme of Compstat 86.

In addition to the above mentioned topics, a certain interest has also been expressed in the aspects of C. S. that involve technological developments such as: Data Base Management Systems, Computer Graphics and the Micro-Computer and Networks.

Among the Compstat'86 themes a special mention is deserved by Knowledge Engineering, which includes applicative, technological as well as methodological aspects. The interchange between these elements brings to the attention of researchers the field of "expert systems for statistics" and "statistics for expert systems".

What, perhaps, represents an innovative element in Compstat'86 is the interest in themes of Statistics which traditionally have not been concerned with computational aspects: Computational approach of Inference, Probability Models and the study of the relationship between Information Science and Statistics.

This in fact shows the deep change occurred in the way Computational Statistics is conceived.

N. Victor, in a lecture entitled "Computational Statistics-Tool or Science?",given at the 32th Meeting of the Italian Statistical Society, defined Computational Statistics as "the examination of possible solutions and simplifications of the statistician's problems by the application of a computer and/or the methods of Computer science," and concludes that it "is not an independent Science, rather an important area of statistics and an indispensable tool for the statistician".

In the light of the above considerations, it is our opinion that Informatics has deeply affected statistical thinking, so that Computational Statistics, more than just being a fresch chapter in Statistics, should be seen as a new way of making Statistics in the computer age.

The interest in the statistical metalanguages, that is peculiar of recent softwares, is just in this trend because it is aimed at the development of new forms of communication between statisticians and, at the same time, of a new way to produce Statistics.
In other words, it deals with an attempt to formalize statistical thinking by substituting the traditional mathematical expression of statistical problems with informatics statements.

To the formulation of Compstat'86 scientific programme and to the selection of the papers published in the proceedings the following Colleagues, have contributed: D. Edwards, Y. Escoufier, A. Gale, T. Havránek, H. Lauter, N. Lauro, G. Leti, A. Mineo, K. Mormirovic, P. Naeve, J. Nelder, A. Rizzi, V. Siesto, P. Sint, A. Verbeek, N. Zagouruiko.

To the formulation of Compstat'86 scientific programme an to the for having contributed to the succes of COMPSTAT'86.

The Editors

Contents

CLUSTER ANALYSIS METHODS AND ALGORITHMS

ROBUSTNESS IN MULTIVARIATE ANALYSIS

COMPUTER GRAPHICS AND DATA PRESENTATION

EXPERT SYSTEMS

STATISTICAL SOFTWARE

DATA HANDLING AND COMPUTATIONAL METHODS OF CLINICAL TRIALS

ECONOMETRIC COMPUTING

Information Science and Statistics

Information Science and Statistics

By E. B. Andersen, Copenhagen

1. Introduction.

The interface between information science and statistics is more often implicit that explicit. It is true that the rapid growth of computer technology and the development of a better understanding of how to handle large amounts of data, has shaped the development of statistics. It is equally true, however, that the way the development of computer technology has influenced the development of statistics seldom in the past was subject to scientific studies on the part of the statisticians. It is of interest, therefore, to discuss the extent to which the development of computer technology has influenced the de facto use of statistical methods.

It is always tempting to speak out of ones own experience, and I shall to some extent do so. Let me, therefore, begin by shortly describing the development of computer technology as I have experienced it. When I started as a graduate student in 1960, there was only one computer in Copenhagen. A large villa was necessary to house it, and it had a very limited capacity. Two years later, we had four computers, of which the three new ones took up the space of ordinary closets. Today we work on personal computers which, I guess, individually are more powerful than our first computer, and where the central unit take up the space of a small briefcase.

It is impossible on a few pages to cover all the impacts on statistics of these developments. What I shall try do do, is to briefly discuss four issues: The extensive use of software packages. The consequences of being able to perform very complicated calculations very fast. The possibilities and limitations of interactive computing. And finally the use of graphical methods.

Compstat 1986

2. Software packages.

No other development has influenced our day to day life as statisticians as much as the availability of software packages containing standard statistical techniques. The existence of such packages allow us to carry out in seconds what we used to spend days to do. In addition, there will be no computational errors as long as the data input is correct. Finally we can in seconds explore the fit of an alternative model to our data. There is a risk, however, connected with the use of packages. Once the data are read into the computer, we will tend to explore fits by alternative models within a certain class of models. If we decide e.g. to analyse the data by a regression model, it is not likely that we change to some type of time series model. This typical behaviour has to do with the difference between playing with data on a data screen and making calculations by hand, It is difficult to get the same feeling for the true nature of a data set by reading output from a computer programme as by writing down figures on a piece of paper. This does not mean that we shall go back to paper and pencil, but that we have to develop methods for displaying the basic data structure often during the analysis of the data in order to evaluate the fit of the model on various stages of the analysis. One way to obtain this goal is probably through a combination of techniques for graphical data display and the analysis of residuals. Analysis of residuals has become more and more important as a method for evaluating the goodness of fit of a model. Hence, if the residuals can be displayed in informative graphical ways on the data screen, there should be a good chance of arriving at a satisfactory description of the data. It is accordingly important that a statistical package is very flexibel both as regards the variety of alternative statistical models included in the package and as regards the easyness with which one can change from one type of model to another. In addition it is important that the package contains powerful methods for evaluating the fit of the models by means of an analysis of the residuals properly displayed on the data screen.

3. Complicated calculations.

Some years ago, the development of statistical methods depended on the extent to which there was an explicit solution to the likelihood equations or other estimation equations. Such explicit solutions were necessary for practical as well as for theoretical reasons. With fast computers available, which can compute iterative solutions almost as fast as explicit solutions, there is no computational necessity connected with solvability of likelihood equations. Neither is solvability necessary for theoretical reasons. Asymptotic properties of suggested estimators can be discussed based on an equation, of which the estimator is a solution as well as based on an explicit expression for the estimator. A striking example is log-linear models for contingency tables. Once the Deming-Stephan iterative procedure was implemented in standard computer programmes, the time needed for the analysis of a contingency table no longer depended on what time was required for the actual computations of the estimates. Whether a log-linear model required iterative solutions of the likelihood equations or a direct solution was available, the estimates were obtained by the computer in seconds. From my own work, I would like to call attention to latent structure models for categorical data. Such models are essentially of two types, depending on the character of the latent variable. If the latent variable is assumed to be discrete, we apply so-called latent class models. In the sixties much efforts were devoted to the discussion of these models. The widespread use of latent class models did not take place, however, until the EM-algorithm was developed in the late seventies. (cf. Goodman (1974) and Dempster, Laird and Rubin (1977)). The EM-algorithm is at the same time a computer algorithm and a statistical method and is thus an example of a concrete interface between information science and statistics. The algorithm is essentially a way of solving a set of likelihood equations, which does not possess a set of explicit solutions, but it is also a methodology for obtaining maximum likelihood estimators in case of incomplete data.

If the latent variable is continuous a typical likelihood equation has the following form (cf. Andersen and Madsen (1977))

$$\sum_{t=0}^{k} y_t A_{\varphi,\mu}/B_{\varphi} = 0,$$

where

$$A_{\varphi,\mu} = \int \frac{\theta-\mu}{\sigma^2} e^{\theta t} \prod_j (1+e^{\theta+\varepsilon_j})^{-1} \varphi(\theta|\mu,\sigma^2)d\theta,$$

$$B_{\varphi} = \int e^{\theta t} \prod_j (1+e^{\theta+\varepsilon_j})^{-1} \varphi(\theta|\mu,\sigma^2)d\theta$$

and $\varphi(\theta|\mu,\sigma^2)$ is the density of a normal distribution with mean value μ and variance σ^2. Hence any solution, iterative or not, depend on the numerical computation of the integrals $A_{\varphi,\mu}$ and B_{φ}. For many years it was considered to be impossible from a practical point of view to work with such models. As computers became faster and more efficient, it turned out, however, that numerical integration could be performed fast and reliable enough to permit applicable solutions of likelihood equations as those shown above. As a consequence there are now computer programmes available for latent structure analysis with a continuous latent variable. These programmes has been used for may succesful fits of real data sets.

It was also thought for some time that the success of the analysis depended on the efficiency of the chosen quadrature formula. It seems, however, that modern computers are so fast, that the choice of quadrature formula does not seem to be as critical a factor as first believed. If the methods just described are to be implemented on personal computers speed will be a factor, however, and the choice of quadrature formula could again be critical.

It should finally be stressed that the availability of software packages for numerical methods, like the NAG library, means that even complicated mathematical calculations are carried by simply calling the appropriate programme from the package available at the local computer. Examples of numerical methods widely used in statistical analyses are numerical optimization, numerical integration, various matrix operations and eigenvalue/eigenvector calculations.

4. Interactive statistical analysis.

One of the main differences between the major statistical programme systems, is the degree to which they are designed for interactive analysis. My first teacher in computer science used to say (in 1959) that the ultimative computer should have an input device and a red and green lamp. The red lamp should flash for a "no" and the green for a "yes". He did not even allow for a yellow flash for "don't know'. As we know now the development has taken a completely different course. Instead of black box computers, we have very open computer systems, where the operator can change the course of the computations as often as he please based on what intermediate results he wants to inspect. This way of interactive work on the computer is extremely well geared to the needs of modern statistics. It should be kept in mind, however, that interactive statistical analyses require a high degree of training on the part of the statistician. If this means that only qualified statisticians performs statistical analysis, this is definitely as it should be. But unfortunately computer work is fun. Often the datacollector does not, therefore, want to leave the fun to the statistician and starts joggling with the data himself. This is quite in order, as long as he or she does not claim that a full statistical analyses has been performed, and starts to draw his or her own conclusions from the "analysis".

5. Graphics.

Anybody who has spend a short while in front of a demonstration stand for computer equipment, software or hardware, cannot fail to notice that graphics is in. Years ago the standard output from statistical analyses were lots of numbers, often badly displayed with too many decimals and tables cut meaningless to pieces by the available paper format. The introduction of computerized statistical analyses in the sixties even made matters worse. It was as if all good rules for the display of numbers in tables were suddenly forgotten. The picture we carry most vividly in our memory is computer outputs with more than ten decimals in the numbers, most of them of no consequence for the conclusions to be drawn.

Today results from a statistical analysis is presented in the form of chic and colourful diagrams and smart tables neatly arranged on the output device.

That the use of graphical statistical methods is one of the central issues in the interface between information science and statistics was emphasizes by Tukey (1983). Tukey has been a clear advocate for new graphical methods which are not illustrations of the results of the statistical analysis, as we traditionally use them, but independent ways of performing a data analysis or a data exploration. The Tukey point of view has been very succesful, but to a large extent the traditional illustrations of analysis results are still the major parts of the graphical software we are offered. The box plot for example has not been able to replace the clumsy histogrammes or the old fashioned box diagrammes, where boxes were piled on top of each other. Future developments will show whether graphics will be able to completely replace the display of numbers on data screens, such that tables and numerical indicators, like estimators and test statistics, are merely intermediate results, which are not part of the displayed output. If the development takes this direction, there are good reasons to sound a warning.

As mentioned earlier it is extremely important that the statistician arrives at a correct understanding of the data structure, or as we traditionally say, gives a correct statistical description of the data. This requires, however, that he must inspect the basic data structure with different model descriptions in mind. It is further important that this inspection takes place when the data is still on as raw form as possible. Hence a warning is called for against computer software, by which it is too easy to transform the basic data and perform various statistically analyses automatically.

It is particularly dangerous if a statistical package is constructed in such a way, that different models are accepted or rejected through automatic data checks during the execution of a subprogramme.

The problem connected with the use of graphics brings to our final attention the key issue of model fitting. Whatever we do in terms os statistical analysis the basic problem will always be to fit a model. A good model should satisfy two criterions, which will often be in conflict. On the one hand the model should reflect our prior knowledge and the problem the data are collected to throw light on. On the other hand the model must fit the data. As mentioned above the only satisfactory method for judging the fit is an analysis of residuals, with residuals that refer to the basic data if at all possible. What we first of all should expect from the computer software is, therefore, adequate graphical ways of displaying the residuals on the data screen such that we can determine whether the chosen model fits our data.

References.

Andersen, E.B. and Madsen, M. (1977): Estimating the parameters of the latent population distribution. Psychometrika. 42, 357-374.

Dempster, A.P., Laird, N.M and Rubin, D.B. (1977): Maximum likelihood from incomplete data via the EM-algorithm. Journ. Royal Statist. Soc. B., 39, 1-22.

Goodman, L.A. (1974): Exploratory latent structure analysis using both identifiable and unidentifiable models, Biometrika. 61, 215-231.

Tukey, J.W. (1983): Another look at the future. In: Heiner, K.W. et al. (eds.) Computer Science and Statistics. Proceedings of the 14th Symposium on the Interface. New York: Springer Verlag.

Catastrophe Theory as a Tool for Statistical Analysis of Systems

By M. Novàk, Praha

Summary:

In the contemporary technology, science and industry the still more and more complicated and expensive systems are used. Their increasing complexity causes naturaly the still harder requirements on the respective system analysis and optimized synthesis. This leads to need to improve the conventional procedure of the system synthesis by the determination of allowable deviations and tolerances of the system parameters and system functions. Because for many systems the deterministic approach to this problem is not suitable, the methods of statistical analysis of system tolerances are used, often with very good results. However, there exist an interesting group of systems, expecially the systems with multivaluability and hysteresis, where the conventional statistical methods based on Monte Carlo approach are not advantageous. In this contribution there will be shown, how in such cases the problem of the tolerance analysis can be solved by modelling, based on some results from the catastrophe theory.

Key words:
Systems
Statistical Analysis
Parameter Tolerances
Hysteresis
Catastrophe Theory

Compstat 1986

1. Introduction

When dealing with the design or with the application of any technical system, which is not quite simple and trivial, one has to face to the problem to determine the allowable deviations of the system functions and the optimal vector of system parameter tolerances. The need for the solution of this problem is stimulated with various serious economic and maitenance aspects, between them the tendency to reach the minimal production cost, the maximal system operation reliability are the most frequent. Suppose that we are dealing with an system S, consisting of N components represented by the system parameters x_i, i = 1, 2 ... N, which are arranged in certain structure so, that the κ system functions F_k, k = 1, 2 ... κ , expressing the considered system properties, fullfil the condition

$$F - F_0 \leqq \Delta F, \tag{/1/}$$

where $F = \{F_k\}_\kappa$ is the set of the actual system functions, $F_0 = \{F_{0k}\}_\kappa$ is the set of the nominal system functions and $\Delta F = \{\Delta F_k\}_\kappa$ is the set of the allowable system functions deviations. The set F depends besides the system structure also on the vector $X = \{x_i\}_N$ of the system parameters. Suppose, that the system structure is invariant. The system with the nominal properties F_0 is then determined by the nominal parameter vector $X_0 = \{x_i\}_N$. When realizing the system /especially if it is producted in large series/, the actual values of the parameter vectors X differ from X_0 according some probability density $\phi(X)$. These can have various character according the kind of manufacturing technology. Though the analysis of the actual probability density $\phi(X)$ is usually quite cumbersome, one can presume here for simplicity, that $\phi(X)$ has normal character. All the points X, representing the possible acceptable systems, where /1/ holds, form in the N-dimmensional space $\{X\}_N$ the so called region of acceptability R_A. All the points X, representing the actually producted systems are involved in the region of deviations R_E, being distributed around X_0 according to to $\phi(X)$. The determination of R_A and the search of the optimal possitioning of R_E with respect to them is therefore the basis for all the system optimization tasks.

2. The multivaluable systems and systems with hysteresis

For conventional systems, the surfaces F over the parameter space $\{X\}_N$ are continuous, smooth and univaluable. This gives the possibility to apply the various statistical approaches for the analysis of R_A and for the optimal positioning of R_E /see e. g. Géher /1984/, Elias /1975/, Styblinski /1980/, Novák /1982/, Pavlík /1979/ /. However for certain important systems, the surfaces F over $\{X\}_N$ have either discrete, or multivaluable character. This cancels down the possibility of the application of standard Monte Carlo approaches to the analysis of R_A. More over: quite often one faces to the systems with hysteresis, i. e. to the systems, where the value $F(X_m)$ in certain point X_m of the space $\{X\}_N$ depends on the direction by which the vector X approaches the point X_m. To such class of systems many important technical, biological and social systems belong. Though the accuracy of the models of the systems in living nature is often problematic, we shall restrict the later discussion to technical systems with hysteresis only. They can be of various physical background /e. g. magnetic, electronic, mechanical etc./, however the effect of systems internal inertia causes the hysteresis phenomena.

Let the system having the system function F(X) be considered, which has over certain region of bifurcations R_B of $\{X\}_N$ multivaluable character. Outside R_B, for every X only one F(X) corresponds. The values F(X) corresponding to $X \in R_B$ depend on the way by which the vector X moves outside R_B to certain position inside it. Quite often, the change of the systems character from the univaluable to multivaluable one when breaking the boundaries of R_B is very sudden - one speaks over the catastrophic change or simply about the catastrophe. In 1972 René Thom has presented a new approach to the analysis of such systems, which later on bas lead to the forming of the so called catastrophe theory. Dealing with such system, the determination of R_B´s boundary is extremly important. This is truth especially for systems, which operation is based on their system function multivaluability, like are e. g. the logic electronic functional blocks. Here the R_B can be identical with R_A.

The main part of the methods for the determination of R_A known from tolerance theory is based on Monte Carlo approach /see e. g. Géher /1984/, Elias /1975/, Styblinski /1980/, Novák /1982/, Pavlík /1979/ /, where one randomly generates the values of X in certain testing region R_P according to some proper approximation of $\phi(X)$ and tests wheather the condition /1/ holds. If the discrepancy of the F values spread inside R_B is comparable or larger then ΔF, this procedure is not useful. One of its several possible modification can be e. g. constructed so, that to every statistically generated X_ν one or more points $X_{\nu n}$ lying in its close neighbourhood are found and from the eventually discrepancies of $F(X_{\nu n})$ values compared with ΔF the question of competency of X_ν inside R_B is solved. Evidently this /and also other similar/ procedures are complicated and computer capacity consuming. Therefore the procedure based on the straightforward R_B modelling by the use of catastrophe theory can give us some serious advantages.

Thom has found, that every catastrophe can be composed from the set of so called elementary catastrophes, which are of seven fundamental types. If one succeeds in searching for the corresponding catastrophe set and for their parameters, one can form a model suitable for straightforward determination of R_B /and R_A/. In certain cases, which are often important for technical praxis, one can recognize from physical base of the system under consideration which of the elementary catastrophes is good for the modelling /see e. g. Novák, Šiška, Šimek /1982/, Novák, Šimek, Šiška /1984/, Novák, Šimek /1985/, Šimek /1985/ /. If this is not evident, one can construct the necessary statistical approximating strategies for operation in the space of catastrophes set. Though such procedures can be also quite laborious, one can reach a significant gain in computer time compared with the pure statistical approach. One of the reasons for this comes from the fact, that once the fundamental parameters of the suitable set of elementary catastrophes is determined, one has the complete information about R_B /or R_A/ over the whole space $\{X\}_N$, which is not truth in the previous case.

Literature:

Elias, N. J. /1975/ New Statistical Methods for Assigning Device Tolerances. IEEE Proceedings of the International Conference on Circuits and Systems ISCAS, Boston, USA.

Géher, K. /1984/ Design of Electronic Circuit for Optimal Production Yield. Periodica Polytechnica, Electrical Engineering, 28, No 1, 67 - 82.

Novák, M. /1982/ Statistical Approach to System Parameter Synthesis. COMPSTAT 82, Toulouse, France.

Novák, M., Šimek, M., Šiška, J. /1982/ Tolerances in Nonlinear Logic Circuits. 7th Colloquium on Microwave Communication, Budapest, 6. - 10. 9.

Novák, M., Šimek, M., Šiška, J. /1984/ Application of the Theory of Catastrophes for the Analysis of Logic Network Tolerances. 5th International Symposium on Network Theory /ETAN/, Sarajevo, September 4. - 7.

Novák, M., Šimek, M. /1985/ Statistical Design of Functional Blocks with Hysteresis. Proceeding of the ECCTD 85. Prague, 2. - 6. 9.

Pavlík, P. /1979/ Some Improvements of Monte Carlo Tolerance Simulation. Software for Computer Control SOCOCO, Prague.

Styblinski, M. A. /1980/ Tolerance Analysis and Optimization of Electronic Circuits Statistical Approach. European Conference on Circuit Theory and Design, Warszaw, Poland, 2. - 5. 9., 97 - 114.

Šimek, M. /1985/ The Analysis of the Multistable Functional Blocks. Phd. Dissertation, General Computing Center, Czechoslovak Academy of Sciences, Prague.

New Procedures for Generating Optimal Experimental Designs

By H. A. Yonchev, Sofia

Summary

The procedure CDOP for generating of D-optimal experimental designs is presented. An empirical comparison of CDOP with some of existing algorithms for the computer generation of exact D-optimal designs is carried out. Among algorithms considered were those due to Fedorov, Wynn, Mitchell, and Johnson and Nachtsheim. It is shown that by the use of CDOP efficient designs can be obtained on convex design spaces.

Key words: D-optimality; Computer search; optimal designs; exchange algorithm.

1. Introduction

Many efforts have been made in the recent years to develop efficient methods of using computer to aid in the constructing of optimal experimental designs. Sophisticated computer search routines are worked out to find designs satisfying different optimality criteria.

We are especially interested in procedures for finding designs which are good with respect to the concept of D-optimality. This very important concept is based on the efficiency of the design to estimate the parameters of an assumed model. It's theory has been considered by Kiefer (1959), Wynn (1970), Fedorov (1972), and several other authors. A review on D-optimality for regression designs is made by St. John and Draper (1975).

To discuss the algorithms for constructing D-optimal designs we first introduce our notation for regression model and the assumptions about the design problem.

We assume that the linear regression model is expressed in matrix notation as

$$Y = Fb + \varepsilon, \qquad (1)$$

where Y is an $N \times 1$ vector of observations; F is an $N \times k$ expanded design matrix with rows $f^T(x_i)$ depending on the assumed model;

Compstat 1986

x_i is m×1 vector of predictor variables; b is a k×1 vector of model coefficients; ε is a N×1 vector of uncorrelated random variables with mean zero and variance σ_ε^2. With no loss of generality we assume σ^2 equals 1.

The covariance matrix of the estimated parameters b is $\sigma_\varepsilon^2(F^TF)^{-1}$. Then the variance of prediction at a point x is

$$d(x) = f^T(x)(F^TF)^{-1}f(x) \quad (2)$$

The maximum values of the variance function d_{max} in the experimental space χ are used for selecting the design points. The efficiency of an experimental design is a function of d_{max} of det (F^TF). A design with matrix F* is said to be D-optimal with respect to the model (1) if it maximizes the determinant of the information matrix F^TF over the factor space χ

$$|F^{*T}F^*| = \max_{\chi}|F^TF| \quad (3)$$

2. Generating D-optimal designs on convex and finite design spaces

The vast majority of algorithms for constructing D-optimal designs are the so called "exchange algorithms". They are based on the algorithm developed by Fedorov (1972). This algorithm begins with nonsingular N-point design. Then a design point is deleted and another point $x \in \chi$ is added in such a way that the resulting encrease in the det(F^TF) is maximal. The exchange of points continues until no further significant encreases in det(F^TF) are obtained. Single point exchange algorithms are worked out also by Wynn (1972), Van-Schalkwyk (1971), Vuchkov (1978) and others. The algorithm of Mitchell (1974) called DETMAX allows an user-selected number of points to be added and deleted at each iteration. Galil and Kiefer (1980) presented an improved version of DETMAX wich is called Modified DETMAX (MD-algorithm). Cook and Nachtsheim (1980) carried out an empirical comparison of the algorithms of Fedorov, Wynn-Mitchell, Mitchell, Van-Schalkwyk and of some combined algorithms on convex (continuous) design space.

3. The computer algorithm CDOP

To show the peculiarities of our algorithm for generating D-optimal designs we describe here it's version CDOP used for generating exact D-optimal designs on convex design space. Other versions used for developing sequentially generated D-optimal designs are also worked out.

The N-point exact D-optimal designs are very convenient for practical use becouse the experimenter is able to set prelemi-nary the desired number of observations N taking into account technological or economic considerations.

An N-point design with matrix F_N is called exact D-optimal design if it satisfies the condition (3) written in the form

$$|F_N^{*T}F_N^*| = \max_X |F_N^T F_N| \qquad (4)$$

The main stages of the algorithm CDOP are as follows:

1. For given N and specified model (1) an initial N-point design ξ_0 with nonsingular information matrix $F_N^T F_N = M_0$ is constructed. The value of $\det(M_0)$ is computed and saved.

2. From each point of ξ_0 considered as an origin, S equally distributed random vectors with a given lenth H are generated in χ. The end points of the vectors together with the initial design form a set cf $h = N(s+1)$ points, caled "candidate list".

3. Using all pcints of the "candidate list" an expanded mattix F_h is formed.

4. The information matrix $M_h = F_h^T F_h$, it's inverse $M_h^{-1} = (F_h^T F_h)^{-1}$ and the value of $\det M_h$ are computed.

5. In each point of the candidate list x_i the value of variance cf prediction $d_h(x_i)$ is calculated

$$d_h(x_i) = x_i^T (F_h^T F_h)^{-1} x_i, \quad i = 1,h, \qquad (5)$$

where $x_i^T = \mathbf{f}^T(\mathbf{x}_i)$ is used for simplicity.

6. The point with minimum value cf d_h denoted X_{min} is found. Then it is deleted from the candidate list wich includes now L-1 pcints.

To update the matrix M_h^{-1} and the values of $\det M_h$ and d_h in the case of h-1 points are used the well known formulae.

$$\det M_{h-1} = \det M_h (1 - X^T_{min} M_h^{-1} X_{min}) \tag{6}$$

$$M^{-1}_{h-1} = M_h^{-1} + (M_h^{-1} X_{min})(M_h^{-1} X_{min})^T / (1 - X^T_{min} M_h^{-1} X_{min}) \tag{7}$$

$$d_{h-1} = d_h + (X_i^T M_h^{-1} X_{min})^2 / (1 - X^T_{min} M_h^{-1} X_{min}) \tag{8}$$

7. The procedure is repeated from point 3 for the candidate list constisting of h-1 points. At each iteration a point with minimum value of $d(x_i)$ is deleted until N-points are left in the candidate list.

8. The last N-points are considered as new design ξ_1. It's information matrix determinant $\det M_{h-N}$ is calculated at $(h-N)^{-th}$ iteration and compared with the value of the determinant of initial design $\det M_o$. If $\det M_{h-N} \geq \det M_o$ the design ξ_1 is set to be a new initial design. We call the search for determining a new initial design "an excursion" according to Mitchell (1974).

9. If $\det M_{h-N} < \det M_o$ the excursion is unsuccessful. A new set of h points is generated from the initial design ξ_o. Then a new excursion is performed.

10. The sequence of excursions terminates whenever $(\det M_{h-N} - \det M_o)/\det M_o < \varepsilon$, where ε is a little positive number. According to Mitchell (1974) we call the sequence of excursions starting from an initial design a "try".

Procedures of the above type (and also all exchange procedures) are not guaranteed to find the global optimum. It has to be run several times with different starting designs. The simplicity of CDOP algorithm makes it very fast and more tries can be run to obtain the best design. We don't need also the special techniques of avoiding singularities of F^TF on excursions used by Mitchell (1974). As a result we could treat problems with n = 10 for quadratic polynomials while for example DETMAX and MD-algorithm were limited to $m \leq 7$.

The computer experiments using FORTRAN-77 program were conducted on VAX 11/750 computer.

4. Examples

We now discuss briefly the results of implementing CDOP in examples compared with some of the best known exhange procedures. We used the examples given by Johnson and Nachtsheim (1983) which are conducted on VAX11/850 computer on convex design space from random initial designs. The first 6 test models of Johnson and Nachtsheim (1983) were used for searching for D-optimal designs on the convex design space $\chi = \{X/-1\le x\le 1/\}$.

The results are given in Table 1, where are compared the algorithms of Fedorov (F), it's modified version (MF) developed by Cook and Nachtsheim, the k-exchange algorithm of Johnson and Nachtsheim (KE), Mitchell (M), Wynn-Mitchell (WM) and our CDOP. Four variables are recorded: the central processing time in seconds (excluding input and output time), the number of the test model, the number of design points N, and the value of $\det(F_N^T F_N)$. The CP-time given for CDOP is an average value obtained from 5 tries with different initial random designs.

Table 1

Algorithm	N=8, Model No 1		N=9, Model No 2		N=9, Model No 3	
	Det.	Time	Det.	Time	Det.	Time
F	.617E-7	3,5	.902	68	.975E-2	26
MF	.644E-7	5,4	.902	35	.975E-2	12
KE	.604E-7	2,5	.902	18	.975E-2	7
M	.631E-7	6,1	.902	17	.975E-2	71
WM	.631E-7	1,6	.902	9	.933E-2	4
CDOP	.624E-7	1,5	.902	5	.218E-1	3
	N=10,Model No 4		N=14,Model No 5		No=15,Model No 6	
	Det.	Time	Det.	Time	Det.	Time
F	.946E-2	29	.728	142	.682E-5	138
MF	.946E-2	25	.728	40	.682E-5	43
KE	.942E-2	8	.728	17	.682E-5	18
M	.944E-2	32	.728	19	.682E-5	15
WM	.941E-2	5	.728	13	.678E-5	16
CDOP	.941E-2	3	.728	8	.675E-5	8

The comparison of algorithms shows that CDOP is nearly as effective as the algorithm of Fedorov (F) being at the same time as a rule faster than (WM).

The relative efficiency of CDOP grows with the encrease of the number of design points N. An illustration is the five factor example for second order polynomial model and N = 21, k = 21. In that case the algorithm of Mitchell found a value of $\det(F_N^{*T} F_N^*) = 0{,}698E\text{-}7$. The algorithm of Galil and Kiefer (1980) achieved $\det(F_N^{*T}F_N^*) = 0{,}789E\text{-}7$ while CDOP found a design with $\det(F_N^{*T}F_N^*) = 0{,}101E\text{-}6$.

The algorithm CDOP is reasonable in terms of both time and relative efficiency. It can be very useful for generating optimal designs on convex spaces in the case of more complicated models and design spaces, for which the set of candidate points is not known or their number is very high.

References

Cook, R.D., and Nachtsheim, C.J. (1980), A comparison of algorithms for constructing exact D-optimal designs, Technometrics, 22, 315-324.

Fedorov, V.V. (1972), Theory of optimal experiments, translated and edited by W.J. Studden and E.M. Klimko, New York: Academic Press.

Galil, Z., and Kiefer, J. (1980), Time and space-saving computer methods, related to Mitchell's DETMAX, for finding D-optimum Designs, Technometrics, 21, 301-313.

Johnson, E.M., and Nachtsheim, C.J. (1983), Some guidlines for constructing exact D-optimal designs on convex design spaces, Technometrics, 25, 271-277.

Kiefer, J. (1959), Optimum experimental designs. J. Roy. Statist. Soc., 8, 21, 273-319.

Mitchell, T.J. (1974), An algorighm for the construction of D-optimal experimental designs, Technomterics, 16, 203-210.

St. John, R.C. and Draper, N.R. (1975). D-optimality for regression designs: a review. Technometrics, 17, 15-24.

Van Shalkwyk, D.J. (1971). On the design of mixture experiments. Ph. D. thesis, University of London.

Vuchkov, I.N., (1978), Optimal design of experiments. Technika, Sofia.

Wynn, H.P. (1970). The sequential generation of D-optimal experimental designs, Ann. Math. Statist., 41, 1655-1664.

Probabilistic Models in Exploratory Data Analysis

Some Guidelines for Principal Component Analysis

By P. Besse, H. Caussinus, L. Ferre, and J. Fine, Toulouse

Summary. - Principal Component Analysis is most often used as a tool of Exploratory Data Analysis to generate graphical displays. The fixed effect model is then a convenient framework to set up a check list of questions to be adressed to the user, in order to help him to perform the analysis as suitably as possible with respect to his goals and his data. The paper attempts to develop this point of view by integrating several previous discussions into this framework and giving some new developments.

Key-words. - Principal Component Analysis, metric choice, transformations of data.

1 - Introduction.

Principal Component Analysis (PCA) is a prevalent technique to analyze data which are collected in an nxp matrix. In an exploratory data analysis context, PCA is expected to summarize the data and to provide graphical displays. However, to achieve these objectives, a large variety of PCA's have been proposed according to the type of data to be processed. For example :

- The raw data can be centered, or standardized ; this is discussed in all text books.
- Different metrics can be chosen to measure distances between rows or columns. For instance, Correspondence Analysis (CA) is a PCA with a special metric which is often effective for processing frequency data (BENZECRI, 1973 ; GREENACRE, 1984). Several modifications or extensions have been proposed : ESCOUFIER (1985) gives a review.
- Non linear transformations of the data turn out to be useful in many cases for reducing dimensionality : de LEEUW et al.(1981) introduce B-spline transformations, WINSBERG and RAMSAY (1983) use monotone splines, AITCHISON (1983) proposes to take the logarithm of proportions when analyzing compositional data, etc.

In this paper, we discuss a general framework which permits

Compstat 1986

us to give some simple guidelines for processing data by means of a suitable variety of PCA. This question has been raised in CAUSSINUS (1985a,b) who noted that, although several models (i.e. several ways of setting up the problem) lead to the same processing of the data (viz. PCA), one of them (the fixed effect model) provides a better rationale for the present purpose. This will be the starting point for our developments herein.

Beforehand, let us stress three points :

- The model which we put forward is very old : see e.g. YOUNG (1940) for arguments in its favour.

- This model is expected to be roughly adequate, but it was mainly chosen for its simplicity : complex situations analyzed by trained statisticians using sophisticated techniques are out of our scope.

- There is a strong similarity with the approach of generalized linear models (Mc CULLAGH and NELDER, 1983) in regression since the model which we use permits a clear separation of the questions concerning the means (structural parameters) and the variances (modelization of the noise) : see section 3.

2 - The basic model.

2.1. General framework.

The nxp matrix to be analyzed is denoted by X, its elements are X_i^j $i = 1,\ldots,n$, $j = 1,\ldots,p$. The rows (columns) of X are px1 (nx1) vectors or matrices $X_i(X^j)$. Distances between rows (i.e. in $\mathbb{R}^p$) are measured by a symmetric positive definite pxp matrix M. The associated norm of X_i is $\|X_i\|_M = ({}^tX_i M X_i)^{1/2}$. Distances between columns are measured by means of an nxn matrix which will be herein a diagonal matrix D of positive weights w_i, $i=1,\ldots,n$ (a more general case is considered in CAUSSINUS (1985a) : it allows us to better account for the exchangeability of rows and columns which appears in many problems). $\mathcal{E}_q$ is the set of all linear subspaces of $\mathbb{R}^p$ whose dimension is less than or equal to a given integer q. The model is defined by the following assumptions :

(i) the X_i's are n independent random vectors

(ii) $\mathbb{E}(X_i) = x_i$

(iii) $Var(X_i) = \sigma^2/w_i\ \Gamma$ (Γ is a pxp symmetric positive definite matrix assumed to be known as well as the positive numbers w_i).

(iv) there exists $E_q \in \mathcal{E}_q$ such that all the x_i's belong to E_q.

Assumption (iv) can be replaced by the more general one (iv') for which the subspace E_q is replaced by a linear manifold F_q.

We shall consider the model (i)-(iv), but only very small modifications are required to deal with (iv') instead of (iv) : it is enough to center the columns to their weighted means. In turn (iv') may be easily replaced by more general assumptions (see e.g. CAUSSINUS, 1985a).

The unknown parameters are E_q, the n vectors x_i belonging to E_q and possibly σ. Least squares estimates of E_q and x_i are obtained by solving the minimization problem :

$$\min_{E_q \in \mathcal{E}_q;\, x_i \in E_q} \quad \Sigma_{i=1}^{n} w_i \| X_i - x_i \|_M^2 \qquad (1)$$

When X_i has a normal distribution, this produces Maximum Likelihood estimates if $M = \Gamma^{-1}$. Otherwise we shall see below (section 4) that $M = \Gamma^{-1}$ turns out to present an optimal property analogous to the Gauss-Markov theorem for linear models.

We denote by Π_E^M the M-orthogonal projector from $\mathbb{R}^p$ onto the subspace E. The minimum in (1) is reached for $\hat{E}_q$ and $\hat{x}_i$ such that :

. $\hat{E}_q$ is the subspace of $\mathbb{R}^p$ spanned by q M-normed eigenvectors of ${}^tXDXM$ associated with the q largest eigenvalues,

. $\hat{x}_i = \Pi_{\hat{E}_q}^M (X_i) \qquad i=1,\ldots,n.$

2.2. Comments on the model.

Without entering into a long discussion, for which we refer to CAUSSINUS (1985b), we need to emphasize some points.

The model in 2.1. is to be considered as a "leading case" (MALLOWS and TUKEY, 1982) or a "virtual model" (BARRA, 1985). It is introduced to make clear the situations in which the statistical analysis will be consistent. And indeed when we display the rows of X on an Euclidean graphic, we assume implicitly that the points which are plotted represent the relevant part of the data while other dimensions contain only random noise It is then natural to suppose that this relevant part is a parameter x_i belonging to a subspace (or a linear manifold) whose dimension is the dimension of the graphical display, while the noise is assumed to be a random error. Thus we can say that our model is the natural model for graphical displays.

The present model now allows us to tackle some questions step by step. From each step we shall try to throw a new light on the problems : which data to analyze (transformations) and how to do it (choice of M and weights) ?

This will provide a check list for PCA's users.

3 - A check list for PCA.

3.1. Invariance.

It is well admitted that invariance with respect to measurement units suggests standardization. But other kinds of invariance have to be considered. With a table of frequency data, Benzecri's principle of distributional equivalence (BENZECRI, 1973) is one of the mainstays of CA (it states that the analysis should be invariant when pooling two proportional rows). For other considerations about this principle see ESCOFIER (1978). Sometimes, the analysis of positive data should be invariant under multiplication of the rows and/or columns by any positive number : KAZMIERCZAK (1985) gives interesting examples and proposes a PCA of the logarithms after a double centering of rows and columns, with $w_i = 1$ and $M = I_p$ (I_p is the unit pxp matrix).

3.2. Structural parameters (means).

For simplicity q should be low while ensuring a reasonable fit to the data. Some authors have suggested transformations to meet this requirement. For instance, AITCHISON (1983) has pointed out that a logarithmic transformation may be valuable in the case of compositional data (a row of X is a vector of proportions adding to 1). Unfortunately it is difficult to propose systematic ways to explore the general problem which shall rest mostly on empirical grounds and on the knowledge of the subject under study. So very general guidelines are almost impossible to provide.

On the other hand, people often expect that vectors x_i nearly occupy an ellipsoid in E_q. This would be the case if these vectors were a sample from a normal distribution : this is not required here and is indeed inconsistent with most real data. However, a too low value of q could be less misleading with such an ellipsoid.

3.3. Variances.

It will be recommended to set $M = \Gamma^{-1}$ because least squares estimators are then more efficient (see section 4). It remains now to discuss the part of the model concerning Γ. This is done in this section in the light of some examples. Another view concerning the choice of Γ will be presented in section 6.

a) In many cases one may assume $w_i = 1$ and $\Gamma = I_p$ or at least Γ diagonal (independence of the columns) : in this latter case the columns are often standardized to come back to the former, but it should be pointed out that this procedure is risky since the X_i^j ($i=1,\ldots,n$) have different expectations.

b) Let us consider a contingency table whose entries are N_i^j, $i=1,\ldots,n$, $j=1,\ldots,p$. We suppose that the row totals N_i^+ are fixed. We analyze the matrix X of the row profiles, defined by $X_i^j = N_i^j / N_i^+$. We have $\mathrm{Cov}(X_i^j, X_i^h) = x_i^j(\delta^{jh} - x_i^h)/N_i^+ \quad j,h=1,\dots,p$ where $x_i^j = \mathbb{E}(X_i^j)$ and δ^{jh} is the Kronecker symbol.

Assumption (iii) of the model fails to be true, but if we take $x^j = \Sigma_{i=1}^n X_i^j N_i^+/N_+^+ = N_+^j/N_+^+$ as an approximation of x_i^j the model holds approximately with

. $w_i = N_i^+$

. the elements of Γ estimated by $(N_+^j/N_+^+)(\delta^{jh} - N_+^h/N_+^+)$

Another difficulty arises from the fact that Γ is now singular, but it is easy to check that Γ^{-1} may be replaced by the generalized inverse $\Gamma^- = \mathrm{diag}(N_+^+/N_+^1, \dots, N_+^+/N_+^p)$. This leads to CA which is once again supported by a new argument.

c) Let the p elements of X_i be the <u>proportions of a mixture</u> ($X_i^+ = 1$). We assume that the ith unit is characterized by $x_i^j = \mathbb{E}(X_i^j)$, $j=1,\dots,p$. There is however a random noise and X_i follows a Dirichlet distribution with parameters $(\alpha_i^1, \dots, \alpha_i^p)$ (note that x_i is assumed to be fixed, and thus the difference from AITCHISON, 1983). If we assume $\alpha_i^+ = \alpha^+$ independent of i, we have :

$x_i^j = \alpha_i^j/\alpha^+$ and $\mathrm{Cov}(X_i^j, X_i^h) = \sigma^2 \; x_i^j(\delta^{jh} - x_i^h)$ with $\sigma^2 = (1+\alpha^+)^{-1}$.

The arguments above (b) can be reproduced with very slight modifications and, on the present grounds, CA of the matrix X can be recommended.

d) Coming back to <u>frequency data</u> (b), we may not be content with the approximation of the covariance structure Γ, mainly if rows and columns are far from being independent. It is clear that our difficulties arise basically from the dependence of the variances on the means. Now, for count data it may be assumed that the random variables N_i^j have independent Poisson distributions : if we take $X_i^j = (N_i^j)^{1/2}$ the variance of X_i^j is approximately constant whatever $\mathbb{E}(X_i^j)$ may be. Thus the PCA of X should be performed with $\Gamma = I_p$ and $w_i = 1$. DOMENGES and VOLLE (1979) have recommended the use of $(N_i^j)^{1/2}$ on different grounds : their arguments are "descriptive" and would rather belong to our section 3.2.

e) The proposal above can be extended to different kinds of data which would present other relations between mean and variance : pick out transformations to stabilize the variance from the usual stock.

4 - A Gauss-Markov type property for PCA.

If the objective is to approximate the display of the vectors x_i in E_q by the display of the vectors $\hat{x}_i$ in $\hat{E}_q$, we can measure the efficiency of the estimators resulting from (1) by a quantity such as $f(M,A) = \mathbb{E}(\Sigma_{i=1}^{n} \|\Pi_{\hat{E}_q}^{M}(X_i) - x_i\|_A^2)$ where A is a given symmetric positive matrix.

Under suitable assumptions (it is enough for example that $X_i - x_i$ be bounded) we can use perturbation theory (KATO, 1966) to expand f(M, A) in successive powers of σ. We get :

$$f(M,A) = \sigma^2[q\,tr(\Gamma A) + (n-q)\,tr(\Pi_{E_q}^{M}\Gamma^t\Pi_{E_q}^{M} A)] + O(\sigma^3) \quad (2)$$

and the following property :

For small σ, whatever the values of A and n, f(M, A) reaches its minimum for $M=\Gamma^{-1}$.

The proof and other developments are given in BESSE et al. (1986)

5 - Illustrative example.

This part is devoted to two illustrations of the previous sections. HAMOUDA (1982) has used CA to analyze a survey concerning offences in Koweit classified according to the nationality of the accused and the nature (robbery, etc.) of the offence. This results in a 15x11 contingency table. An advantage of CA is to display row (and column) profiles which show the important features of the data (the marginal fact that Koweit natives are more numerous is obvious and irrelevant). On the other hand it is clear that these data can be considered as 15x11 independent Poisson variates N_i^j. Then section 3.3 suggests that dealing with the matrix of elements $(N_i^j)^{1/2}$ may result in a more efficient processing as far as the quality of estimates is concerned. The first factor of this analysis is (as it might be expected) a "size effect" : it corresponds to the "trivial factor" of CA. The projections of the profiles $((N_i^j/N_i^+)^{1/2})_{j=1,\ldots,11}$ on the subspace of the second and third factors show a configuration looking very much like the one of CA (factors 1 and 2). The fact that we have obtained similar graphical displays by means of two varieties of PCA resting on rather different grounds should support the value of these graphics.

The same example has served to illustrate the behaviour of f(M, A) in a case where σ is not too small. We have generated sixty 15x11 tables where each entry is a Poisson variate whose parameter is the corresponding entry of the given table. Expectations being taken over these 60 realizations, we have computed $f(M, I_{11})$ for some values of M, including I_{11}. The comparaison with the leading term in (2) is fairly satisfactory.

6 - Another approach to the choice of Γ.

The purpose of this part is a little different : we try now to remove some well known or obvious features from the data. This is achieved by considering particular choices of the error structure and thus particular choices of metrics, allowing us to transfer trivial parts of the data correlation structure from the mean to the error. This approach is illustrated in the case of longitudinal data : n functions or trajectories observed at p argument values t_j and, without loss of generality, assumed to be centered. In that context, each error is a noise process with covariance matrix $\sigma^2\Gamma$.

Consider the simple problem of removing a constant component. Of course, since a metric change in a PCA is equivalent to a linear transformation of the centered data, that can be performed by considering the

$$\{X(t_1), X(t_1)-X(t_2), \ldots, X(t_{p-1})-X(t_p)\}$$

columns sequence instead of the $\{X(t_j)\}$ one. But a more general approach, allowing us to remove any shape of components with any kind of boundary conditions, can be developped as an application of BESSE and RAMSAY (1986)'s work. This leads us to consider the error process as a shifted brownian motion (BESSE, 1985) instead of a white noise which is the classical use when $\Gamma = \sigma^2 I_p$. Then the metric to be used is given by the matrix Γ_1^{-1} where $\Gamma_1 = [1+\min(t_i, t_j)]_{1\leqslant i,j\leqslant p}$. Thus in this case the first axes of PCA provide discrimination between trajectory variations instead of absolute values.

A more complete example dealing with periodic components can be found in BESSE and RAMSAY (1986) and, in a similar context, BESSE (1986) proposes the same kind of approach to filter singularities of noisy data.

7 - Conclusion.

A help system (which could be an "expert system") for exploratory data analysis should force the user to give all relevant information about his data (see : concept of "idéomorphisme" in BARRA and BECKER, 1985). Around the algorithmic core of singular value decomposition, our heuristic approach to PCA (others would have said Factor Analysis) is expected to enables statisticians to ask useful questions, whose answers will serve to guide the user towards proper choices of transformations, metrics, etc.,(although it is to be stressed that we do not intend to force him to use dogmatically a unique analysis).

References.

AITCHISON, J. (1983), Principal Component Analysis of compositional data, Biometrika, 70, 1, 57-65.

BARRA, J.R. (1985), Méthodes statistiques en psychiatrie - Modèles virtuels. In "Model Choice", Proceed. of the 4th Franco-Belgian Meet. of Statisticians. Publi. des Fac. Univers. Saint-Louis, Bruxelles.

BARRA J.R., BECKER, M. (1985), Conception d'un logiciel d'assistance - Intelligence en analyse factorielle canonique, 4èmes Journées Internat. An. Donnée. et Inform., Versailles.

BENZECRI, J.P. (1973), L'analyse des données ; II. L'analyse des correspondances, Dunod, Paris.

BESSE, Ph. (1985), PCA of longitudinal data as a fixed effect model, Fourth European Meeting of the Psychometric and the Classification Societies, Cambridge.

BESSE, Ph. (1986), Choice of a metric in PCA or choice of a model : the case of discrete event data, COMPSTAT 86, Rome.

BESSE, Ph., CAUSSINUS, H., FERRE, L., FINE J. (1986), Une propriété du type Gauss-Markov en ACP ; unpublished manuscript.

BESSE, Ph., RAMSAY, J.O. (1986), Principal Component Analysis of sampled functions, Psychometrika, to appear.

CAUSSINUS, H. (1985a), Models and uses of Principal Component Analysis, Multidimensional Data Analysis Workshop, Cambridge.

CAUSSINUS, H. (1985b), Quelques réflexions sur la part des modèles probabilistes en Analyse des Données, 4èmes Journées Internat. An.Donn. et Inform., Versailles.

DOMENGES, D. VOLLE, M. (1979), Analyse factorielle sphérique : une exploration, Annales de l'INSEE, 35, 3-84.

ESCOFIER, B. (1978), Analyse factorielle et distances répondant au principe d'équivalence distributionnelle, Rev.Stat.App., XXVI, 4, 29-37.

ESCOUFIER, Y. (1985),L'analyse des correspondances : ses propriétés et ses extensions. Bull.I.I.S., Actes 45ème session, LI, 4, 28-3, 1-16.

GREENACRE, M.J. (1984), Theory and applications of correspondence analysis, Academic Press, London.

HAMOUDA, K.I. (1982),Répartition des crimes et délits au Koweit d'après la nationalité des auteurs, Cah.An.Donn.,VII,1,117-123.

KATO, T. (1966), Perturbation theory for linear operators, Springer-Verlag, New-York, 1rst ed.

KAZMIERCZAK, J.B. (1985), Analyse logarithmique : deux exemples d'application, Rev.Stat.App., XXXIII, 1, 13-24.

LEEUW, J. de, VAN RIJKERVORSEL, J., VAN der WOUDEN, H. (1981), Non linear principal components analysis with B-splines, Methods of Operations Research, 33, 379-393.

MALLOWS, C.L., TUKEY, J.W. (1982), An overview of techniques of data analysis, emphasizing its exploratory aspects, In "Some Recent Advances in Statistics", J. Tiago de Oliviera et al. ed., 11-172, Academic Press, London.

MC CULLAGH, P., NELDER, J.A., (1983), Generalized Linear Models, Chapman and Hall, London.

WINSBERG, S., RAMSAY, J.O. (1983), Monotone spline transformation for dimension reduction, Psychometrika, 50, 1, 91-119.

YOUNG, G. (1940), Maximum likelihood estimation and factor analysis, Psychometrika, 6, 1, 49-53.

Comparison of Least Squares with Least Absolute Deviation Forecasting Using Simulation Techniques

By W. C. Conley, Wisconsin

SUMMARY

Linear least squares forecasting is used on two data sets and these resulting equations are compared, along with their larger total absolute deviations, to regular least absolute deviation (LAD) equation fits done with multi stage Monte Carlo optimization (MSMCO). Then two nonlinear models are fitted using LAD with MSMCO. One is a regular LAD fit and the other uses mini max LAD curve fitting.

KEY WORDS
LEAST SQUARES LEAST ABSOLUTE DEVIATION COMPARISON
MULTI STAGE MONTE CARLO OPTIMIZATION

INTRODUCTION

The least squares theory of curve fitting and forecasting is quite well developed. It has been one of the best application areas of applied statistics and computer programming. The computer can implement the theory and speed the calculation and testing of the betas B_i's in most any setting as long as the normal equations are linear in the beta's.

However new computer techniques like Multi Stage Monte Carlo simulation (MSMCO) can fit curves to data even when the normal equations are nonlinear in the betas.

Additionally, because multi stage simulation is a general purpose algorithm it can fit least absolute deviation curves to data, somewhat overcoming the problem of a least squares curve being distorted by one or two outlier points whose squares of Y_i-Y overly weight their importance. Using the multi stage algorithm with least absolute deviation curve fitting, each data point contributes an $|Y_i-Y|$ term, so each point is treated equally tending to produce a more representative curve fit in many cases.

Compstat 1986

The paper presents examples of a variety of types of curve fits using least squares and the least absolute deviation L.A.D. technique (in its regular and mini max forms) with the multi stage Monte Carlo algorithm. Error terms for the least squares and L.A.D. curves are compared. As the curve fitting work is new, the references Clark (1984), Conley (1983), Conley (1984), Conley (1985), Sylwestrowicz and Parkinson (1984) and Woodside (1984) consist of a variety of multi stage applications and related topics to show the generality of the technique.

LEAST SQUARES AND REGULAR LAD

Consider the data set

X_1	X_2	X_3	X_4	Y
100	100	150	250	213,250
600	425	258	317	623,926
800	199	480	750	892,275
300	299	752	400	829,601
500	500	600	400	858,200
395	800	200	100	533,385
750	250	500	550	600,000

Fitting $Y=B_0+B_1X_1+B_2X_2+B_2X_3+B_4X_4$ to the data using least squares yields $Y=-170{,}000-216X_1+787X_2+490X_3+1091X_4$ with a total absolute deviation of 356,199. The following multi stage Monte Carlo program finds a least absolute deviation equation of $Y=-170{,}000 -79.5479X_1+650.454X_2+599.853X_3+944.725X_4$ (using the least squares B_0) in about three minutes (20,000 samples) of computer time. Its total absolute deviation is 268,395 or about a 25% reduction in total error of the least squares total absolute deviation.

```
1       DIM N(4),U(4),X(4),L(4),A(4),B(4),C(7)
2       DIM D(7,5),E(7)
3       FOR Y=1 TO 7
4       READ D(Y,1),D(Y,2),D(Y,3),D(Y,4),D(Y,5)
5       NEXT Y
10      F=2
11      X=1
12      M=1.0E50
```

```
14      FOR I3=1 TO 4
15      X(I3)=-10000
16      A(I3)=0
17      N(I3)=10000
18      NEXT I3
20      FOR J=1 TO 20
23      Z=1000
25      FOR I=1 TO Z
28      FOR K=1 TO 4
30      IF A(K)/F**J<X(K) THEN 50
40      GO TO 60
50      L(K)=X(K)
55      GO TO 65
60      L(K)=A(K)-N(K)/F**J
65      IF A(K)+N(K)/F**J>N(K) THEN 80
70      GO TO 90
80      U(K)=N(K)-L(K)
85      GO TO 100
90      U(K)=A(K)+N(K)/F**J-L(K)
100     B(K)=L(K)+RND(X)*U(K)
102     NEXT K
103     R=0
104     FOR W=1 TO 7
106     C(W)=D(W,1)*B(1)+D(W,2)*B(2)
107     C(W)=C(W)+C(W,3)*B(3)+D(W,4)*B(4)
108     E(W)=ABS(C(W)-D(W,5)-170000)
110     R=R+E(W)
112     NEXT W
120     IF R < M THEN 150
140     GO TO 175
150     A(1)=B(1),A(2)=B(2)
155     A(3)=B(3),A(4)=B(4)
160     M=R
175     NEXT I
180     NEXT J
185     PRINT A(1),A(2),A(3),A(4),M
200     DATA 100,100,150,250,213250
210     DATA 600,425,258,317,623926
```

```
220        DATA 800,199,480,750,892275
230        DATA 300,299,752,400,829601
240        DATA 500,500,600,400,858200
250        DATA 395,800,200,100,533385
260        DATA 750,250,500,550,600000
270        END
```

Now consider the data set

X_1	X_2	X_3	X_4	Y
100	100	150	250	500,000
600	425	258	317	623,926
800	199	480	750	892,275
300	299	752	400	829,601
500	500	600	400	858,200
395	800	200	100	533,385
750	250	500	550	600,000

The least squares fit of $Y=B_0+B_1X_1+B_2X_2+B_3X_3+B_4X_4$ yielded $Y=153{,}621-564X_1+607X_2+207X_3+1277X_4$ with a total absolute deviation of 303,920. The multi stage BASIC program was rerun after adjusting for the new data and the least squares B_0 and yielded a LAD equation $Y=153{,}621-511.762X_1+512.470X_2+238.649X_3+1242.040X_4$ with a total absolute deviation of 268,849. This represents a reduction in error of over 12% for a three minute computer run.

A MINI MAX LAD CURVE FIT

Regular LAD (least absolute deviation) fits equations to data by minimizing the sum of the absolute deviations, whereas mini max LAD fits curves using MSMCO by minimizing the maximum absolute deviation hence forcing the curve fitting process to deal with all data especially outlier points. Mini max LAD could be looked at as a compromise between least squares and regular LAD made available by the MSMCO algorithm. Consider the data set.

X_1	X_2	X_3	Y
82	51	35	235
20	55	91	198
66	49	46	219
20	90	15	185
95	15	81	238

10	25	85	136
66	19	68	194
75	20	30	179
70	80	10	239
14	90	85	233
38	75	50	219
90	85	75	322
10	15	20	67
35	50	60	188
70	50	30	214
40	65	10	176

This time a nonlinear curve fit of the functional form $Y=B_1X_1+B_2X_2+B_3X_3(X_1+X_2)^{B_4}+(X_1+X_3)^{B_5}+(X_2+X_3)^{B_6}+(X_1+X_2+X_3)^{B_7}$ using mini max LAD gave us $Y=.423468X_1+.499068X_2+.002639X_3+(X_1+X_2)^{.727445}+(X_1+X_3)^{.605490}+(X_2+X_3)^{.155586}+(X_1+X_2+X_3)^{.934257}$ with total absolute deviation of 5.85 and an individual maximum deviation of .64792

A NONLINEAR REGULAR LAD EQUATION FIT

Using MSMCO and regular least absolute deviation, fit the family of functions $Y=B_1X_1+B_2X_2+B_3X_3+B_4X_4+(X_1X_2)^{B_5}+(X_1X_3)^{B_6}+(X_1X_4)^{B_7}+(X_2X_3)^{B_8}+(X_2X_4)^{B_9}+(X_3X_4)^{B_{10}}+(X_1+X_2+X_3)^{B_{11}}+(X_1+X_2+X_4)^{B_{12}}+(X_1+X_3+X_4)^{B_{13}}+(X_2+X_3+X_4)^{B_{14}}+(X_1+X_2+X_3+X_4)^{B_{15}}$ to the following data

X_1	X_2	X_3	X_4	Y
50	50	50	50	309
200	200	200	200	943
400	400	400	400	1684
600	600	600	600	2380
850	850	850	850	3215
250	500	700	800	2165
800	100	600	750	2225
900	750	80	825	2755
900	800	700	210	2516
300	150	750	875	1958
200	650	100	780	1905
150	800	800	100	1654

700	100	50	680	1755
800	125	900	75	1768
680	900	150	250	2127
57	190	40	750	1215
130	200	720	50	992
60	800	100	200	1281
900	150	200	250	1702
450	200	700	150	1450
250	350	450	500	1601
800	650	600	700	2724
240	300	390	400	1408
255	700	250	600	1921
280	490	275	500	1670

The result was B_1=689.6616, B_2=495.823, B_3=201.4014, B_4=663.9407, B_5=.2893040, B_6=.2957349, B_7=.1313781, B_8=.2717300, B_9=.3960642, B_{10}=.0926007, B_{11}=.7412842, B_{12}=.6141448, B_{13}=.6231946, B_{14}= .4314253, B_{15}=.7613076 with a total absolute deviation of 21.48 and an average absolute deviation of (21.48/25).

REFERENCES

Clark, K. (1984), Operations research and policy problems: simplified computer optimization techniques for policy analysis, Int. J. on Policy and Inform, 8,84-94.

Conley, W. C. (1983), An electrical resistance network, Int. J. of Math. Educ. in Sci and Technol., 14,287-291.

Conley, W. C. (1984), An RC switching network with fixed components, Int. J. Electronics, 57,433-438.

Conley, W. C. (1985),Solving systems of equations with multi stage Monte Carlo optimization, Int. J. Systems Sci., 16,1239-1248.

Conley, W.C. (1985) Die Losung nichtlinearer Gleichungssysteme: Ein Beispiel mit Kapazitaten,ntz Archiv Zeitschrift fur Informationstechnik, 7,13-16.

Conley, W. C. (1985), Linear and nonlinear Kirchoff's laws problems, IETE Technical Review of India, 2,248-250.

Sylwestrowicz, J. D. and Parkinson, D., (1984), An experiment with Conley's multi stage Monte Carlo optimization program, Queen Mary College of the University of London Technical Report No. 233.

Woodside, W. (1984), A multiple commodity inventory problem with storage constraint, Int. J. of Math. Educ. in Sci and Technol., 15,1-6.

Modeling the Reliability of Fault-Tolerant Software Systems*

By T. M. Gerig and J. R. Cook, Raleigh

Software Reliability: With the advent of highly complex computer programs whose correct execution are essential to the proper functioning of a critical system, the concept of reliability has been extended to software. This is necessary for complex software because it is impossible to verify that it will execute correctly under all conceivable inputs. By testing a software product extensively and attempting to correct the errors that are discovered, confidence that it will execute correctly in a give situation is increased. The "time to failure" between corrections (measured, for example, in numbers of executions or CPU cycles) may be used to gauge the increasing reliability of the product. Indeed, contracts for the purchase of custom software frequently include specifications on the minimum allowable mean time to failure.

Fault-Tolerent Systems: The engineering concept of fault-tolerances of hardware systems has also been extended to software systems. The mechanism for introducing fault-tolerance into software is through the use of redundant components. The system is equipped with two or more components developed independently to perform the same computation. One approach for connecting the components is called Recovery Block Programming.

Recovery Block (RB) Programming is used in applications where a correct solution can be recognized. In its simplest form, several programs written to perform the same task are grouped to form a single software subsystem.

* This work was partially supported by the US - Spain Joint Committee for Science and Technological Cooperation.

Compstat 1986

A problem submitted to the subsystem is solved by the first program and the solution is checked for correctness. If the resulting solution is found to be correct, the subsystem reports to the host system accordingly. If not, the problem is passed to the second program and the cycle is repeated. The subsystem reports a failure only if none of the programs are able to correctly solve the problem. Under this setup, there can be only two outcomes: the subsystem returns a correct solution or it reports a failure. It is possible that the acceptance procedure may itself be flawed so that it will occasionally incorrectly classify an outcome. In this case, there is a chance that the subsystem will return an incorrect solution or incorrectly report a failure.

A major difference between hardware and software redundancy is the viability of the assumption that the failure of components occurs independently. Customarily, reliability computations for hardware systems hinge on the assumption that if two components possess probabilities of failure on a given trial of, say, $1-R_1$ and $1-R_2$, then the probability that both components fail on that trial is equal to the product, $(1-R_1)(1-R_2)$. Early attempts at computing software system reliability were based on this assumption. Since then, it has been demonstrated by Vouk (1986) that this assumption cannot be assumed to hold for software systems.

<u>Latent Class (LC) Model</u>: The purpose of this paper is to demonstrate the usefulness of the Latent Class model (Lazarsfeld and Henry (1968)) for this problem. Serving as a less restrictive alternative to the assumption of independence, it provides a probabilistic framework for computing the reliability of Recovery Block subsystems. With certain applications, it may be reasonable to hypothesize that the space of input problems can be grouped into classes on the basis of their difficulty. It is expected that a problem drawn from a class of higher difficulty will, for a given program, more often result in an incorrect solution. These classes are termed latent because they are assumed to be present but are not apparent. In general, it may be reasonable to assume that problems fall into latent classes according to their nature, type or difficulty with respect to one or more subtasks. A particular program may be successful in finding solutions to problems from one such class but not from another.

Specifically, let E_i be the event that program i obtains the correct answer to a randomly chosen question, $i=1,2,\ldots,n$, and L be the random variable (or latent variable) which indicates from which of the K latent classes the problem was drawn $(k=1,2,\ldots,K)$. Let $p_i^k = P(E_i|L=k)$ be the probability that program i obtains a correct answer given that the problem came from latent class k. The marginal distribution of the latent variable, L, is $v^k = P(L=k)$, $k=1,2,\ldots,K$, $\Sigma_{k=1}^K v^k = 1$. The LC model makes the assumption that for problems from a given latent class, E_i and E_j are independent, i.e. $P(E_i \text{ and } E_j|L=k) = p_i^k p_j^k$.

Reliability of Recovery Block Systems: Two descriptive measures that can be used to compare RB systems are overall system reliability and, given that a correct solution is reported by the system, the expected number of executions required to obtain that solution. Under the LC model, the reliability to an RB system composed of n programs is calculated as

$$R = P(\text{System obtains correct solution}) =$$

$$1 - \Sigma_{k=1}^K v^k [\pi_{i=1}^n (1 - p_i^k)] \quad (1)$$

and, given that the system obtains the correct solution, the expected number of executions, M, (for the given ordering of programs in the system) is given by

$$E(M) = \Sigma_{m=1}^n m q_m / \Sigma_{m=1}^n q_m, \quad (2)$$

where

$$q_1 = \Sigma_{k=1}^K v^k p_1^k,$$

$$q_m = \Sigma_{k=1}^K v^k p_m^k [\pi_{i=1}^{m-1} (1 - p_i^k)], \ m>1$$

and it is understood that the p_1^k, $p_2^k, \ldots,$ p_n^k are listed in the same order that the programs appear in the system.

In applications, the parameters, $\{p_i^k\}$ may be assumed known or estimated experimentally. The distribution for the latent class indicator variable,

$\{v^k\}$, reflects the mix of problems that are anticipated. In practice, this distribution may be assumed known, may be experimentally determined or may be chosen to be the "least favorable" distribution so as to yield worst-case values for R and E(M).

<u>RB System Optimization</u>: If N programs are available for inclusion in an n program RB system, the most reliable system among the $\binom{N}{n}$ possible systems will depend on $\{p_i^k\}$ and $\{v^k\}$. For a given set of n programs, their ordering can be selected to minimize the E(M) criterion. Algorithms have been developed for obtaining the optimum system with respect to both of these criteria. The minimax system is obtained as that which has the greatest value of

$$\text{Min}\{1 - \pi_{i=1}^{n}(1 - p_i^k),\ k=1,2,\ldots,K\}. \tag{3}$$

<u>Example</u>: An experiment to test the effectiveness of fault-tolerant software systems was performed in 1983 and reported by Scott (1983) and Wiggs (1984). Fifteen students were hired to write programs to solve a complex network routing problem. Each was provided with the same detailed instructions for writing and debugging their program. The students worked independently and were told that their monetary rewards would be greater for producing a more reliable program. One hundred network routing problems were generated to provide representative coverage of the space of potential inputs. Each of the 100 problems was submitted to each of the 15 programs and a record kept of whether a correct or incorrect solution was obtained. The resulting data form a 15 by 100 matrix of zeroes and ones. The EM algorithm (Dempster, Laird and Rubin (1977)) was used to obtain maximum likelihood estimates of the parameters of the LC model with n=3 classes. They are listed in Table I.

Based on criterion (1) and using $\{\hat{p}_i^k\}$ and $\{\hat{v}^k\}$, the most reliable three program system among the $\binom{15}{3}$=455 possible is {6,12,15} with reliability R=0.964. By contrast, the optimum system based on the conventional independence assumption is found to be {1,6,10}. This system consists of those three programs which have the highest individual overall reliability on the 100 test problems or the largest value of $r = \Sigma_{k=1}^{K} v^k p_i^k$.

The reliability of this system under the LC model is R=0.908 which is significantly lower than the R=0.964 obtained for the optimum under the LC model. By taking $\{\hat{v}^k\}=\{1,0,0\}$ which is the least favorable choice of problem mix for each of these systems, a lower bound for the system reliability of R=0.855 for {6,12,15} and of R=0.602 for {1,6,10} are obtained. The reason for this large difference is that under the independence model, the program choice is based solely on overall individual reliability thereby allowing the possibility of having no program which performs well on problems from a particular latent class. In this case, the independence model selects three programs whose reliabilities, r, with problems from class I are 0.175, 0.347 and 0.261 in favor of two others with corresponding reliabilities of 0.434 and 0.607. Of course, there are choices of $\{\hat{v}^k\}$ for which the reliability of system

TABLE I. MAXIMUM LIKELIHOOD ESTIMATES OF PARAMETERS

Program i	$\hat{p}_i^1$	$\hat{p}_i^2$	$\hat{p}_i^3$
1	0.175	1.000	0.958
2	0.000	0.375	0.983
3	0.262	0.792	0.855
4	0.174	0.264	0.924
5	0.0451	0.897	1.000
6	0.347	0.884	0.967
7	0.132	0.931	1.000
8	0.0429	0.0472	0.805
9	0.176	0.999	0.917
10	0.261	0.933	1.000
11	0.0868	0.371	0.880
12	0.434	0.752	0.859
13	0.173	0.346	1.000
14	0.0868	0.0463	0.222
15	0.607	0.719	0.900
	$\hat{v}^1$	$\hat{v}^2$	$\hat{v}^3$
	0.230	0.290	0.480

{1,6,10} would be higher. Indeed, $\{\tilde{v}^k\}=\{0,1,0\}$ yields R=0.992 for {6,12,15} and R=1.000 for {1,6,10}.

The order within the optimum system that minimizes E(M) given by (2) over the 3! possible arrangements is 6-15-12.

Based upon $\{\tilde{p}_i^k\}$, the system that has the greatest value of (3) and, therefore that best protects against the unfavorable choices of $\{v^k\}$, turns out in this case to be the same as the maximum reliability system , {6,12,15}. The lower bound for the reliability of this minimax system is R=0.855.

It is interesting to compare the reliability computations based on the LC model with those based on the independence model. For example, for {6,12,15} the LC model and independence model yield R=0.964 and R=0.988, respectively. While for {1,6,10}, they yield R=0.908 and R=0.992, respectively. It is possible that computations based on the independence model may under- or overestimate the reliability based on the LC model.

REFERENCES

Dempster, A. P., Laird, N. and Rubin, D. B. (1977) Maximum likelihood from incomplete data via the EM algorithm. Journal of the Royal Society, Series B, 39, pp 1-22.

Lazarsfeld, P. F. and Henry, N. W. (1968) Latent structure analysis. Houghton Mifflin, Boston.

Scott, R. K. (1983) Data domain modeling of fault tolerant software reliability, Ph. D. Dissertation, North Carolina State University.

Vouk, Mladen A., McAllister, David F. and Tai, K. C. (1986) An experimental evaluation of the effectiveness of random testing of fault-tolerent software. Proceedings of the IEEE Workshop on Software Testing, July 1986, Banff, Canada.

Wiggs, J. E. (1984) Experimental validation of fault-tolerant software reliability models, M. Sc. Thesis, North Carolina State University.

Computerized Exploratory Screening of Large-Dimensional Contingency Tables

By S. Kreiner, Copenhagen

This paper is concerned with programs and strategies for the initial analysis of large-dimensional contingency tables, i.e. contingency tables which include so many categorical variables that it is not technically possible to undertake a unified statistical analysis of the full table.

Keywords: Contingency tables, exploratory analysis, log linear models, graphical models, collapsibility

1. Introduction.

Available programs for the analysis of contingency tables are seriously limited. One can, dependent on economic and time considerations, analyze at most 7-10 dimensional contingency tables with the standard programs now available. These of course cannot adequately address the analytic problems in most large scale survey studies, where the number of variables typically far exceeds these limits.

The practical solution to this problem is that analyses are conducted in marginal contingency tables of a manageable size.

There is of course nothing wrong with this type of marginal analytic procedures, but problems arise, however, if generalizations are made from the marginal analyses to conclusions regarding the model for the full table. Simpson's paradox (Dawid (1977)) is a part of the problem. As we for practical reasons are required to do so, it follows that in the course of the analysis great weight should be placed on a running check of the model for the complete set of variables. This means that model checking must assume a central position in any strategy for the analysis of large dimensional contingency tables.

2. The frame of reference of the analysis.

This paper is concerned with the initial mainly exploratory analysis or screening of contingency tables. Before the strategy and the programs are presented it is however necessary to give a short review of the elements of the frame of reference of the complete analysis of which the initial screening is but only the first step.

1) The analysis discussed here will be performed within the framework of recursive log linear models.

   We presume, that is, that the set of response variables consists of several recursive or causal levels where the variables at a given level are to be taken as dependent on the variables at the lower levels as well as on the set of explanatory variable for the complete system of response variables.

   A recursive log linear model is defined as the product of ordinary log linear models for each separate level of response variables, where the variables of the lower levels are treated as explanatory variables and the variables of the higher levels are ignored.

   The product model may but need not be log linear with a structure that reflects the structure of the models of the separate levels (Asmussen & Edwards (1982)).

2) The special sub-class of graphical log linear models defined by conditional independencies of certain variables without any assumptions on the order of interactions of the model (Darroch, Lauritzen & Speed (1980), Edwards & Kreiner (1983)) plays an important part in the strategy of the initial screening of the contingency table. It is not presumed, however, that the analysis is restricted to this class of models.

Compstat 1986

3) Collapsibility of contingency tables, that is the property that the analysis concerning specific interaction parameters may be performed in certain collapsed tables, have been defined in several ways (Whittemore (1978), Asmussen & Edwards (1983), Davis(1986), Kreiner(1986)). We will discuss collapsibility **onto** a marginal table **over** the variables not included in the marginal table **with respect to** certain interaction parameters of interest and will distinguish between **weak collapsibility**, that is the property that disappearance of the parameter of interest in the full model implies the disappearance of the parameter in the collapsed table, and **strong collapsibility**, that is the property that the parameters of interest in the model of the full table is exactly the same as the corresponding parameters in the model of the collapsed table and that the estimates of thĕse parameters and the likelihood ratio test statistics for hypotheses concerning these parameters are the same for the collapsed as for the full table.

   Collapsibility properties are closely connected to the cut-sets of the interaction graph (general cut-sets in the case of weak collapsibility and clique cut-sets for strong collapsibility). Collapsibility may therefore be determined by graph theoretical analysis of the interaction graph corresponding to the full table.

4) Several strategies for the analysis of multidimensional contingency tables have been suggested. Almost all however refer to a general analysis in two steps

   1) Choice of or initial screening for a base model.

   2) Stepwise model selection starting from and improving the base model.

   Most attention have of course been given to the second step. The first step is often done with in a rather haphazard way, as efficient and quick strategies for the second step will remedy even very badly fitting base models. (That different model selection srategies may lead to different models is quite another matter which won't be discussed here).

5) Cox (1977), Cox & Snell (1981) and Chatfield (1985) discuss the initial exploratory data analysis, distinguishing between **specific** analysis concerned with the primary aspects of direct scientific concern and **exploratory** analysis concerned with the secondary aspects of the statistical model. The secondary aspects complete the statistical model and indicate appropriate statistical procedures for the specific analysis.

   In the case of log linear models for large dimensional contingency tables, the distinction between primary and secondary aspects becomes a question of distinction between the interaction parameters of interest and the secondary parameters, the appearance or disappearance of which indicates (the) appropriate statistical procedures by determining collapsibility properties w.r.t. the primary interaction parameters.

3. An integrated strategy for the analysis of large-dimensional contingency tables.

The traditional strategies for analysis of contingency tables are clearly inadequate for the large-dimensional table. They do not explicitely address the problem of collapsibility, which is alpha and omega for the large-dimensional table, or do it only vaguely, stating that it would be an advantage to use the collapsibility properties, and they make no distinction between primary and secondary parameters. It may in fact be argued, that step-wise model selection procedures which are aimed at fitting parsimonious models are more exploratory than anything else.

In the case of large-dimensional tables we have to be a little more explicit about what we do and why we do it.

We can't possibly look at all interactions and therefore have to put some kind of order of priority on the parameters and we have to decide which collapsed tables to analyze. While this paper is primarily concerned with the strategy for the initial data analysis it should be understood, that this strategy is only the first part or should only be the first part of an integrated strategy covering the whole of the analysis. This strategy contains in addition to an exploratory screening aimed at the construction of a base model (the subject of this paper) further exploratory step-wise model selection procedures and specific phases, where conclusions are drawn concerning the primary parameters. Interspersed are intermediate steps where the model is analysed with respect to collapsibility and the results hereof checked (recognizing the fact, that the analysis may run completely astray if erroneous collapsibility properties are applied).

4. The strategy and programs for exploratory screening of large -dimensional contingency tables.

Results connecting collapsibility properties of the statistical model and cut sets of the interaction graph constitute the main feature of a series of programs which may be used (separately or consequtively) for initial exploratory screening of large dimensional contingency tables.

The strategy automized by these programs consists of five phases:

1) A preliminary analysis of all or some of the two-way tables.
2) An analysis of a selection of three-way tables to disclose hidden interactions (that is, interactions among variables where no evidence against marginal independence were found).
3) Analysis of three-way tables to check for conditional independence (given one third variable) of variables for which interaction were implied by the first two phases.
4) Construction of a graphical base model defined by conditional independence of pairs of variables for which no evidence was found against conditional independence in at least one three-way table.
5) Graph theoretical analysis of the interaction graph corresponding to this model aimed at determining
   5a) the cliques of the graph (corresponding to the generating sets of the base model)
   5b) the minimal cut sets of the graph for given pairs of variables corresponding to the minimal w(eakly)-collapsed tables
   5c) the minimal complete cut sets for given pairs of variables corresponding to the minimal s-collapsed tables (under preparation).

The results of the last phase of this analysis are then used to indicate the proper checks of the validity of the base model (looking at the conditional independencies in w-collapsed tables) and the s-collapsed marginal tables where step-wise model search is in order (Kreiner(1986)).

5. An example.

The strategy will be illustrated with data collected in a national survey of health and self care behaviour in Denmark. A total of 22 variables organized in four recursive levels of response variables and one level of explanatory levels were included in the analysis as shown in Figure 1.

Only the last part of the output summarizing the results of the analysis is presented (Figure 2).

For level 1, the graphical base model implied is defined by the generating sets/cliques

[ABU] [AEU] [BCSTU] [CDEFGHIJKLMNOPQRSTUV].

The interaction graph for this model is shown in Figure 3.

The primary research questions of this investigation were concerned with the interaction between social network (variables E,F,M,N,O,P and Q) and health (variables A and B).

Consider for instance the AE-interaction. Collapsibility properties w.r.t. the AE-interaction are as follows.

The table is first of all weakly collapsibel onto the marginal ABEU table w.r.t. the AE-interaction as conditional independence given the rest of the variables infers conditional independence given only B and U.

Evidence against conditional independence of A and E may thus be sought for in the ABEU table. The test performed in this table will however differ from the test which conceivably could be performed in the full table. For this not to be the case we have to have strong collapsibility which presupposes that the boundaries of the connected subsets of variables/vertices not included in the collapsed table are complete. For the ABEU table this means that EBU should be complete, which is not the case.

Strong collapsibility is however given for the 7-dimensional ABCESTU table, where the boundary of the variables not included is the complete subset, CESTU.

The validity of these results depends however on the appropriateness of the base model generated by the screening programs. That is, on whether the results on conditional independence of A respectively B and certain variables at the lower recursive levels are valid.

Before the tests for and estimates of the AE-interaction are calculated these results should be checked, at least in 5-dimensional weakly collapsed tables (including the EBU variables) or even better in the 8-dimensional strongly collapsed tables including the BCESTU variables.

6. Concluding remarks.

By incorporating results on correspondance between properties of log linear models and interaction graphs we have developed strategies and computer programs for the exploratory analysis of large-dimensional contingency tables. This paper have only presented a short review of the considerations which have motivated the development of these programs. Details concerning the execution and application of these programs will be presented in research reports from the Danish Institute for Educational Research.

The programs are written in ALGOL and PL/I. Interested readers may contact the author.

References

Asmussen, Søren and Edwards, David (1983). Collapsibility and Response Variables in Contingency Tables. Biometrika 70, 567-578.

Chatfield, C.(1985). The Initial Examination of Data (with discussion). J.R.Stat.Soc. A 148, 175-214.

Cox, D.R. (1977). The Role of Significance Tests. Scand.J.Stat. 4, 49-70.

Cox, D.R. and Snell, E.J. (1981). Applied Statistics. London: Chapman and Hall.

Darroch, J.N. and Lauritzen, S.L. and Speed, T.P. (1980). Markov Fields and Log Linear Interactions Model for Contingency Tables. Ann.Statist. 8, 522-539.

Davis, L.J. (1986a). Whittemore's Notion of Collapsibility in Multidimensional Contingency Tables. (To appear in Comm.Stat. A - Theory Methods).

Davis, L.J. (1986b). Collapsibility of Likelihood Ratio Tests in Multidimensional Contingency Tables. (To appear).

Dawid, A.P. (1979). Conditional Independence in Statistical Theory (with discussion). J.R.Statist.Soc.B 44, 1-31.

Edwards, D.G. and Kreiner S. (1983). The Analysis of Contingency Tables by Graphical Methods. Biometrika 70, 553-565.

Kreiner, S. (1986). Analysis of Multidimensional Contingency Tables by Exact Conditional Tests: Techniques and Strategies. (To appear in Scand. J. Stat.).

Whittemore, A.S. (1978). Collapsibility of Multidimensional Contingency Tables. J.R.Statist. Soc. B 40, 328-340.

```
RESPONS VARIABLES
=================
LEVEL :  1
-----------
  A: 289 - ILLN. EPISODES   4 CATEGORIES
  B: 262 - PERCEIV. HEALTH  2 CATEGORIES
LEVEL :  2
-----------
  C: 191 - DIF. DAILY ROUT  3 CATEGORIES
  D: 190 - STRESSFUL PROBL  2 CATEGORIES
  E: 298 - SAT. W. NETWORK  3 CATEGORIES
  F: 297 - SAT. W. CONTACT  3 CATEGORIES
  G: 192 - STRESS REDUCTIO  8 CATEGORIES
  H: 150 - ALCOHOL CONSUM   3 CATEGORIES
  I: 149 - TOBACCO CONSUM.  4 CATEGORIES
  J: 299 - ALIENATION       3 CATEGORIES
  K: 308 - LOCUS OF CONTRO  3 CATEGORIES
  L: 311 - ATT. - MED CARE  3 CATEGORIES
LEVEL :  3
-----------
  M: 163 - CONTACT-FAMILY   4 CATEGORIES
  N: 165 - CONTACT-FRIENDS  4 CATEGORIES
  O: 164 - DIST. - FAMILY   4 CATEGORIES
  P: 312 - SOMEONE AS.RESP  3 CATEGORIES
  Q: 272 - MARITAL STATUS   3 CATEGORIES
LEVEL :  4
-----------
  R: 287 - SOCIAL CLASS     8 CATEGORIES
  S: 285 - INCOME           4 CATEGORIES
  T: 292 - EMPLOYMENT       5 CATEGORIES
EXPLANATORY VARIABLES:
======================
  U: 264 - AGE              6 CATEGORIES
  V: 265 - GENDER           2 CATEGORIES
```

Figure 1. The 22 variables of the survey of health and self care in Denmark. The Four recursive levels of the response variables were defined as part of the input to the SCREENING programs.

GRAPH AND CLIQUES

	A	B	C	D	E	F	G	H	I	J	K	L	M	N	O	P	Q	R	S	T	U	V
A:ILLN. EPISODES	*	+	-	-	+	-			-			-		-	-						+	
B:PERCEIV. HEALTH	+	*	+			-	-			-						-	-	-	+	+	+	
C:DIF. DAILY ROUT	-	+	*	+	+	-	+			+			-			-				-	-	+
D:STRESSFUL PROBL	-		+	*	-		-															+
E:SAT. W. NETWORK	+		+	-	*	+				+	+	-		-		-	+					
F:SAT. W. CONTACT	-	-	-		+	*		+		-		-	-	+		-	-	+	-	+	+	-
G:STRESS REDUCTIO		-	+	-			*	+	-			-								-	+	-
H:ALCOHOL CONSUM						+	+	*	-			-		-		-		+		-	-	+
I:TOBACCO CONSUM.	-						-	-	*								+			-	+	
J:ALIENATION		-	+		+	-				*		+						-				
K:LOCUS OF CONTRO					+						*					-						+
L:ATT. - MED CARE	-				-	-	-	-		+		*		-	-		-	+	-	-	+	
M:CONTACT-FAMILY			-			-							*	+	+	+	-		-			
N:CONTACT-FRIENDS	-				-	+		-				-	+	*	-	-	+		-		+	
O:DIST. - FAMILY	-											-	+	-	*			+	-			
P:SOMEONE AS.RESP		-	-		-	-		-			-		+	-		*	+					+
Q:MARITAL STATUS		-			+	-			+			-	-	+		+	*	-	+	-	+	-
R:SOCIAL CLASS		-				+		+		-		+			+		-	*	+	-	-	-
S:INCOME		+				-						-	-	-	-		+	+	*	-	+	
T:EMPLOYMENT		+	-			+	-	-	-			-					-	-	-	*	+	+
U:AGE	+	+	-			+	+	-	+			+		+			+	-	+	+	*	*
V:GENDER			+	+		-	-	+			+					+	-	-		+	*	*

\+ : CONDITIONAL DEPENDENCE
\- : CONDITIONAL INDEPENDENCE
* : NO TESTS PERFORMED
' ': MARGINAL AND CONDITIONAL INDEPENDENCE

Figure 2. Summary of results concerning of the inital screening of the interactions between the 22 variables of the survey on health and self care behaviour.

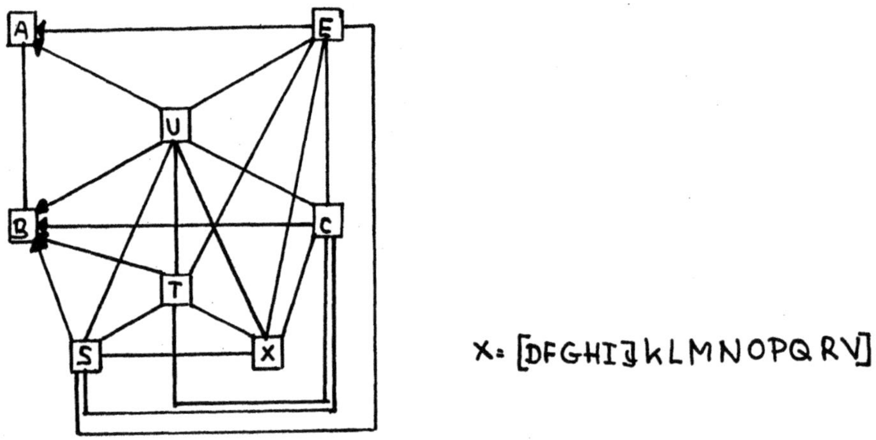

Figure 3. Interaction graph of the model for level 1.

A Fast Algorithm for Some Exploratory Methods in Categorical Data Problems

By T. Rudas, Budapest

Two methods for categorical data problems, a discriminant analysis procedure and a procedure for the reduction of dimensionality can be summarized as an attempt to give a good forecast knowing an individuals categories on a subset of the variables for the category of a fixed variable and for the categories of the rest of the variables, respectively. The most simple algorithm checks all subsets that are to be considered and chooses the best one. This needs several re-orderings of the observed frequencies. In this paper an eager and fast algorithm is proposed that skips the cells of the contingency table in descending order of their observed frequencies and puts them into a set as long as they can be correctly forecasted together based on one of the subsets of the variables considered. Such a maximal subset of cells defines a subset of the variables that is optimal in some cases and is 'not very bad' at least.

Some key words: discriminant analysis, reduction of dimensionality, categorical data, eager algorithm.

0. Introduction

The author of the present paper suggested two methods for categorical data problems that can mostly have exploratory applications. A version of the discriminant analysis procedure (Rudas, 1984) and a method for the reduction of dimensionality for categorical data problems (Rudas, 1985a). These procedures will be briefly outlined in Sections 1. and 2.

The present paper is concerned with an algorithm to carry out the computations that, for both procedures require the finding of certain subsets of the variables with a maximum property. In this respect both procedures belong to a general class of exploratory methods, that among others contains cluster analysis methods (cf. Rudas, 1985b). The algorithm suggested in this paper makes use of the fact that a subset of the variables with the desired maximum property defines a subset of the cells of the table (and the set of variables is essentially defined by the set of cells)

Compstat 1986

and the latter is in some cases much more easy and fast to find. In its most simple application the procedure leads to subsets that are optimal if a certain condition is fulfilled. More sophisticated application yields the optimal subset.

1. Discriminant analysis

Let $V=\{v_1,\dots,v_q\}$ be a set of categorical variables. The set of all possible observations on V is a contingency table denoted by C. Variable v_i has n_i categories that can be denoted by $1,\dots,n_i$. If H is a subset of V then H defines a contingency table which is a marginal of C, C_H. The (marginal) cells of C_H are denoted by c_H. Suppose variable v_q represents groups that are to be discriminated by the classifying variables $v_1,\dots,v_{q-1}$. If S is a subset of the classifying variables and an individual belongs to $c_S \in C_S$ then it can be classified into that category j of v_q for which the conditional probability

$$P(v_q=j \mid S=c_S) \quad , \qquad 1 \leqslant j \leqslant n_q \tag{1}$$

is maximal.

A forecast is said to be incorrect if the individual does not belong to the category j of v_q yielded by the above rule. It is easy to see that this rule minimizes the probability of incorrect forecast. The probability of correct forecast that can be achieved by a set S is denoted by A(S). If $S_1 \subseteq S_2$ then

$$A(S_1) \leqslant A(S_2) \quad ,$$

i.e. the forecast based on a set can not be less accurate than a forecast based on one of its subsets.

In practical applications the probabilities can be replaced by the observed frequencies. The aim is to find a subset with great A(S) and small number of variables in it or few cells in C_S. Sometimes it may be useful to apply a stepwise procedure instead of investigating all subsets of the classifying variables.

The forecasting rule based on (1) may not be unique. In such cases randomization is to be used. This procedure minimizes the risk if the loss function is the increase in the probability of incorrect forecast due to using, instead of the optimal forecasting rule, the rule deduced from the sample.

2. Reduction of dimensionality via subset selection

The aim here is to substitute V with one of its subsets in the sense that if $S \subset V$ and an individual is known to belong to $c_S \in C_S$ we would like to forecast the cell of C where it really belongs to. The forecast will be that cell of C for which the conditional probability

$$P(V=c \mid S=c_S) \qquad (2)$$

is maximal. For a given subset in this procedure forecasting is made into the categories of the rest of the variables, rather than into the category of a given variable.

If for a given subset the probability of correct forecast is high enough, it can be used as a substitute for the whole set, as the information contained in the q groupings, to a large extent, can be gained by using only the variables in S.

Similarly to the discriminant analysis procedure the forecasting rule based on (2) yields the smallest probability of incorrect forecast; as a subset of S can not yield more accurate forecast than S, stepwise procedures can be applied; in practical applications the probability in (2) is to be replaced by the observed frequency and randomization is to be used if necessary, that, again minimizes the risk.

The results referred to in this and the previous Sections are proved in Rudas (1984, 1985a).

3. The algorithm

Both procedures can be formally described as follows: Having selected $S \subset V$ C_S is the source space and C_T is the target space, where $T=\{v_q\}$ for the discriminant analysis procedure and $T=V \setminus S$ for the reduction of dimensionality. A forecasting rule is a function

$$r : C_S \rightarrow C_T$$

such that the conditional probability

$$P(T=r(c_S) \mid S=c_S) \qquad (3)$$

is maximal for all $c_S \in C_S$. It is easy to see that (1) and (2) are special cases of (3). For (2) this follows from the fact that the conditional probability (2) is zero if the projection of c to C_S is not equal to c_S.

For a given $S \subset V$ the probability of correct forecast is

$$A(S) = \sum_{c \in H} P(c)$$

where H is the set of those cells of C that are correctly forecasted i.e. $c \in H$ iff $T(c)=r(S(c))$, where for $U \subset V$ and $c \in C$, $U(c)$ is the projection of c to C_U. (Note that if there are more rules maximizing (3), H depends on r, $H=H(S,r)$, but $A(S)$ does not.)

Given a class of subsets of V, W the subset $S \in W$ is sought for that maximizes A(S) over W. The most natural algorithm to find this optimal S is to try all $S \in W$ and find r then compute A(S). To find the maximal conditional probability in (3) several reorderings of the observed frequencies are necessary.

An other algorithm that is to be outlined here tries to find directly a subset of the cells that is of maximal frequency and can be forecasted together using a subset of the variables $S \in W$. If $G \subset C$ and $S \subset V$ then G is said to be forecastable by S, which will be denoted by $G \in F(S)$ if the following condition holds for any $c_1, c_2 \in G$:

$$\text{if} \quad S(c_1) = S(c_2) \quad \text{then} \quad T(c_1) = T(c_2)$$

It is easy to see that if $G \in F(S)$ then there exists a forecasting rule r such that $G \subseteq H(S,r)$.

The algorithm is eager in the sense that it puts the cells of the table, starting from the one with the greatest probability moving toward the one with the smallest probability (or frequency) in the same set G as long as there is at least one $S \in W$ for which $G \in F(S)$. Suppose there are p cells with positive observed frequencies:

$$f(c_1) \geq f(c_2) \geq \ldots \geq f(c_p) > 0$$

then

step 1.
$G_1 = \{c_1\}$, $W_1 = W$
step i. (repeat $i=2,\ldots,p$)
a./ decide if there is a subset $S \in W_{i-1}$ such that

$G_{i-1} \cup \{c_i\} \in F(S)$.
if yes then go to b./
if not then go to c./

b./ $G_i = G_{i-1} \cup \{c_i\}$
$W_i = W_{i-1} \setminus \{S:\ S \in W_{i-1},\ G_{i-1} \cup \{c_i\} \notin F(S)\}$
go to step i+1.

c./ $G_i = G_{i-1}$
$W_i = W_{i-1}$
go to step i+1.

The algorith terminates with G_p and W_p for which $G_p \in F(S)$ for all $S \in W_p$.

4. Properties, uses and modifications

It is clear that there may be a subset $S' \in W$ such that

$$A(S') > \sum_{c \in G_p} f(c) \ .$$

This may be caused by two types of 'errors'. If in step r b./ was executed, the cell included in G_r, c_r

(i) ruled out S' for $G_{r-1} \cup \{c_r\}$ was not forecastable by S', or

(ii) made the later inclusion of some cells of H(S') impossible.

To improve the results one may re-analyze the data by starting the algorithm instead of c_1 with c_m and performing p-m+1 steps. It is clear, however, that starting with a very large m will not improve the results for

$$\sum_{i \geq m} f(c_i) \geq \sum_{c_i \in G_p} f(c_i)$$

if the procedure is being started at m, that gives an upper limit for the starting point if it is different from 1.

More generally one may try to leave out certain cells from G_p which will allow some others to enter, by performing c./ instead of b./ in certain steps.

It is also possible to try all subsets E of C if there is an $S \in W$ such that $E \in F(S)$. Not all subsets are worth trying for

(iii) if $E \subseteq E'$ then $\sum_{c \in E} f(c) \leq \sum_{c \in E'} f(c)$.

(iv) if $E \notin F(S)$ and $E \subseteq E'$ then $E' \notin F(S)$.

(v) if for $c \in E' \setminus E$ $f(c)=0$ then $\sum_{c \in E} f(c) = \sum_{c \in E'} f(c)$ i.e. only the subsets of p cells are to be investigated, not that of q.

It is not difficult to give, based on (i) and (ii) above a condition that assures that every $S \in W_p$ is optimal. Unfortunately the checking of this condition would require a very long computing. Heuristically one may say that if most of the observations are in the first few cells then the subsets in W_p are close to the optimum.

The algorithm considered in this paper can be used either instead of the stepwise subset selection procedure or combined with it. In the second case W is the class of subsets considered in the given step of the stepwise procedure.

When the number of subsets in W is large but p is small it may be worth investigating all subsets of C with the restrictions (iii)-(v) above. This procedure may be much faster than investigating all $S \in W$ which would require several re-orderings of the data. If not all subsets of C are investigated, the procedure will rapidly lead to a 'not very bad' subset. (A similar price is being paid by any application of stepwise procedures.)

5. References

Rudas, T. (1984) Stepwise discriminant analysis procedure for categorical variables, in Havranek, Sidak, Novak (eds.) COMPSTAT 1984, 389-394. Physica Verlag, Wien.

Rudas, T. (1985a) Reduction of dimensionality in categorical data problems via subset selection, unpublished.

Rudas, T. (1985b) Exploratory methods for categorical data problems, 45th Session of ISI, Contributed papers, 102-103, Amsterdam.

Easy to Generate Metrics for Use with Sampled Functions

By S. Winsberg, New York and J. Kruskal, Murray Hill

Generalized PCA for use with sampled functions using metrics designed to "filter" known variation in the data so as to uncover the subtle kinds of variation is discussed. Examples are presented.

Key Words: generalized PCA, filtering, sampled functions.

Consider data that arise from sampling n functions of time, which may be ordinary or periodic. Each function is sampled at p times t_j, yielding a row p-vector of values, which is referred to as a sampled function. Thus the data form an n by p matrix X of values $x_i(t_j)$.

The obvious variation in the data often may be described as a linear space of functions, which we call s. Additional subtle variation may be suspected, so before analyzing such data, it is sometimes useful to remove the obvious or known variation in order to uncover the subtle or unknown kinds of variation. Consider the data set consisting of (periodic) mean monthly temperatures of thirty-two French cities. The data look like sampled cosine functions with peaks at July. The obvious variation consists of the mean and some multiple of the cosine, i.e., $s = \{1, \cos\}$, so we may want to remove this variation in order to study the residual, Res. The goal then is: to filter out some known kinds of variation.

Consider those methods for analyzing X, such as principal components analysis (PCA), cluster analysis, and quadratic discriminant analysis, that are based on a quadratic metric between p-dimensional row vectors. To accomplish our goal we incorporate a "filter" into the metric so that it ignores the obvious variation. Besse and Ramsay (1982) focused on filtering for such methods, and developed a family of

Compstat 1986

quadratic metrics which accomplish it. They have emphasized PCA, but much of their work can be applied to other metric based methods.

Conceptually, the B-R approach is based on interpolation or smoothing of the data, though the interpolation/smoothing need not actually be carried out. When dealing with sampled functions of time, one may desire to use interpolation and/or smoothing for its own sake. The B-R approach starts by choosing a differential operator that annihilates the functions to be removed. In the temperature example above, the appropriate operator is $D + (1/w^2)D^3$ where D is d/dt. Now working in norms instead of metrics, the B-R approach is to calculate and use the norm defined by these steps: (i) interpolate or smooth the sampled function; (ii) apply the operator; (iii) take the usual function space norm, L_2.

We, Kruskal and Winsberg, were stimulated by the B-R work but concerned about its mathematical complexity, which involves consideration of a reproducing kernel p associated with the differential operator. We have developed two simpler approaches for filtering which do not use interpolation/smoothing. The first K-W approach is the discrete analogue of the B-R approach, and filters by the difference operator that corresponds to the B-R differential operator. The second K-W approach filters by projection onto the space orthogonal to the functions to be removed.

The first or difference operator approach may be illustrated as follows: to annihilate $s = \{1, \cos\}$, the differential operator is $D + (1/w^2)D^3$ which annihilates the larger space $\{\sin; cos, 1\}$. We use the filter $F = \Delta + (1/w^2)\,\Delta^3$ where for periodic and nonperiodic functions respectively when $p = 4$

$$\Delta = \begin{matrix} -1 & 0 & 0 & 1 \\ 1 & -1 & 0 & 0 \\ 0 & 1 & -1 & 0 \\ 0 & 0 & 1 & -1 \end{matrix} \qquad \Delta = \begin{matrix} -1 & 0 & 0 \\ 1 & -1 & 0 \\ 0 & 1 & -1 \\ 0 & 0 & 1 \end{matrix}$$

In the second or projection operator approach one selects any basis for s. These functions need not be orthogonal. Form the matrix K whose rows consist of the p values of the null fractions. The filter is then defined as $F = I - K^0 = I - K^+K$. One advantage of the projection operator approach is that it is not necessary to find an operator that annihilates s, nor is it necessary to filter out extra functions, as the operator approach sometimes requires. The practice is then to do PCA of the filtered data matrix XF based on the usual identity (I) metric, or equivalently to do PCA on the data matrix X based on the metric $M = FF'$.

Generalized PCA with respect to a metric M ($M = I$ yields ordinary PCA) may be stated as follows: given X (I by J), of full rank, with $I \geq J$, find A and B that minimize $\|X - AB'\|$ subject to (i) A is I by R, B is J by R, with $R \leq J$; (ii) B is M orthogonal (i.e. $B'MB = I$); and (iii) the columns of A are in order of decreasing magnitude. (If M is singular, the answer is not unique.) To find A and B', let $Y = YE\Lambda$, and take the SVD, PDQ' of Y. Then $A = PD$ and $B' = Q'\Lambda^{-1}E'$.

In the projection operator approach, annihilation leaves precisely Res. However in both the differential and difference operator approach it leaves not Res but the image of Res under F.

Example:

To demonstrate how filtering works in practice, we turn to the temperature data presenting the results for the projection operator approach only. As can be seen from Figure 1 once the obvious constant and cosinusoidal components are removed, three interesting phenomena may be noticed. The first eigenfunction corresponds to a warm spring and is characteristic of the inland cities as opposed to coastal cities as can be seen from Figure 2a. The second eigenfunction corresponds to an early spring and is characteristic of cities lying in a narrow latitude band i.e. Embrun, Grenoble,

Toulouse, Bordeau and Biaritz, as opposed to another lattitude band further north containing Brest, Nancy, Lille, Rouen, and Paris (see Figure 2b). The third eigenfunction corresponds to a sharp dip in temperature from November to December. This dip separates Embrun the only city in the group at a high altitude from the rest (see Figure 2). In conclusion filtering appears to be useful when trying to uncover subtle variation in a set of data consisting of sampled smooth functions.

References

Besse, P. and Ramsay, J. O. Principal components analysis of interpolated functions. Psychometrika, in press.

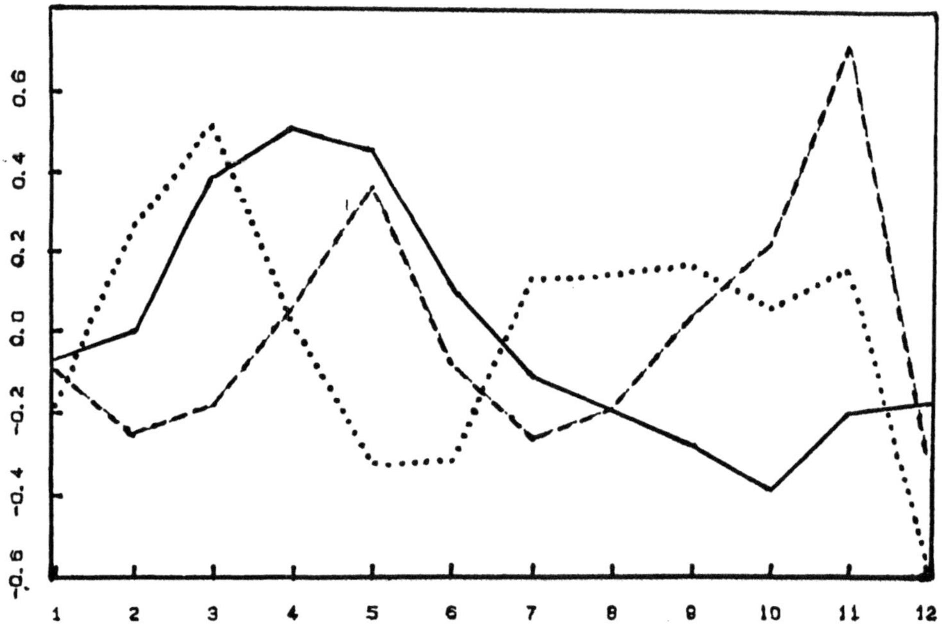

FIGURE 1 First three eigenfunctions for PCA analysis of French cities with filtering s=(cos,1).

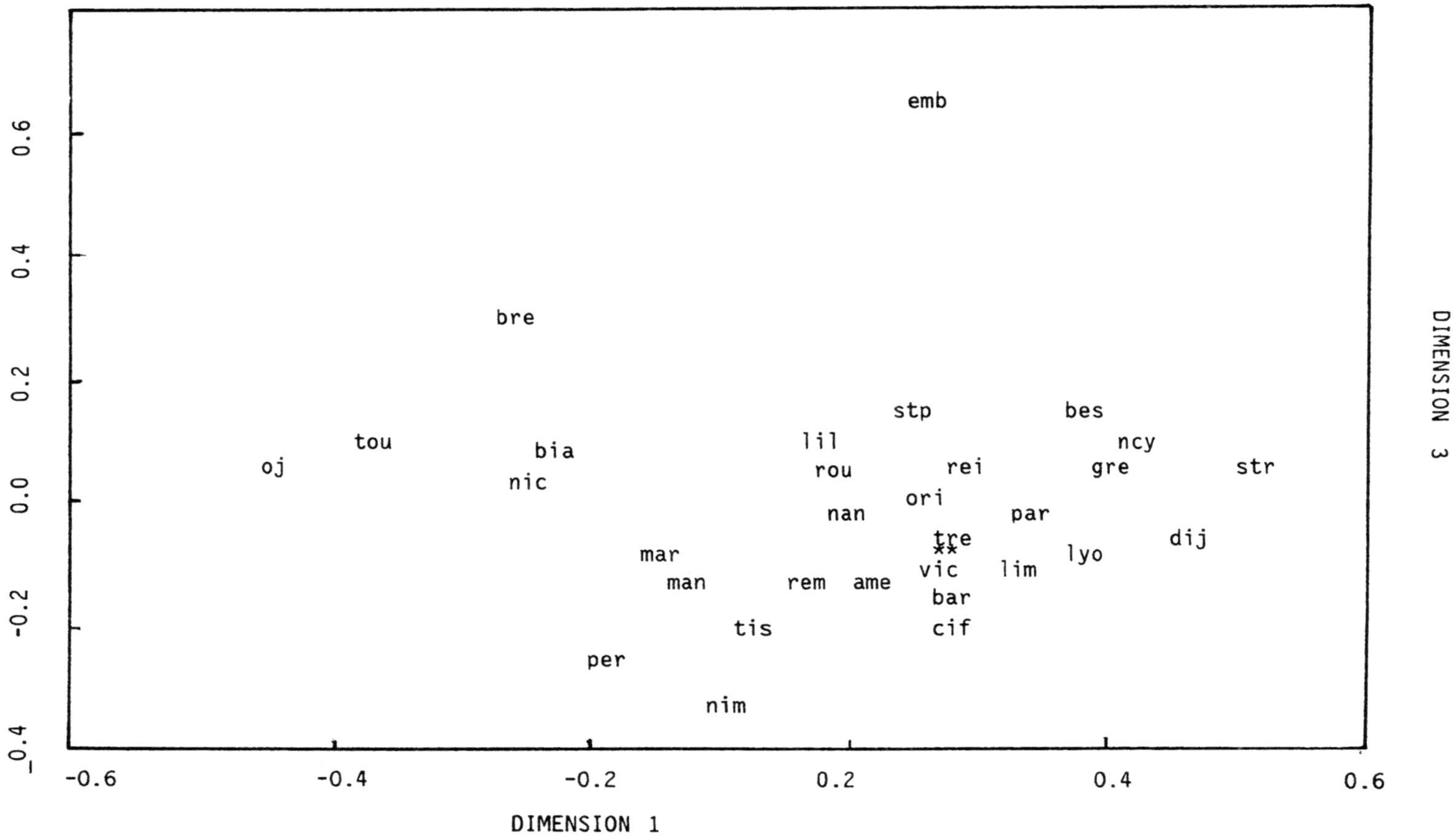

Figure 2a configuration of Frech cities with filtering s=(cos,1); dimension 1 versus dimension 3

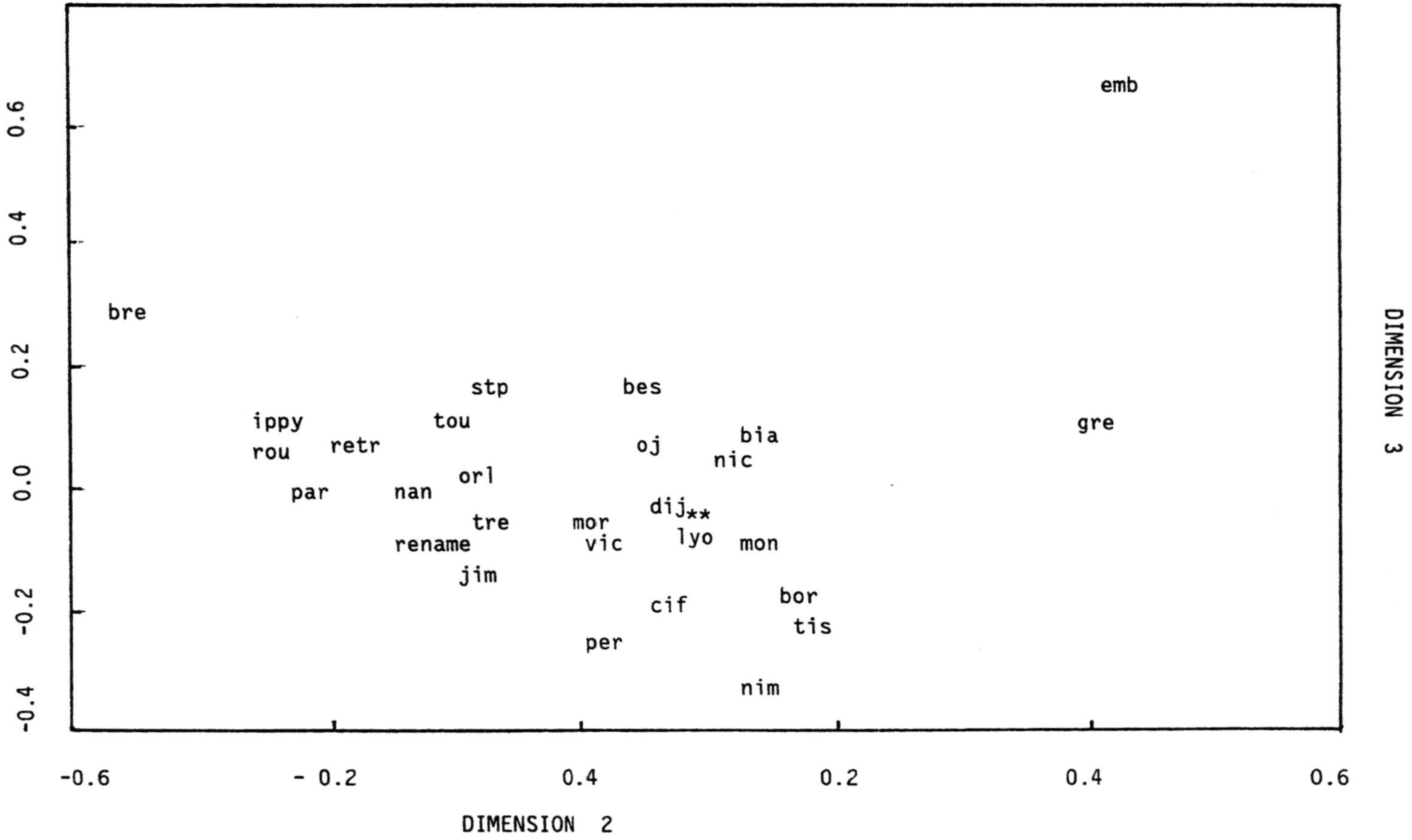

Figure 2b configuration of French cities with filtering of (cos,1);dimension 2 versus dimension 3

Computational Approach of Inference

Sequential Inference Processes in Computational Aspects

By C. Asano, K. Kurihara, and Z. Geng, Fukuoka

[Keywords]
successive inference, pooling data, sequential test and estimation, evaluation and choice of statistical processes, NISAN system

1. Introduction and Summary

Since Wald's stimulative work [11], a number of literature related to several kinds of sequential plans have been presented until today, although there exist difficulty and complication on seeking mathematical properties and concretely numerical evaluation.

In view of statistical computing, however, there exist much more possibility to propose extensively their proposals and also another aspects to be considered. Actually it was surveyed in the recent 300 papers concerning sequential inference that many plans proposed in about 90 papers are possible to be examined by using NISAN system.

The purpose of the present paper is to demonstrate some extensive results of the recent papers and to propose some new sequential plans of another types by using only few commands involved in NISAN system.

2. Applied knowledge-commands in NISAN system

NISAN system has many knowledge commands with several operands, in which the following commands are mainly useful to evaluate properties of sequential inference plans constructed by research workers. SQTP command gives the power of test, average sample number, the variance and expected cost for one-by-one, group- and partial- sampling rules of sequential plans. SQTE command gives estimated values of parameters after a preliminary test process. SQDIST is to give a conditional distribution of a sequential test statistic at the final stage of a closed plan.

3. Classification of sequential inferences

In view of successive plans, there exist three categories [3], depending on the aim of actual situations of statistical inferences. The first is sequential test plans via identical tests, e.g. repeated significance test and SPRT like Wald [11]. The second is given as sequential test plans via

Compstat 1986

difference tests with a similar purpose, e.g. Elder-Muse [5]. The third is given as sequential test plans via different tests with different purposes, e.g. a test process of a t-test after a preliminary F-test, although these examples are rather simple two-stage plans. On the other hand, if the final stage is to estimate a parameter, such plans tend to a kind of sequential estimation, like Asano [2].

Every sequential plan proposed stands naturally on a special background with distribution theory, sampling rule, estimated parameter, pooling of data, open type or closed type, and the decision rule, and more sophisticated plans will be suggestively considered to insist on the superiority. Thus mathematical difficulty arises concretely from numerical evaluation of statistical properties of such plans.

NISAN system has a high flexibility for the requests and is able to give quickly numerical evaluation with good precision.

4. Sequential test plans via identical tests

Plans in this category are most popular, in which a statistical test is repeated by adding samples at every stage, when any decision is reserved.

4.1 Repeated significance tests

Pocock [10] proposed an N-stage group repeated significance test of Type 1 with sample size n for a normal mean test of overall α=5%, and asymptotically recommended $\alpha_1=\alpha_2$=2.9% and 1.6% for N=2 and 5, respectively, to obtain an overall α=5% of t-test. For small sample size, however, actual t-test depends on n, i.e. degrees of freedom. SQDIST command is able to give easily such nominal probabilities. Table 4.1. shows the values of final α_N in case of every α_i=2.9% or 1.6% as Pocock suggested. Table 4.2 shows the values of final α_N in case of every α_i=1% for overall α=5%.

Gould-Pecore [6] has investigated about Type 2 at Pocock's case, and

Table 4.1 The final α_N in case of every α_i=2.9% or 1.6% as Pocock suggested for overall α=5% of t-test.

N \ n	5	20	100
2	0.027	0.029	0.030
5	0.015	0.016	0.016

Table 4.2 The final α_N in case of every α_i=1.0% for overall α=5% of t-test.

N \ n	5	20
3	0.040	0.041
5	0.033	0.034
7	0.025	0.027

Table 4.3 The final α_N in case of every α_i=3.1% or 2.4% as Gould-Pecore suggested for overall α=5% of t-test.

N \ n	5	20	100
2	0.028	0.031	0.032
3	0.019	0.023	0.024

gave α_i=3.1%, 2.4%, as they gave, respectively, and α_A=50%, for overall α=5%. Table 4.3 shows final values α_N corrected for overall α=5%, where Types 1 and 2 are shown in the following figures.

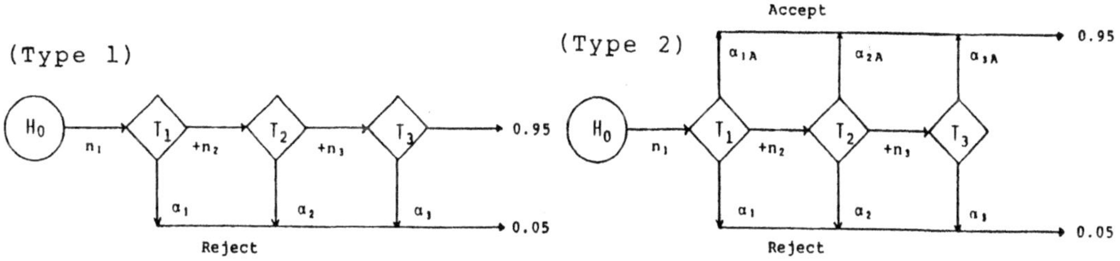

Figure 4.1 Type 1 and Type 2 plans.

4.2 Direct method

Aroian [1] presented a method of numerical evaluation of OC and ASN for normal mean test of a closed type, n_0=7 and 14, basing on one-by-one sampling SPRT, and showed some numerical illustrations. Although his mathematical formulations are rather simple, the numerical evaluation is so complicated that his work seems to be very laborious.

Regarding such kinds of investigation, more extensive studies are able to be proceeded with OC, ASN, S.D.(ASN), and E(cost) for one-by-one-, group-, partial- sampling rules and sample- and sampling-cost, by applying MIXRAN and SQTP commands, where the MIXRAN is for a mixed distribution of several distributions. Table 4.4 shows properties of Aroian's case for a mixed distribution, where p is a mixed rate and * denoted Aroian's illustrated values, which coincide completely with values in Table 4.4.

Table 4.4 The OC, ASN and S.D. of Aroian's case for a mixed distribution.

p	OC	ASN	SD
1.0*	0.898*	4.391*	1.981
0.9	0.834	4.541	2.016
0.8	0.754	4.685	2.051
0.7	0.671	4.763	2.060
0.6	0.592	4.892	2.074
0.5	0.491	4.927	2.050
0.4	0.412	4.851	2.067
0.3	0.324	4.783	2.072
0.2	0.243	4.707	2.049
0.1	0.162	4.559	2.011
0.0*	0.102*	4.391*	1.986

Table 4.5 The OC and ASN for group sampling in Meeker's case.

$p_{1\cdot}$	$p_{\cdot 1}$	p_{11}	OC	ASN
0.10	0.30	0.03	0.830	31.130
0.20	0.30	0.06	0.860	21.710
0.30	0.30	0.09	0.856	18.760
0.40	0.30	0.12	0.857	17.425
0.50	0.30	0.15	0.844	16.880

Meeker [9] studied sequential independence test for 2×2 contingency table on a closed type, basing on one-by-one SPRT. For this matter, Table 4.5 shows more extensive results for a group-sampling, n=5, in Meeker's case.

4.3 Sequential robust test

Lim and Fund [8] proposed a sequential a-trimmed t-test, basing on a closed one-by-one SPRT, where $0 \leq a < 0.5$. In this case, their plan is able to modify to a closed group-sampling repeated significance test with an overall α. Table 4.6 demonstrates only OC for various trim-rates and mixed rates, basing on two stage group-sampling repeated significance test.

Table 4.6 The OC for trim-rates a and mixed rates p, basing on 2 stage group RST.

p \ a	0.0	0.1	0.2
1.00	0.049	0.049	0.051
0.95	0.041	0.047	0.050
0.90	0.040	0.046	0.047
0.85	0.040	0.041	0.045
0.80	0.039	0.037	0.044
0.75	0.038	0.033	0.042
0.70	0.036	0.029	0.041

4.4 Sequential correlation coefficient test

Wolde-Tsadik [12] proposed a sequential test plan for a correlation coefficient, which applied a sequential binomial test based on SPRT, and discussed on OC and ASN by Wald's approximation.

Regarding his plan, an N-stage group-sampling SPRT plan may be proposed with more merit. Figure 4.2 demonstrates only both curves of OC and ASN for $N=2$, $n_1=n_2=20$, and $N=3$, $n_1=20$, $n_2=n_3=10$ by the SQTP.

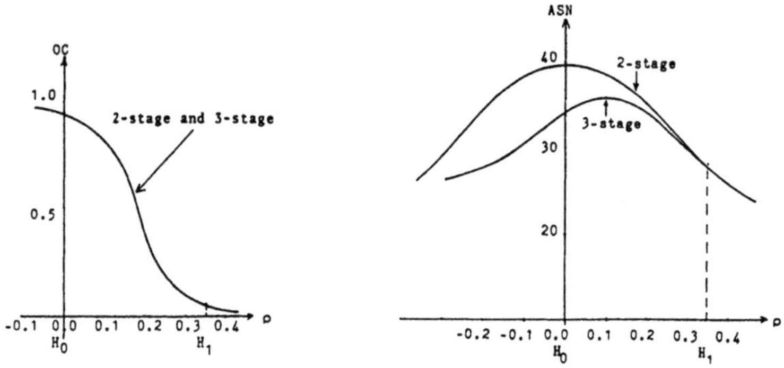

Figure 4.2 The OC and ASN curves for 2, 3 stage group-sampling SPRT.

5. Sequential test plans via different tests with a similar purpose

In view of practical and reasonable reasons, there are some sequential plans which compose of two or three different tests, e.g. both nonparametric and parametric tests, with the similar purpose, e.g. location tests.

Elder and Muse [5] discussed a two stage test plan, including both tests by attributes and by variables, where the exact formulation and the evaluation are rather complicated.

In this case, DEFTP and POWER commands are easily applicable to investigate statistical properties of the plan. Figure 5.1 demonstrates only an OC curve with n=5 and $\alpha_1=\alpha_2=5\%$ in their case.

Another plan is shown in Figure 5.2, where median-, t- and Wilcoxon-tests in order are for location. The DEFTP and POWER commands show an overall $\alpha=6.80\%$, i.e. 2.36% for median test, 3.18% for t-test, 1.26% for Wilcoxon test, to an exponential distribution with mean zero and variance unity, where N=3, $n_1=10$, $n_2=n_3=0$ and $\alpha_1=\alpha_2=\alpha_3=5\%$.

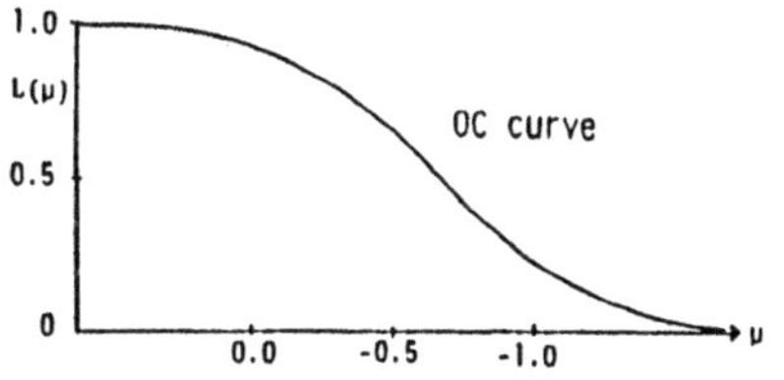

Figure 5.1 The OC curve for Elder-Muse plan.

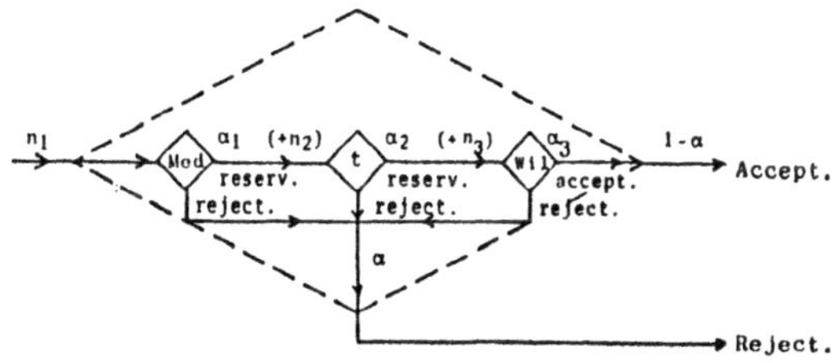

Figure 5.2 A kind of test processes for location.

6. Sequential test plans via different tests with different purposes

One of the simple plans is a double inspection plan with χ^2-test for dispersion and normal test for mean. The two-stage plan and statistical performance are shown in Figure 6.1.

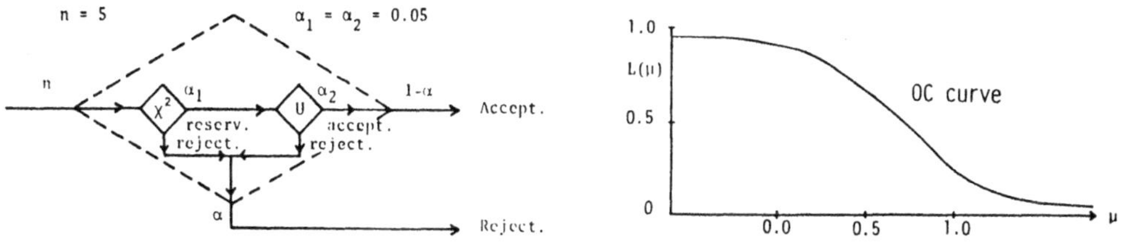

Figure 6.1 A two-stage plan and OC curve.

7. Parameter estimation after a sequential test plans

Plans in the present section have been discussed by many authors, e.g. Kitagawa [7], Bancroft [4], Asano [2].

Such generalized plans are related to pooling methodology of data to estimate parameters, and SQTE command in NISAN system evaluates the properties of estimation based on this plans. Figures 7.1 demonstrates the procedure of the present plan and the operational output, respectively.

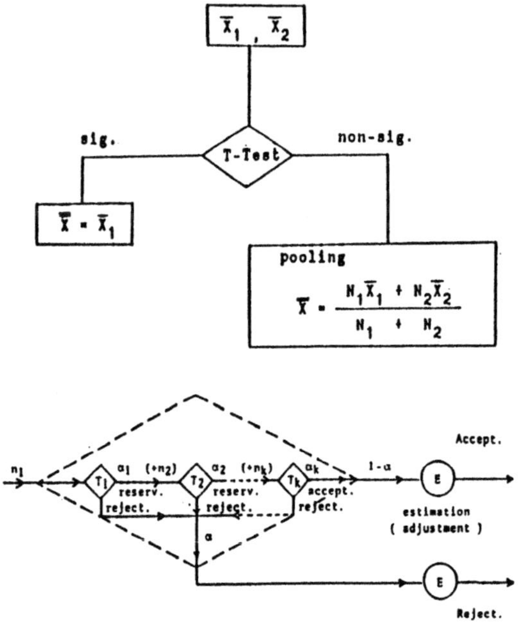


```
NISAN
SQTE TEST(T T T) POP(2) TYPE(NOR) ALPHA(2.21 2.21 2.21)
    I(5) ADD(5 5) PARA(0.0 0.2 1.0 1.0) SIM(5000)

                 1-TEST    2-TEST    3-TEST
 TEST NAME          T         T         T
 SAMPLE SIZE        5        10        15
(REJECT)
 ALPHA           0.0221    0.0221    0.0221
 SIDE               2         2         2

 NUMBER OF POPULATIONS                              2
 TYPE OF DISTRIBUTION                          NORMAL
 NUMBER OF SIMULATION                            5000

 CASE 1)              1-POP.        2-POP.
  1-PARA   :           0.00          0.20
  2-PARA   :           1.00          1.00

NO.  REJECT    RATE     ACCEPT    RATE
T 1     132   0.026          -       -
T 2     126   0.026          -       -
T 3     125   0.026       4617   0.974

 PROB. OF ACCEPT       0.923
 MEAN SAMPLE SIZE     14.610
 S.D. SAMPLE SIZE      1.766

          MEAN                                  0.065
          VARIANCE                              0.064
          M.S.E.                                0.069
          RELATIVE RISK                         0.842
NISAN
```

Figure 7.1 The procedure and computer output.

References

[1] Aroian L.A. (1968): Sequential analysis, Direct method, Technometrics, 10, 125-32.

[2] Asano Ch. (1960): Estimations after preliminary tests of significance and their applications to biometrical researches, Bull. Math. Statist., 9, 1-23.

[3] Asano Ch. Jojima K., Kurihara K. and Ichimura M. (1984): Developmental studies and evaluation of sampling inspection plans on the basis of a software system, Frontiers in Statist. Quality Control, 107-20, Physica-Verlag.

[4] Bancroft T.A. and Han C.P. (1977): Inference based on conditional specification : A note and a bibliography, Intern. Statist. Review, 45, 117-27.

[5] Elder R.S. and Muse H.D. (1982): An approximate method for evaluating mixed sampling plans, Technometrics, 24, 207-11.

[6] Gould A.L. and Pecore V.J. (1982): Group sequential methods for clinical trials allowing early acceptance of H_0 and incorporating costs, Biometrika, 69, 75-80.

[7] Kitagawa T. (1950): Successive process of statistical inferences, (1), Mem. Fac. Sci. Kyushu Univ., Ser. A. 5, 139-80.

[8] Lim T.K. and Fung K.Y. (1980): Sequential trimmed t tests, Biometrika, 67, 181-6.

[9] Meeker W.Q. (1978): Sequential test of independent for 2×2 contigency tables, Biometrika, 65, 85-90.

[10] Pocock S.J. (1977): Group sequential methods in the design and analysis of clinical trials, Biometrika, 64, 191-9.

[11] Wald A. (1947): Sequential Analysis, John Wiley & Sons.

[12] Wolde-Tsadik G.(1976): A generalization of an SPRT for the correlation coefficient, J. Amer. Statist. Assoc., 71, 709-10.

Variance Components as a Method for Routine Regression Analysis of Survey Data

By. N. Longford, Lancaster

ABSTRACT

We discuss a general modelling framework for variance component analysis of data from hierarchical structures. Areas of application include large scale surveys, small area statistics, longitudinal analysis, multivariate data and repeated measurements and experiments. The framework is also applicable for variance heterogeneity modelling.

Keywords: variance components, hierarchical data, survey data, statistical modelling, maximum likelihood.

1. INTRODUCTION

Historically development of variance component models has been associated with analysis of animal and agricultural experiments (Patterson, Thompson (1971), Harville (1977)). The formulation of the general EM algorithm by Dempster, Laird and Rubin (1977) has generated a lot of interest in other fields of application, especially survey data analysis; Aitkin, Anderson and Hinde (1981), Anderson and Aitkin (1985), Dempster, Rubin and Tsutakawa (1981), Mason, Wong and Entwisle (1984). More recently Goldstein (1986) and Longford (1986b) have independently constructed algorithms for maximum likelihood estimation in variance component models with hierarchically nested random effects. These algorithms are not based on EM methodology, and use iterative procedures which converge rapidly and do not require inversion of large matrices.

In description of our modelling framework we will refer to the interactive software produced by the author (Longford 1986a), and based on the Fisher scoring method described in Longford (1986b).

Compstat 1986

We assume that we have observations on N elementary units (individuals), which are grouped into n_2 units (clusters) of level 2. These clusters can be further grouped into n_3 units of level 3, and so on . . ., units of level m-1 grouped into n_m units of the highest level of the nesting hierarchy m. For example, in a nationwide household survey we may have individuals as elementary observations within families (level 2), families within interviewers (level 3), and interviewers within regions (level 4). We may have explanatory variables for the elementary units (age, sex, occupation), for families (housing type, financial and socio-economic status), or for the higher levels of the nesting hierarchy. Of course the response variable has to be a variable defined for elementary units.

2. THE MODEL

In the variance component approach we model two kinds of features of the data. The first kind refers to the systematic explanation of the variability and describes the relationship of the y-variate on the explanatory variables for the 'typical', or average elementary unit in the sample/population. The parametric representation of this relationship is in terms of a linear regression formula, in complete analogy with the ordinary least squares models.

The other feature of interest in data with hierarchical structure is the variability of this relationship among the units on the various levels of the hierarchy. While in OLS models variability is completely described by the variance of an observation, in variance component models every level of the hierarchy is associated with a variance. For every level this variance may be constant, or in general a non-negative quadratic function of some of the explanatory variables. The general parametric description consists of the variance attributable to the elementary units (analogous to the OLS variance) and variance matrices corresponding to the levels of the nesting hierarchy.

As an example of a variance component model with 3 levels of hierarchy we consider the model

$$y_{ijk} = \beta_0 + \beta_1 x_{ijk}^{(1)} + \beta_2 x_{ijk}^{(2)} + \beta_3 x_{ijk}^{(3)}$$

$$+ \xi_{3k}^{(0)} + \xi_{3k}^{(1)} x_{ijk}^{(1)} + \xi_{3k}^{(2)} x_{ijk}^{(2)}$$

$$+ \xi_{2jk}^{(0)} + \xi_{2jk}^{(1)} x_{ijk}^{(1)} \qquad + \xi_{2jk}^{(3)} x_{ijk}^{(3)}$$

$$+ \epsilon_{ijk}, \tag{2.1}$$

where i is the identification for the elementary units (individuals), j for the units of level 2 (say, towns), and k for the units of level 3 (countries); the β's are unknown constants, and the ξ's are normally distributed random variables - they are further specified in the next paragraph.

The first line of (2.1) is the fixed or systematic part of the model describing features present in the data in all the units. The second line models the variability of the relationship of the y-variate on the explanatory variables from country to country, the third line describes the variability among the towns of a country. The vectors $\underline{\xi}_{2jk} = (\xi_{2jk}^{(0)}, \xi_{2jk}^{(1)}, \xi_{2jk}^{(3)})$, $k = 1, \ldots, n_3$, $j = 1, \ldots, n_{3k}$, and $\underline{\xi}_{3k} = (\xi_{3k}^{(0)}, \xi_{3k}^{(1)}, \xi_{3k}^{(2)})$, $k = 1, \ldots, n_3$ are assumed to be two independent random samples from $N(\underline{0},\Theta_2)$ and $N(\underline{0},\Theta_3)$, respectively, and they are independent of the random sample of $\langle\epsilon_{ijk}\rangle \sim N(\underline{0},\sigma_e^2)$. The variance matrices Θ_2 and Θ_3 are unknown positive semi-definite matrices.

The variance of an observation consists of three additive components:

$$\mathrm{var}(y_{ijk}) = \sigma_e^2 + \underline{z}_{ijk}^{(2)\,T}\Theta_2\,\underline{z}_{ijk}^{(2)} + \underline{z}_{ijk}^{(3)\,T}\Theta_3\,\underline{z}_{ijk}^{(3)}$$

$$= \sigma_e^2 + \sigma_2^2(\underline{z}_{ijk}^{(2)}) + \sigma_3^2(\underline{z}_{ijk}^{(3)}) , \tag{2.2}$$

where $\underline{z}_{ijk}^{(2)} = (1, x_{ijk}^{(1)}, x_{ijk}^{(3)})$ and $\underline{z}_{ijk}^{(3)} = (1, x_{ijk}^{(1)}, x_{ijk}^{(2)})$. The covariance of two observations in the same unit of level 2 is $\sigma_2^2 + \sigma_3^2$, the covariance of two observations in different units of level 2 but in the same unit of level 3 is σ_3^2.

3. RANDOM EFFECTS ANOVA AND RANDOM PARAMETER MODELS

The variance (2.2) is constant only if var $(\xi_{3k}^{(1)})$ = var $(\xi_{3k}^{(2)}) = 0$ and var $(\xi_{2jk}^{(1)})$ = var $(\xi_{2jk}^{(3)}) = 0$, in which case (2.1) collapses to the random effects ANOVA model

$$y_{ijk} = \beta_0 + \beta_1 x_{ijk}^{(1)} + \beta_2 x_{ijk}^{(2)} + \beta_3 x_{ijk}^{(3)} + \epsilon_{ijk} + \xi_{2jk} + \xi_{3k} \tag{3.1}$$

In general the variance component model (2.1) involves variance heterogeneity. The interpretation of variable relationships in (2.1) is suitable only if the x-variables involved in the town-level random part $(\underline{z}^{(2)}\underline{\xi}_2^T)$ are not constant within all the towns, and the variables involved in the country-level random part $\underline{z}^{(3)}\underline{\xi}_3^T$ are not constant within all the countries; otherwise these variable regression slopes (random slopes) cannot be identified. However the variance heterogeneity interpretation is always applicable, even if, say, $x^{(1)}$ in (2.1) is defined for countries. As a special case we have a simple random parameter model

$$y_j = \beta_0 + \xi_j^{(0)} + (\beta_1 + \xi_j^{(1)})\, x_j^{(1)}, \tag{3.2}$$

which is obtained from (2.1) by having only 1 country, and every town j represented by a single individual; indices k=1 and i=1 are left out. The model (3.2) does not involve any hierarchies but it models the variance of y_j as a quadratic function of $x^{(1)}$:

$$\text{var}\,(y_j) = \sigma_0^2 + \sigma_{12} x_j^{(1)} + \sigma_1^2\,[x_j^{(1)}]^2.$$

4. MODEL SPECIFICATION

Model specification and selection in the variance component approach is substantially more complex than in the OLS framework. Firstly, we select the variables to be included in the fixed part of the model - this is the direct analogue of the OLS model selection. Secondly, for every level $\ell = 2, \ldots, m$ of the hierarchy we sub-select from the fixed part the variables to be included in the random part of that model. It is clear that even with a small number of variables and small number of levels of nesting the variety of possible models and numbers of parameters involved in them increase very rapidly, and fitting all possible models is not a feasible proposition. As a 'rule of thumb' we may presume that if a variable is important or dominant in description of systematic relationships in the data (fixed part) then the corresponding slope may also vary among units of the various levels, and so the variable should be considered for inclusion in the random part on all or at least some of the levels.

In the matrices Θ_2 and Θ_3 the diagonal elements (variances) are of primary interest, the covariances are usually treated as nuisance parameters. In a complex model retaining of all the covariances may lead to a large number of parameters; on the other hand all the covariances cannot be constrained to 0, otherwise certain invariance properties of the model are violated. In the VARCL software discussed in Section 5 all the intercept-by-slope covariances are estimated as free parameters subject to constraints of non-negative definiteness, and all the slope-by-slope covariances are constrained to 0.

5. SOFTWARE

Details of a Fisher scoring algorithm for maximum likelihood estimation in variance component models are given in Longford (1986b). The interactive software VARCL based on this algorithm has been used in Aitkin and Longford (1986).

At the beginning of a session the user has to specify a 'maximal' model - maximal lists of variables for the fixed part and random parts of the models considered for fitting. The the data are input and certain within level-2-unit cross-products are calculated. These statistics are a set of sufficient statistics for all the submodels of the maximal model; in the iterations of the computational procedure data on individuals are not used at all. This also enables the user to dump, and later restore, a session of VARCL. The dump-files are usu-ally much smaller than the original data files and restoring a dump-file saves the user repeating the initial specifications and input of data. An arbitrary number of submodels of the maximal model can be fitted within a session; alteration of the parts of the model is carried out by menu-driven conversation.

The output from a model fit contains:

1. Estimates of the fixed effects parameters and their standard errors.

2. Estimates of variances, their square roots and st.errors of these.
3. Estimates of covariances (slope * intercept) and st. errors.
4. Value of the deviance (-2*log-likelihood).
5. Optionally, conditional means of the random effects (ξ's).

The present implementation of the software can handle data with up to 3 levels of nesting, up to 200.000 units of input data and calculated variables, 200.000 crossproducts (fixed-effects * random-effects * groups), 5000 units on levels 2 and 3, 40 fixed-effect parameters, 10 variances.

A separate program has been developed for random effects ANOVA with up to 9 levels of hierarchy.

REFERENCES

Aitkin, M., Anderson, D., and Hinde, J. (1981), Statistical modelling of data on teaching styles (with Discussion). JRSS A 144, 419 - 461.

Aitkin, M., and Longford, N. (1986), Statistical modelling issues in school effectiveness studies. JRSS A 149, in press.

Anderson, D., and Aitkin, M. (1985), Variance component models with binary response: interviewer variability. JRSS B 47, 203 - 210.

Dempster, A.P., Laird, N.M., and Rubin, D.B. (1977), Maximum likelihood for incomplete data via the EM algorithm. JRSS B 39, 1 - 38.

Dempster, A.P., Rubin, D.B., and Tsutakawa, R.K. (1981), Estimation in covariance components models. JASA 76, 341 - 353.

Goldstein, H. (1986), Multilevel mixed linear model analysis using iterative generalized least squares. Biometrika 73, in press.

Harville, D.A. (1977), Maximum likelihood approaches to variance component estimation and to related problems. JASA 72, 320 - 340.

Longford, N.T. (1986a) VARCL Manual. Centre for Applied Statistics, Lancaster University, Lancaster, U.K.

Longford, N.T. (1986b) A fast scoring algorithm for maximum likelihood estimation in unbalanced mixed models with nested random effects. Submitted to Biometrika.

Mason, W.M., Wong, G.Y., and Entwisle, B. (1984) The multilevel linear model: A better way to do contextual analysis. Sociological Methodology, Jossey Press, London.

Patterson, H.D., and Thompson, R. (1971) Recovery of inter-block information when block sizes are unequal. Biometrika 58, 545 - 554.

On the Assessment of Case Influence in Generalized Linear Models

By G. U. H. Seeber, Innsbruck

SUMMARY: This paper deals with likelihood distances as proposed by Cook and Weisberg (1982) as a general approach to the assessment of case influence. Approximate solutions are given in the framework of generalized linear models. An example demonstrates their implementation into GLIM.

KEYWORDS: Regression diagnostics, likelihood distances, generalized linear models, GLIM.

1. INTRODUCTION

Statistical modelling is iterative in nature: a model is formulated, fitted to data, and critically evaluated, eventually leading to the formulation of a new model. During the third model criticism step questions typically to be asked include: Are there any observations not well explained by the model, pointing out general inadequacies ?, or: Is the result of a fit unduly influenced by a single observation, i.e. are conclusions drawn from a fit substantially affected by the deletion of one data point ? Methods of residual analysis are in common use to treat questions like the first. One approach to case analysis, referred to by the second question, uses the following procedure: Compute estimates from slightly perturbed (e.g. by deletion) data and compare the result with the original estimates by means of some distance function. For linear models Cook distances are accepted as standard measures which are easy to interpret and not costly to compute.

Section 2 introduces likelihood distances as a generally applicable distance function. These reduce to a (known) monotone function of Cook distances for linear models. Calculations may be time consuming with nonlinear models, since an iterative fit is

Compstat 1986

required for each case deleted. Section 3 provides approximate solutions for generalized linear models. An example is given in section 4.

2. LIKELIHOOD DISTANCES

The likelihood function contains information on both the stochastic mechanism producing the data and the observed values. Given the data its logarithm is considered as a measure of evidence given to parameter values by the data. This view is reflected by the maximum likelihood principle according to which the most supported parameter should be chosen as an estimate. In the context of case analysis we are faced with the problem of comparing a full data estimate $\hat{\theta}$ with an estimate $\hat{\theta}_{(i)}$ obtained by the deletion of the i-th case. It seems natural to choose

$$LD_i = -2\cdot\log(L(\hat{\theta}_{(i)})/L(\hat{\theta})) = 2\ (l(\hat{\theta})-l(\hat{\theta}_{(i)}))$$

where L and l are the likelihood function and its logarithm, respectively. These quantities have been proposed by Cook and Weisberg (1982) as a general approach to the assessment of case influence and have been named likelihood distances. The formal similarity to the likelihood ratio statistic is obvious. Based on its asymptotic chi-squared distribution likelihood distances have an immediate interpretation in terms of asymptotic confidence regions.

If $\theta = (\theta_1,\theta_2)$ contains a nuisance parameter θ_2 the definition is easily modified using the profile likelihood function

$$L_{\theta 2}(\theta_1) = \sup_{\theta 2}\ L(\theta_1,\theta_2).$$

This is of particular interest when dealing with the scale parameter in generalized linear models.

3. APPROXIMATIONS

Let Y have the log-likelihood function

$$l(\theta,\phi) = \sum_i [y_i\cdot\theta_i - b(\theta_i)]\ /\ a(\phi) + c(y_i,\phi)$$

with canonical parameter θ, scale parameter ϕ, and some specific functions b, a, and c. Assuming a generalized linear model, $\mu_i = E(Y_i) = b'(\theta_i)$ is related to a linear predictor $\eta_i = \sum_j \beta_j \cdot x_{ij}$ by the link function $\eta_i = g(\mu_i)$. Maximum likelihood estimates for the regression parameter β are obtained by an iteratively weighted least squares algorithm. For the details see McCullagh and Nelder (1983).

The following procedure yields one step approximations to parameter estimates. Let $f: R^k \longrightarrow U \subset R^n: \beta \longmapsto E_\beta(Y)$ be the map carrying over regression parameters to expectations of Y. The image U of f forms a smooth k-dimensional submanifold.If $\hat{\beta}$ is the ML-estimate, the vector $f(\hat{\beta})$ of fitted values is the orthogonal projection of the vector y of observed values onto the tangent hyperplane $H_{\hat{\beta}} = \{[f(\hat{\beta}) - Df(\hat{\beta}) \cdot \hat{\beta}] + Df(\hat{\beta}) \cdot \beta;\ \beta \in R^k\}$ on U at $f(\hat{\beta})$. Orthogonality is understood with respect to the scalar product defined by the (diagonal) variance/covariance matrix of Y. Related to this projection is the linear regression problem with dependent variable $y - f(\hat{\beta}) + Df(\hat{\beta}) \cdot \hat{\beta}$, a design matrix equal to the Jacobian Df evaluated at $\hat{\beta}$, and weights equal to the estimated variances of the Y_i's. Well known diagnostic tools applied to this new problem lead to generalizations of linear regression diagnostics to nonlinear situations. Moolgavkar, Lustbader and Venzon (1984) used this approach to construct a hat matrix, Cook distances, and related quantities for exponential families. Here, the linear problem is used to obtain approximations to the $\hat{\beta}_{(i)}$'s, either by use of an updating formula or by n non-iterative fits.

Calculations may furthermore be simplified, if models with canonical links (i.e. canonical parameters equal linear predictors, $\theta_i = \eta_i$) are used. Then $f(\beta) = g^{-1}(\eta)$, and $Df(\hat{\beta}) \cdot \beta = Dg^{-1}(X \cdot \hat{\beta}) \cdot X \cdot \beta$. Now, the above mentioned linear regression problem may be linearily transformed using $Dg(f(\hat{\beta}))$ as transformation matrix. This results in the weighted least squares problem with adjusted dependent variable $X \cdot \hat{\beta} + Dg(f(\hat{\beta})) \cdot (y - f(\hat{\beta})) = \hat{\eta} + Dg(\hat{\mu}) \cdot (y - \hat{\mu})$, design matrix X, and the same weights as above. But this is identical to performing one further iteration by the IWLS algorithm. So a single step with starting values obtained by the original full data fit and observation i deleted yields the same approximations as we had before. In GLIM this is easily accomplished by the use of the $weight and the $recycle directives as demonstrated in the next section.

4. SHIP DAMAGE INCIDENTS DATA

McCullagh and Nelder (1983; p. 137) reproduce data on the number of reported damage incidents (count) and aggregate months of service (serv) by ship type (ship), year of construction (cons) and period of operation (perd). They fitted a model including main effects and the logarithm of service period as an offset with log link and Poisson errors. To account for over-dispersion they assumed that Var(count) = $\sigma^2 \cdot$E(count) for some $\sigma^2 > 1$. Parameter estimates obtained by the sequence of GLIM directives

```
$ca lser=%log(serv)  $off lser  $link l  $err p
$scale 0  $fit ship+cons+perd  $dis e
```

coincide with those given by McCullagh and Nelder with the exception $\hat{\sigma}^2$ = 1.548 (estimated by the mean deviance). This leads to slightly different results for the standardized residuals sres

```
$ca sres=(%yv-%fv) / %sqr(%sc*%fv)
```

as well. To compute exact estimates for case %i being deleted use

```
$ca we=1  : we(%i)=0  $we we  $fit ship+cons+perd $
```

then twice the negative log-likelihood may be computed by

```
$extract %pe
$ca off=%pe(1)+%pe(2)*(ship==2)+ ... +%pe(9)*(perd==2)+lser
$off off  $fit nul-1
$ca ldis(%i)=%dv-%d  $
```

where nul is a vector containing zeroes only and the deviance of the full data model was saved into %d. Now, assume that vector fv contains the fitted values under the original model. Since the canonical link is in use, approximations to the parameter estimates are obtained by recycling just one step:

```
$ca we=1  : we(%i)=0  $we we  $ca %fv=fv
$recycle 1 1  $fit ship+cons+perd $
```

no	LD	app.LD	SRES
1	0.0021	0.0021	-0.3682
2	0.0012	0.0012	-0.3243
3	0.0204	0.0206	-0.2665
4	0.1001	0.1025	-0.4645
5	0.0064	0.0063	0.1143
6	0.0226	0.0229	-0.0826
8	0.7850	0.7277	0.6679
9	2.7397	3.0535	-0.4971
10	1.7810	1.6853	0.7837
11	0.7419	0.7215	0.3127
12	1.9856	2.0479	-0.4910
13	0.5022	0.5124	-0.6896
14	0.3863	0.3815	0.2680
16	0.9122	0.8455	0.4487
17	0.0000	0.0001	0.0168
18	0.0130	0.0127	0.3198
19	0.1989	0.2203	-0.9174
20	0.0550	0.0581	-0.4098
21	3.5756	2.3128	2.9958
22	2.7277	4.1981	-1.1732
24	0.0299	0.0290	0.5251
25	0.0132	0.0135	-0.4983
26	0.0050	0.0051	-0.3906
27	0.0689	0.0718	-0.7563
28	0.0642	0.0669	-0.7484
29	0.0588	0.0565	0.5761
30	6.2161	4.0844	1.5684
32	2.7923	3.5877	-0.9721
33	0.0006	0.0006	-0.2579
35	0.7794	0.7158	1.4254
36	1.0707	0.9587	1.8967
37	0.0679	0.0699	-0.3280
38	0.3038	0.3203	-0.8835
40	0.3038	0.3203	-0.8859

Table 1: Case number, exact and approximate likelihood distances, and standardized residuals for ship damage incidents data.

Directives may conveniently be arranged into macros to loop through the cases, but care has to be taken to the correct setting of weight and offset variables.

Table 1 gives exact and approximate likelihood distances and standardized residuals. It is not surprising that approximations are the poorer the larger the distances are. As a result of the analysis attention is drawn to case no. 30, which potentially hides a difference between ships of type A and D. This case deleted, the main effect on the level corresponding to type D is nearly ten times larger than for the full data model (-0.74, s.e. 0.52).

REFERENCES:

Cook R.D., S. Weisberg (1982): Residuals and Influence in Regression. New York, London: Chapman and Hall.

McCullagh P., J.A. Nelder (1983): Generalized Linear Models. London, New York: Chapman and Hall.

Moolgavkar S.H., E.D. Lustbader, D.J. Venzon (1984): A Geometric Approach to Nonlinear Regression Diagnostics with Application to Matched Case Control Studies. Annals of Staistics 12, 816-826.

Numerical Aspects of Statistical Computations

Algoritmic Development in Variable Selection Procedures

By J. Antoch, Prague

Summary : Robust procedure for variable selection in linear model closely related to the all possible model approach is described. Method combines idea of α-acceptability and principle of coherence and use the class of γ-trimmed least squares estimators as the basis for evaluation of estimates.

Key words : Linear model, variable selection, γ-trimmed least squares estimator, α-acceptability, principle of coherence.

Motto : All models are wrong, but some of them are useful (G.Box).

1. Introduction

Issues of model (variable) selection arise in many statistical problems which deal with a large number of observed variables and there is often desire to find a relatively simple class of functions that model some underlying phenomenon. Whenever dealing with this problem one should always keep in mind following two simple, nevertheless key, views.

On the one hand, model (variable) selection is only a part of the more general problem of analyzing the structure and character of the data and by far not the first one. Therefore, variable selection ought to be performed only in the later stage of data analysis and to follow iterative application of the procedures checking for influential cases, providing residual diagnostic, transformations, detecting possible collinearity etc. Literature provides us a host of material concerning this problem. On the other hand, the criterion used to decide on the appropriate model(s) should be related to the intended use.

Through this paper we shall assume that we have data on a large number of potential variables which include all relevant ones and, possibly, some other extraneous ones which could be removed without no serious harm done. For simplicity, linear model is used for description of the situation.

In spite of preliminary analysis one can still have a suspicion that several outliers may occur in avaiable data. It is

Compstat 1986

well known that if classical least squares are used to fit the models, detailed subset selection analysis may be only meaningless excercise in such a case. Therefore, proposed procedure is based on a class of robust L-estimators, spec. γ-trimmed least squares estimator, which are used for fitting in every submodel. As was shown in numerical studies, these estimators behave quite analogously as the classical least squares when the distribution of errors is normal, but their performance is much better when outliers occur. Besides, our approach combines the idea of α-acceptability for determination of a class of reasonable submodels with the fast searching algorithm based on a principle of coherence enabling us to select as much as suitable submodels. All ideas, although presented for robust variable selection in linear model, can be modified for other purposes, too.

2. Robust α-acceptable estimates in linear model

Let us assume model

$$\underset{\sim}{Y} = \underset{\sim}{X}\underset{\sim}{\beta} + \underset{\sim}{e} , \tag{2.1}$$

where $\underset{\sim}{X}=(x_{ij})_{i=1,\dots,n}^{j=0,\dots,p}$ is a full rank design matrix, $\underset{\sim}{Y}=(Y_1,\dots,Y_n)'$ is an observable random variable, $\underset{\sim}{\beta}=(\beta_0,\dots,\beta_p)'$ is a vector of unknown regression coefficients and $\underset{\sim}{e}=(e_1,\dots,e_n)'$ is a vector of i.i.d. errors. Moreover, let us suppose $x_{i0}=1$, $i=1,\dots,n$ and let $\underset{\sim}{x}_i'$ denotes i-th row of the matrix $\underset{\sim}{X}$. For the model selection we shall assume that last q variables will be eliminated (otherwise we renumerate them). In such a case following partition of $\underset{\sim}{X}$, resp. $\underset{\sim}{\beta}$, is suitable

$$\underset{\sim}{X} = ({}_r\underset{\sim}{X} | {}_q\underset{\sim}{X}), \text{ resp. } \underset{\sim}{\beta} = \begin{pmatrix} {}_r\underset{\sim}{\beta} \\ {}_q\underset{\sim}{\beta} \end{pmatrix} , \tag{2.2}$$

where ${}_r\underset{\sim}{X}$ is nxr matrix, $r=p+1-q, 1\leq q\leq p$, etc., resp. use of the model

$$\underset{\sim}{Y} = {}_r\underset{\sim}{X}\,{}_r\underset{\sim}{\beta} + \underset{\sim}{e} \tag{2.3}$$

is often useful.

As was mentioned before, we are looking for a robust procedure. Therefore, the basic estimator of $\underset{\sim}{\beta}$ will not be the classical least squares estimator, but γ-trimmed least squares estimator (γ-TLSE) as the most important and the most typical representant of the class of robust L-estimators in (2.1). This estimator was suggested by Koenker and Bassett (1978) in the following way. Let us fix γ, $0\leq\gamma<1/2$. Then γ-regression quantile $\hat{\underset{\sim}{\beta}}(\gamma) = (\hat{\beta}_0(\gamma),$

$\dots,\hat{\beta}_p(\gamma))'$ corresponding to (2.1) is defined as that vector $\underset{\sim}{b}\in R_{p+1}$ which solves $\min\{\underset{\sim}{b}\in R_{p+1} \mid \sum_{i=1}^{n} \varphi_\gamma(Y_i-\underset{\sim}{x}_i'\underset{\sim}{b})\}$, where $\varphi_\gamma(x)=x.(\gamma-I[x<0])$, $x\in R_1$. Let $\underset{\sim}{D}$ be diagonal matrix such that

(2.4) $$d_{ii}=\begin{cases} 0 & Y_i<\underset{\sim}{x}_i'\hat{\underset{\sim}{\beta}}(\gamma) \text{ or } Y_i>\underset{\sim}{x}_i'\hat{\beta}(1-\gamma), \\ 1 & \text{otherwise} \end{cases} \qquad i=1,\dots,n.$$

Then γ-TLSE $\hat{\underset{\sim}{\beta}}(\gamma T)=(\hat{\beta}_0(\gamma T),\dots,\hat{\beta}_p(\gamma T))'$ is the ordinary least squares estimator computed from the model $\underset{\sim}{D}\underset{\sim}{Y}=\underset{\sim}{D}\underset{\sim}{X}\underset{\sim}{\beta}+\underset{\sim}{D}\underset{\sim}{e}$, i.e.,

(2.5) $$\hat{\underset{\sim}{\beta}}(\gamma T) = (\underset{\sim}{X}'\underset{\sim}{D}\underset{\sim}{X})^{-}\underset{\sim}{X}'\underset{\sim}{D}\underset{\sim}{Y}\ .$$

Put

(2.6) $$s^{*2}(\gamma) = (1-2\gamma)^{-2}.\{S^{*2}/(n-p-1) + \gamma(1-\gamma)(c_1-c_2)\}\ ,$$

where $S^{*2}=\underset{\sim}{Y}'\underset{\sim}{D}(\underset{\sim}{I}-\underset{\sim}{X}(\underset{\sim}{X}'\underset{\sim}{D}\underset{\sim}{X})^{-}\underset{\sim}{X}')\underset{\sim}{D}\underset{\sim}{Y}$, $c_1=\hat{\beta}_0(\gamma)-\hat{\beta}_0(\gamma T)$, $c_2=\hat{\beta}_0(1-\gamma)-\hat{\beta}_0(\gamma T)$.

Following definition of α-acceptability gives us an idea for evaluation of more simple, but statistically undistinguishable, alternatives to the basic full model estimator.

Definition 2.1. Let $0\leqq\gamma<1/2$ and $0<\alpha<1$ are fixed constants. Suppose that $\hat{\underset{\sim}{\beta}}(\gamma T)$ is γ-TLSE computed from the full model (2.1) and m is the number of observations which have been removed by trimming.

(a) Values of $\underset{\sim}{b}\in R_{p+1}$ such that

(2.7) $$A^{*}(\underset{\sim}{b}) = (\hat{\underset{\sim}{\beta}}(\gamma T)-\underset{\sim}{b})'(\underset{\sim}{X}'\underset{\sim}{D}\underset{\sim}{X})(\hat{\underset{\sim}{\beta}}(\gamma T)-\underset{\sim}{b}) \leqq \leqq (p+1).s^{*2}.F^{1-\alpha}_{p+1,n-m-p-1} = a^{*}(\alpha)\ ,$$

where $F^{1-\alpha}_{k,l}$ is the $(1-\alpha)$-th quantile of F distribution with k and l degrees of freedom, are called α-acceptable.

(b) For any $\underset{\sim}{b}\in R_{p+1}$ the α such that

(2.8) $$A^{*}(\underset{\sim}{b}) = a^{*}(\alpha)$$

is called its acceptance level.

Remarks : (a) For any estimator $\hat{\underset{\sim}{\beta}}$ the value of $A^{*}(\hat{\underset{\sim}{\beta}})$ describes us distance between this estimator and the full model estimator. On the other hand, corresponding acceptance level provides us another look on this idea, i.e., the extreme level on which both these estimators are statistically undistinguishable (from the hypothesis testing point of view).

(b) Note that for any $0<\alpha<1$ fixed, confidence elipsoid

(2.9) $$B^{*}_{\alpha} = \{\underset{\sim}{b}\in R_{p+1} \mid A^{*}(\underset{\sim}{b})\leqq a^{*}(\alpha)\ \}$$

for $\underset{\sim}{\beta}+\underset{\sim}{\delta}(\gamma)$, where $\underset{\sim}{\delta}(\gamma)$ is a vector whose components are all zero except for the first one, has asymptotic confidence coefficient $1-\alpha$. Morover, if $\alpha<\alpha'$ then $B^*_{\alpha'} \subset B^*_{\alpha}$.

(c) For $\gamma=0$, $\hat{\underset{\sim}{\beta}}(OT)$ is classical least squares estimator $\hat{\underset{\sim}{\beta}}(LS)$, no observations are trimmed off and, if $\Lambda(\underset{\sim}{e})\sim N(\underset{\sim}{0},\sigma^2\underset{\sim}{I})$, we shall receive the exact confidence region for $\underset{\sim}{\beta}$, i.e.,

$$(2.10)\qquad B_\alpha = \{\underset{\sim}{b}\in R_{p+1} \mid (\hat{\underset{\sim}{\beta}}(LS)-\underset{\sim}{b})'(\underset{\sim}{X}'\underset{\sim}{X})(\hat{\underset{\sim}{\beta}}(LS)-\underset{\sim}{b}) \le \frac{p+1}{n-p-1} RSS(\hat{\underset{\sim}{\beta}}(LS)) \,.\, F^{1-\tilde{\alpha}}_{p+1,n-p-1}\}.$$

Idea described above enables us to define a class of regression equations, each member of which is not significantly poorer than the equation for the full model (when α is fixed). Among them there is a number of suitable subset candidates. This, by itself, does not solve the problem how to find the "best" equation. Let us notice that from hypothesis testing point of view, there cannot exist a "best" model (equation). Therefore, some additional procedure for choice among avaiable comparable models is necessary.

Moreover, computation of robust estimators, even the simplest ones, is rather costly - in any case substantially more costly than computation of ordinary least squares. So, much more than at some other time, one will appreciate the efficient procedure for model selection which considerably reduce the number of models to be fitted, especially when one works, as we do, with the analog of all possible model method. One such procedure is described in following section.

3. Principle of coherence and basic algorithm for model selection

One suitable approach, based on a general principle of statistical inference, so called coherence, was recently introduced by Edwards and Havránek (1985 b). Their procedure, as well as the idea of α-acceptability, is applicable in wide range of statistical disciplines. In this paper we shall present a short sketch of their approach for model selection from an arbitrary family F of models . In section 4. their procedure will be used for our purposes, i.e., how to choose from the set of α-acceptable submodel estimates.

Principle of coherence was introduced by Gabriel(1969) in a following way :

"...Procedure involving testing a set of models ought not accept a model while rejecting a more general model..."

This principle immediately implies following two rules (we suppose that the coherent procedure is used):

Rule I. If a model is accepted than all models that include it are considered to be accepted (we shall call such models weakly accepted and denote them w-accepted).

Rule II. If a model is rejected than all its submodels are considered to be rejected (we shall call such submodels weakly rejected and denote them w-rejected).

These two rules enable us to reduce drastically the number of models to be fitted because we can concentrate our attention on two suscint classes of them,i.e.,

A = (set of the simplest accepted models in F), resp.

R = (set of the most complex rejected models in F).

All other models can be classified as either w-accepted or w-rejected.

Of course, given test may or may not follow this principle. Nevertheless, when the violation of this principle is only exceptional, one can call this procedure almost coherent and use it instead of the coherent one having in the mind,that it may happen that the solution will be non-unique. In any case, the procedure can be repeated using a different starting point.

The important common feature of considered approach is the fact, that multiple models are selected both to A and R in contrast to commonly used stepwise procedures. If one looks at the models as on the working hypotheses (and we should do so), this fact may be viewed rather advantageous. Instead of one parsimonious model (so much appreciated by non-statisticiens) with possibility to overlook some other, equally plaussible ones, we have a set of models which can be regarded as equivalent alternative interpretations of the data and should be evaluated further only after careful consultations with client. The choice rests with the experimenter, using the context of the problem, to decide which of the adequate equations is most appropriate. In the absence of any other considerations one of the minimal adequate equations might be considered.

Determination of the sets A and R make possible construction of a-dual and r-dual of a set of models.

Definition 3.1. Let F be the set of all possible models and $S \subset F$.

(a) The a-dual of S, denoted $D_a(S)$, is the set of simplest models in F that are not contained in any model in S.

(b) The r-dual of S, denoted $D_r(S)$, is the set of the most complex models in F that do not contain any model in S.

Remarks: (a) If S contains rejected models, than $D_a(S)$ consists of the simplest models in F which could eventually be accepted.

(b) If S contains accepted models, than $D_r(S)$ consists of the most complex models in F which could eventually be rejected.

The basic algorithm for determination of sets A and R follows

Algorithm : (Edwards and Havránek (1985 b)).

(A1) An initial set of models S_0 is tested and the models are assigned coherently to A or R.

(A2) If A is empty, go to (A3). If R is empty, go to (A4). If neither A nor R are empty, choose between (A3) and (A4).

(A3) Test the models in $D_r(A)$-R. If these are all rejected, then stop; otherwise update A and R and go to (A2).

(A4) Test the models in $D_a(R)$-A. If these are all accepted, then stop; otherwise update A and R and go to (A2).

Remarks : (a) Initial set of models S_0 may be chosen arbitrarily, but with respect to the principle of coherence the models in S_0 would be incomparable.

(b) The natural choice in (A2) is to compare the number of models in each set and to choose the smallest one.

(c) If one wish to have some submodel, say M, included in any regression equation, it is sufficient to force $D_r(M)$ into the R without testing.

(d) Efficient algorithm with extensive discussion for computation of a-duals and r-duals, resp. $D_r(A)$-R and $D_a(R)$-A , which can be used whenever the family of models forms a lattice, is described in Edwards and Havránek (1985 b). Program in Pascal is also avaiable.

(e) Let T is the set of models which are not classified to be rejected (w-rejected) or accepted (w-accepted) at the moment on the basis of rules I. and II. and the actual content of A and R.

Then $D_a(R)-A$ and $D_r(A)-R$ represent the "extreme points" of T. This fact explains, with respect to the coherence principle, why one test just these sets in (A3) and (A4).

4. Robust procedure for variable selection in linear model

Through this section we shall assume that $0<\gamma<1/2$ and $0<\alpha<1$ are fixed constants. Morover, in accord with reasons described in section 1. we shall use the class of γ-TLSE both for the full model and submodels. Analogously as in section 2., let us denote by $\hat{\underset{\sim}{\beta}}(\gamma T)$ γ-TLSE computed for the full model (2.1), resp. by ${}_r\hat{\underset{\sim}{\beta}}(\gamma T)$ γ-TLSE for submodel (2.3). It is advantageous to replace ${}_r\hat{\underset{\sim}{\beta}}(\gamma T)$ by ${}_r\hat{\underset{\sim}{\beta}}^*(\gamma T)=({}_r\hat{\underset{\sim}{\beta}}'(\gamma T),{}_q\underset{\sim}{0}')$, where ${}_q\underset{\sim}{0}$ is a $(q\times1)$ vector of zeros, $q=p+1-r$, $1\leqq r\leqq p$, because this enables us decide directly if corresponding submodel is α-acceptable or not.

More precisely, we shall call submodel (2.3) α-acceptable if

$$(4.1)\qquad \frac{(\hat{\underset{\sim}{\beta}}(\gamma T)-{}_r\hat{\underset{\sim}{\beta}}^*(\gamma T))'(\underset{\sim}{X}'\underset{\sim}{X})(\hat{\underset{\sim}{\beta}}(\gamma T)-{}_r\hat{\underset{\sim}{\beta}}^*(\gamma T))}{s^{*2}} \leqq \chi_q^2(1-\alpha) \quad ,$$

where $\chi_q^2(1-\alpha)$ is the $(1-\alpha)$-th quantile of χ^2 distribution with q degrees of freedom. This acceptance rule, which is asymptotically equivalent with Definition 2.1., is in real situations a bit more strict then (2.7). Nevertheless, even for moderate samples the difference is negligible. From the point of view of section 3. this rule is almost coherent.

Let us consider set $AS(\alpha)$ of all α-acceptable submodels ((4.1) was used as an acceptance rule). Important property of this set, justifying the use of it as the fundamental part of our procedure, is given by following fact.

<u>Fact 1 :</u> Let $0<\alpha<1$ is a fixed constant. If the subset model is the "true" model, than probability that is included in $AS(\alpha)$ is greater or equal than $(1-\alpha)$.

There exist at last two great problems concerning the immediate use of $AS(\alpha)$. On the one hand, direct determination of $AS(\alpha)$ is very costly because we must fit all the models and in every case decide according to (4.1). On the other hand, even if one proceed such a way, $AS(\alpha)$ may be still too vast. Hence, this is typical situation where one can effectively use ideas described in section 3.

Now, we can proceed to the short description of our procedure.

(P1) $\hat{\underset{\sim}{\beta}}(\gamma T)$ for the full model is computed.

(P2) The initial set of models S_0 is determined. One recomendable choice is the set consisting of the submodels formed by succesive ommiting of one regressor, except the intercept one, from the full model .

(P3) We proceed according to the algorithm described in section 3., where (4.1) is used as an acceptance rule.

(P4) When set of simplest accepted models A and set of the most complex rejected models R are formed, more detailed analysis is postponed up to the evaluation of results by experimenter.

Remarks : (a) Let us notice that algorithm used does not violate the result of Fact 1.

(b) Aside proposed procedure all ideas used, i.e., robust L-estimators, principle of coherence, framework of α-acceptability and fast searching algorithm offer much more possibilities we might described in this short paper, how interested reader easily find from quoted literature. All of them are of great both practical and theoretical importance.

Acknowledgement : Author is grateful to dr.T.Havránek for providing him manuscript of his work.

References :

Aitkin M.A.(1974). Technometrics (16),221-227.
Antoch et all.(1984). Proc. of COMPSTAT 84,49-54.
Antoch J.(1986). Numerical properties of L-estimators in linear model. To appear in Comp. Stat. Quarterly.
Edwards D. and Havránek T.(1985 a). Biometrika (72),339-351.
Edwards D. and Havránek T.(1985 b). A fast selection procedure for large families of models. Preprint.
Gabriel K.R.(1969). The Ann.of Math. Stat. (40),224-250.
Hocking R.R.(1976). Biometrics (32), 1-49.
Hocking R.R.(1983). Technometrics (25),219-249.
Jurečková J.(1983). Comm. Math. Univ. Carolinae (24),597-607.
Koenker R. and Bassett G.(1978). Econometrica (46),33-50.
McCabe G.P.(1978). Technometrics (20), 131-139.
Ruppert D. and Carroll R.J.(1980). J.A.S.A. (75), 828-838.

Linear Models of Categorical Variables

By D. Denteneer, Voorburg

Linear models of factors play an important role in current data analytic practice. The computational aspects of these models are fairly well established, given the numerous statistical packages that treat this class of models, but the literature on this subject is scarce. Yet the subject is non trivial and has several surprising and elegant results. This paper surveys the main results, including some generalizations.

Keywords
generalized linear models, factors, sum-of-products models, restrictions, normalizations.

1. Introduction

Generalized linear models play an important role in current data analytic practice. The basic structure of such a GLM is

$$\mathbf{y} = \mathbf{X} \cdot \theta . \qquad (1.1)$$

Here $\mathbf{y}$ is the parameter-vector (usually a transformed expectation) to be modelled, $\mathbf{X}$ is the model matrix, generated by a set of predictor variables, and θ is another parameter vector. In this paper we primarily pay attention to predictor variables that are factors (categorical variables), but extensions of the results to variates (numerical variables) are sketched in the last section.

This paper essentially deals with the structure of the model matrix $\mathbf{X}$ for the subclass of all linear models that are 'sum-of-products' (SOP-) models, and with the solution of (1.1) in θ under 'normalizing' restrictions. Standard text books deal with $\mathbf{X}$ by means of a few well chosen examples (see Searle, 1971). For the working statistician generalization is fairly obvious, but for computations we need more generality.

In (1.2) we give an example of such a SOP-model for a 2-by-3-by-2 table of factors A, B, and C:

$$Y_{abc} = \theta_1 + \theta_A|_a + \theta_B|_b + \theta_C|_c + \theta_{AB}|_{ab} + \theta_{AC}|_{ac} + \theta_{BC}|_{bc} \qquad (1.2)$$
$$a=1,..,n_A,\ b=1,..,n_B,\ c=1,..,n_C$$

This is formally described

in GLIM as	A + B + C + A.B + A.C + B.C	
in ECTA as	1 2 0 1 3 0 2 3	
in Fienberg (1977) as	[AB][AC][BC]	
and in this note as	1 + A + B + C + A.B + A.C + B.C	(1.3)

Compstat 1986

The name SOP derives from the GLIM notation, which goes back to Wilkinson & Rogers (1973), where the models are described as a formal sum of formal products of variables.

Model (1.2) is quite special in the sense that it has exactly one y-component for each possible combination of a, b, and c: the parameter to be modelled, that is $\mathbf{y}$, is given in the form of a 'complete' table, or cross-classification. This important form receives much attention in this paper. Generalizations are: to allow the table to be incomplete, to allow multiple data tuples per cell, and to include variates in the model formula. These extensions are discussed in section 5.

Greater generality, proofs, and much more detail can be found in Denteneer and Verbeek (1985).

2. SOP-model matrices

A model formula is denoted by Ω, and each term T in Ω is a product of factors. With a little abuse of language, we may write $T \in \Omega$ for a term T in Ω, and $C \in T$ for a factor C in T. The model formula induces a natural partition on the model matrix $\mathbf{X}$. In particular $\mathbf{X} = (\mathbf{X}_1, \ldots, \mathbf{X}_T, \ldots)$, and $\boldsymbol{\theta} = (\theta_1, \ldots, \theta_T, \ldots)$, and model (1.1) can be written as

$$\mathbf{y} = \sum_{T \in \Omega} \mathbf{X}_T \cdot \theta_T$$

$\mathbf{X}_T$ is called a block of $\mathbf{X}$ associated with T. We mention some well known facts about $\mathbf{X}_T$

THEOREM 2.1 (a) The block $\mathbf{X}_T$ of the model matrix, corresponding to a product of factors T is:

$$\bigotimes_C \begin{cases} \mathbf{I}_C & \text{if } C \in T \\ \boldsymbol{\iota}_C & \text{if } C \notin T \end{cases} \tag{2.1}$$

where $\boldsymbol{\iota}_C$ is the n_C-vector of ones.
(b) $\mathbf{X}_T$ has exactly one '1' in each row.

Observe that $\mathbf{X}_T = \mathbf{I} \otimes \boldsymbol{\iota}$ after a suitable reordering of the factors. Statement (b) immediately follows.

Note the sparseness of $\mathbf{X}_T$: each row contains exactly one '1'. This allows traditional sparse storage schemes to be applied. For each block of the model matrix, we can store the column number in which this '1' occurs, thereby reducing storage requirements by a factor 'number of columns of $\mathbf{X}_T$'. However, the special structure of $\mathbf{X}_T$ allows for even greater savings in storage requirements, brought about by the fact that the rows of $\mathbf{X}_T$ are 'congruent':

THEOREM 2.2 The pattern of the 1-s in any column of $\mathbf{X}_T$ is a copy of the pattern in the first column, shifted downward.

Thus to store a block of the model matrix, we need only store the addresses of the ones in the first column of this block and the size of the 'offsets' of the other columns. An application of this storage scheme to iterative proportional fitting of log linear models gives a very fast algorithm, see Denteneer and Verbeek (1986a).

More properties of SOP-model matrices can be derived; in Denteneer and Verbeek (1985) the kernel $\{\theta: \mathbf{X}.\theta = \mathbf{0}\}$ of the model matrix is derived and an analytical expression is given for its eigenvalues and condition number, based on McIntosh (1982). In this paper we concentrate on another important property of $\mathbf{X}$: its rank deficiency. It is obvious that each block of the model matrix is of full column rank; blocks however, are mutually dependent, as the sum over the columns yields the vector of ones for each block. As a consequence, the solution vector θ in (1.1) is not identified. Linear restrictions on θ are called normalizations if they pinpoint one solution from the solution space and if they do not change the model space $\{\mathbf{X}.\theta: \theta\}$. Sections (3) and (4) discuss two popular ways to normalize the parameter vector.

3. GLIM-normalized model matrices

A common way to normalize the parameter vector is to restrict certain parameters to zero. The GLIM way to normalize the model matrix is an example of this technique. For model (1.2) the GLIM restrictions are

$$\begin{aligned} &\theta_{A|1} = \theta_{B|1} = \theta_{C|1} = 0 \\ &\theta_{AB|i1} = \theta_{AB|1j} = \theta_{AC|i1} = \theta_{AC|1k} = \theta_{BC|j1} = \theta_{BC|1k} = 0 \end{aligned} \tag{3.1}$$

To define the GLIM way to normalize the parameter vector in full generality we need a new definition. A SOP-model is called **pyramidal** if the inclusion of a term T in the model formula Ω implies inclusion of all lower order terms. Thus if $T \in \Omega$ then for each term $S \subset T$ we have that $S \in \Omega$. Note that if $S \subset T$ then $R(\mathbf{X}_S) \subset R(\mathbf{X}_T)$. Thus inclusion of the lower order terms does not alter the model space: it is a matter of parametrization only.

For pyramidal models the GLIM-way to identify the parameter vector boils down to 'set all parameters to zero that have at least one '1' in their index vector'. Setting parameters to zero is equivalent to deletion of these parameters from the parameter vector and deletion of the corresponding 'redundant' columns from the model matrix. The matrix thus obtained will be denoted by $\mathbf{G}$, and will be called the GLIM normalized model matrix. The matrix $\mathbf{G}$ is partitioned in blocks $\mathbf{G}_T$ as $\mathbf{X}$ is in blocks $\mathbf{X}_T$. Stripping the redundant columns from each block $\mathbf{X}_T$ yields $\mathbf{G}_T$. Equivalently, the first column of each identity matrix in

(2.1) must be deleted to obtain G_T.

For non-pyramidal models the following theorem shows how a GLIM normalized model matrix **G** can be obtained from the model matrix **X**. It is remarkable that the redundancy of a column is determined by the position of the first '1' in this column.

THEOREM 3.1 (a) A column from the model matrix is redundant iff there exists a preceding column that has its first '1' in the same row.
(b) Using a packed storage scheme as described in section 2, we can construct **G** from **X** in O(min(n, p.log(p)) time, where p is the number of columns of **X**, and n the number of rows of **X**.

Theorem (3.1.a) has another important consequence in that it tells us that no two columns in the normalized GLIM model matrix have identical first non zero entry. Indeed, this property implies that **G** is lower triangular, apart from a reordering of the columns. Thus to solve a consistent system of linear equations $\mathbf{y} = \mathbf{G} \cdot \boldsymbol{\theta}$, one may apply Gaussian elimination. See theorem 3.2. Theorems 3.1 and 3.2 are examplified in table 3.1.

THEOREM 3.2 (a) The GLIM normalized model matrix is lower triangular, up to a reordering of the columns.
(b) A consistent system of linear equations $\mathbf{y} = \mathbf{G}.\boldsymbol{\theta}$ can be solved in $O(p^2)$ time.

These results apply to consistent systems of linear equations only. In Denteneer and Verbeek (1986b) methods are given to solve inconsistent systems of linear equations $\mathbf{y} = \mathbf{G}\ \boldsymbol{\theta}$. They are based on the Cholesky decomposition of $\mathbf{G}^t\mathbf{G}$ or the QR de composition of **G**. There it is shown that almost analytical expressions can be given for these decompositions of the GLIM normalized model matrix.

Table 3.1 The normalized GLIM model matrix is lower triangular after a permutation of columns, cf. theorem 3.2. The example shows a 2x3x2-table with model formula 1 + A + B + C + A.B.

X, redundant columns are starred

1	*	3	*	5	6	*	8	*	*	*	*	13	14
1	1	.	1	.	.	1	.	1	.	.	.	.	.
1	1	.	1	.	.	.	1	1	.	.	.	.	.
1	1	.	.	1	.	1	.	.	1	.	.	.	.
1	1	.	.	1	.	.	1	.	1	.	.	.	.
1	1	.	.	.	1	1	.	.	.	1	.	.	.
1	1	.	.	.	1	.	1	.	.	1	.	.	.
1	.	1	1	.	.	1	.	.	.	.	1	.	.
1	.	1	1	.	.	.	1	.	.	.	1	.	.
1	.	1	.	1	.	1	.	.	.	.	.	1	.
1	.	1	.	1	.	.	1	.	.	.	.	1	.
1	.	1	.	.	1	1	.	.	.	.	.	.	1
1	.	1	.	.	1	.	1	.	.	.	.	.	1

G, after a reordering of columns

1	8	5	6	3	13	14
1						
1	1					
1	.	1				
1	1	1				
1	.	.	1			
1	1	.	1			
1	.	.	.	1		
1	1	.	.	1		
1	.	1	.	1	1	
1	1	1	.	1	1	
1	.	.	1	1	.	1
1	1	.	1	1	.	1

4. Ecta-normalized model matrices

Another established way to identify the parameter vector is to set certain *sums* of parameters to zero. This is done for instance in BMDP4F, ECTA, and in most analysis of variance packages. The following restrictions would be imposed on (1.2)

$$\begin{aligned} &\theta_{A|+} = \theta_{B|+} = \theta_{C|+} = 0 \\ &\theta_{AB|i+} = \theta_{AB|+j} = \theta_{AC|i+} = \theta_{AC|+k} = \theta_{BC|j+} = \theta_{BC|+k} = 0 \end{aligned} \tag{4.1}$$

Here '+' means summing over the corresponding index. More generally it is required that summing a block of parameters over any index results in zero. These restrictions suffice to identify the parameter vector. They do not reduce the model space $\{\mathbf{X}.\boldsymbol{\theta}: \boldsymbol{\theta}\}$ in a pyramidal model.

Note that (4.1) can be obtained from (3.1) by replacing all indices equal to '1' by a summation over the corresponding index. This observation leads to a formula for the normalized ECTA model matrix; simply replace each identity matrix in (2.1) by $\mathbf{K}_C$, where $\mathbf{K}_C$ is a n_Cx(n_C-1)-matrix defined by

$$\mathbf{K}_C = \begin{pmatrix} \mathbf{I}_{n_C-1} \\ -\boldsymbol{\iota}^t_{n_C-1} \end{pmatrix} \tag{4.2}$$

Note that $\mathbf{K}_C$ is orthogonal to $\boldsymbol{\iota}$, and as a consequence different blocks of $\mathbf{E}$, the normalized ECTA matrix are mutually orthogonal. Thus if each factor has two levels $\mathbf{E}$ is an orthogonal matrix, and $\mathbf{E}^t\mathbf{E}$ is diagonal, this property of $\mathbf{E}$ underlies Yates' adding and subtracting algorithm for solving for the parameter vector in a 2^n-table. For pyramidal models, the case for general numbers of levels is also easily treated. To this end we note that $\mathbf{E}^t\mathbf{E}$ is block-diagonal; formulas (2.1) and (4.2) allow easy computation of these diagonal blocks. Moreover these blocks can be analytically inverted.

The following theorem generalizes this result to non-pyramidal models. It tells us that the solution in $\boldsymbol{\theta}$ of (1.1) under the normalizing ECTA-restrictions can be obtained by solving a lower triangular system, that is related to $\mathbf{X}$ in a surprisingly simple way.

THEOREM 4.1 (generalization of Yates' algorithm) The solution of

$$\mathbf{X}^t.\mathbf{y} = \mathbf{X}^t.\mathbf{X}.\boldsymbol{\theta},$$ normalized by the ECTA restrictions

is equal to the solution of

$$\mathbf{X}^t.\mathbf{y} = \Lambda(\mathbf{X}^t.\mathbf{X}).\boldsymbol{\theta},$$

where $\Lambda(\mathbf{X}^t.\mathbf{X})$ denotes the 'lower triangle' of $\mathbf{X}^t.\mathbf{X}$, or more precisely, the matrix of which the elements on and below the diagonal equal the corresponding elements of $\mathbf{X}^t.\mathbf{X}$, while the elements above the diagonal are zero.

5. Extensions

The form of a complete cross-classification, discussed in sections 1 through 4, is an important but rather special class of data for GLM's. Below we give an example of the more general case. We have included a single variate, U:

$$\Omega = A + B + B.U, \qquad (5.1)$$
$$y_{abi} = \theta_{A|a} + \theta_{B|b} + \theta^{U}_{B|b}.U_{abi} \qquad (a=1,2;\ b=1,2,3;\ i=1..N_{ab}).$$

It is clear that if the number of cells 'missing' in comparison to a complete table is not too large, then the sparse storage scheme as introduced in section 2 may still succesfully be applied, be it that additional information concerning U, y_{abi}, and the vector N_{AB} must be stored.

The rank of the new model matrix is also determined by the data. The following theorem specifies the linear dependencies in the model matrix that can be detected by looking at the model formula, and defines 'intrinsic' aliasing. These dependencies hold for any data set. Moreover, for each model formula there are data sets for which these linear dependencies are the only ones. Linear dependencies due to the data are called 'extrinsic' aliasing.

THEOREM 5.1 (aliasing) Let $\Omega = \Omega_1 +..+ \Omega_k$, where Ω is a SOP-model, and each Ω_i is a SOP-model with the same variate(s) in each term (possibly the variate 1 only). Let $\mathbf{X}_i$ be the model matrix corresponding to Ω_i, moreover, let $\mathbf{Z}_i$ be the model matrix corresponding to Ω_i with the variate(s) singled out. Then a column of $\mathbf{X}_i$ is intrinsically aliased iff the corresponding column of $\mathbf{Z}_i$ is redundant.

In fact this theorem tells us that within groups of terms connected with the same variate(s), intrinsic aliasing is as in the situation for complete tables; it is not possible to say something about the linear dependencies between groups of terms connected with different variates, that is between the Ω_i.

REFERENCES

Baker, R.J. & J.A. Nelder, 1978, The GLIM system, Release 3, Manual. (NAG Central Office, Oxford)

Denteneer, D. & A. Verbeek, 1985, Factorial parameters in linear models, (Internal report, the Netherlands Central Bureau of Statistics, Voorburg)

Denteneer, D, & A. Verbeek, 1986a, A fast algorithm for iterative proportional fitting in log-linear models. CSDA 3, 251-64.

Denteneer, D, & A. Verbeek, 1986b, Decompositions of sum-of-products model matrices (in preparation)

Dixon, W.J. & M.B. Brown (Eds), 1983, BMDP-83, Biomedical Data Programs, P-Series. (BMDP Inc, Berkeley CA)

Fay, R., 1971, ECTA: Everymen's contingency table analysis. (University of Chicago, Chicago)

McIntosh, A, 1982, Fitting Linear Models, An Application of Conjugate Gradient Algorithms. (Springer Verlag. New York, Heidelberg, Berlin)

Searle, S.R., 1971, Linear Models. (Wiley, New York)

Wilkinson, G.N. & C.E. Rogers, 1973, Symbolic description of factorial models for analysis of variance. Applied Statistics 22, pp. 392-399.

A Fast Algorithm for Curve Fitting

By A. A. Georgiev, Wroclaw

SUMMARY. In this article I propose a new algorithm for nonparametric curve estimation using the Fast Fourier Transform. The proposed method is extremelly useful when the calculation of the curve is required at a grid of points, for example in order to plot the estimate. Applications to other curve estimation methods are discussed.

KEYWORDS: curve fitting; nonparametric regression; Fast Fourier Transform; convolution; kernel; smoothing.

1. INTRODUCTION

Recently nonparametric curve fitting (regression estimation) has attracted the attention of many research workers and many methods have been proposed in the literature. A good survey of these methods is given in the article by Collomb (1985). It is known that for each of these estimates how good an estimate is depends on how smooth the underlying regression curve is. Typically optimal estimation involves a parameter which depends on the smoothness of the underlying curve, e.g. the bandwidth for the kernel method (Georgiev, 1984 a, eq. 12). In practice, however, the amount of smoothness is never known. So finding a method of estimation for finite sample size (without any assumption of smoothness of the unknown regression curve) is an important task.

Frequently used method is to choose the bandwidth in kernel methods subjectively between estimates drawn with different bandwidths. In this article I propose an algorithm for calculation of the Priestley-Chao nonparametric curve estimate using the Fast Fourier Transform (FFT).

Compstat 1986

2. PRIESTLEY-CHAO ESTIMATE

Let $Y_1, \ldots, Y_n$ be observations according to the model

$$Y_i = g(x_i) + \varepsilon_i \qquad , \qquad i = 1, \ldots, n,$$

taken at fixed points $x_1, \ldots, x_n$, where ε_i's are independent random variables having zero mean

$$|\varepsilon_i| < \infty \qquad \text{almost surely,} \quad i = 1, \ldots, n.$$

We assume that g is bounded over the interval $[0,1]$ and $0 = x_o < x_1 < \ldots < x_n \leq 1$. The estimate of g examined in this paper was introduced by Priestley and Chao (1972) and is of the following form

$$g_n(x) = \sum_{i=1}^{n} Y_i K[(x - x_i)/h] [(x_i - x_{i-1})/h]$$

where K is a kernel function and $\{h = h(n)\}$ is a positive number sequence (the smoothing parameter or bandwidth). I now present a theorem concerning the consistency properties of g.

THEOREM (Georgiev and Greblicki, 1986). Let K be almost everywhere continuous on R. Let K have a majorant, i.e. let $|K(x)| \leq H(x)$, all $x \in R$, where H is symmetric on $[0,\infty)$, $\int H(y)\,dy < \infty$ and $\sup_x H(x) < \infty$. Let

$$\max_i (x_i - x_{i-1}) \leq \beta/n \quad , \qquad \beta > 0 \quad .$$

If

$$h(n) \to 0 \qquad \text{as} \qquad n \to \infty \quad ,$$

and

$$\sum_{n=1}^{\infty} \exp[-\alpha n h(n)] < \infty \quad , \qquad \text{all} \quad \alpha > \infty \quad ,$$

then

$$g_n(x) \to g(x) \qquad \text{completely as} \quad n \to \infty ,$$

all $x \in (0,1)$ at which g is continuous.

For other consistency results see Georgiev (1985), and Georgiev and Greblicki (1986).

3. THE ALGORITHM

The theoretical results in Section 2 are not directly applicable in the common situation faced by data analyst or scientist with limited prior knowledge. Practice (see Georgiev, 1984b) suggest that the choice of kernel function K is not crucial to the performance of the method and therefore it is quite reasonable to choose a kernel for computational efficiency.

In the Priestley-Chao estimate the choice of bandwidth is left to the statisticians. It is useful to examine estimates with several different bandwidths for the same finite sample $\{(x_1, Y_1), \ldots, (x_n, Y_n)\}$. The character of the estimate $g_n(x)$ mainly depends by the choice of h, which determines how much the data are smoothed. It is highly inefficient to use g_n directly for calculation at a grid of points.

Here I follow the idea of Silverman (1982) under the convolution form of g_n. The Fourier transform of $g_n(x)$ is

$$G_n(s) = (2\pi)^{1/2}\, \tilde{K}(h s)\, U(s)$$

where $\tilde{K}$ is the transform of the kernel function K, and U(s) is the transform of the data

$$U(s) = (2\pi)^{-1/2} \sum_{i=1}^{n} Y_i (x_i - x_{i-1}) \exp(j s x) .$$

The proposed algorithm is:

Step 1 - obtain a discrete approximation of data at a grid of design points;

Step 2 - calculate U(s) applying FFT ;

Step 3 - for given kernel K and bandwidth h calculate $\tilde{K}(h s)$ at a grid of design points ;

Step 4 - $G_n(s)$ is found ;

Step 5 - g_n is calculated at a grid of design points by inverse FFT.

The main advantage of the proposed algorithm over the direct use is in time saving. For the same data and grid of points, but different values of the bandwidth, the calculation begins from Step 3. The algorithm may be used successively after the first calculation in order to obtain various estimates from the same data.

4. COMPUTATIONAL REMARKS

Actualy I implement the proposed algorithm and the simulation results will appear in the forthcoming paper Georgiev (1986). Here I make only a few general remarks.

I use the implementation of FFT by Monro (1976). The main time in the algorithm is consumed for calculation the FFT of the discrete approximation of the data. Generaly the time increases linearly with increasing the number of design points. The effect of bandwidth on execution time is marginal and it depends of the choose of the kernel function K.

5. DISCUSSION

The proposed approach can be used also for calculation of other type of estimates developed by Georgiev and Greblicki (1986), e.g.

$$m_n(x) = \ell_n(x)/f_n(x)$$

where

$$\ell_n(x) = (1/n\,h) \sum_{i=1}^{n} Y_i \, K[(x - x_i)/h] ,$$

$$f_n(x) = (1/n\,h) \sum_{i=1}^{n} K[(x - x_i)/h] .$$

For calculation of $\ell_n(x)$ the above algorithm can be used, and for $f_n(x)$ - the algorithm of Silverman (1982). After that two sequences are divided and the estimates m_n at a grid of points are found.

REFERENCES

Collomb G. (1985), Nonparametric regression: an up-to-date bibliography, *Statistics*, 16, 309 - 324.

Georgiev A.A. (1984a), Kernel estimates of functions and their derivatives with applications, *Statistics and Probability Letters*, 2, 45 - 50.

Georgiev A.A. (1984b), Kernel and nearest neighbor curve estimates: a simulative comparison, *COMPSTAT 84*, short communication.

Georgiev A.A. (1985), Propriétés asymptotiques d'un estimateur fonctionnel non paramétrique, *Comptes Rendus Des Seances De L'Académie Des Sciences Paris*, 300 I, 407 - 410.

Georgiev A.A. (1986), Nonparametric curve estimation using the Fast Fourier Transform, working paper.

Georgiev A.A. and Greblicki W. (1986), Nonparametric function recovery from noisy observations, *Journal of Statistical Planning and Inference*, 13, 1 - 14.

Monro D.M. (1976), Real discrete fast Fourier transform, *Applied Statistics*, 25, 166 - 172.

Priestley M.B. and Chao M.T. (1972), Non-parametric function fitting, *Journal of the Royal Statistical Society, Ser. B*, 34, 385 - 392.

Silverman B.W. (1982), Kernel density estimation using the Fast Fourier Transform, *Applied Statistics*, 31, 93 - 99.

Generalized Multiplicative Models

By M. Green, Lancaster

SUMMARY

An algorithm is described for the Maximum Likelihood fitting of models where y_k has a distribution from the exponential family and $\mu_k = E(y_k)$ has the form,

$$\mu_k = \prod_{j=1}^{m} \mu_k^{(j)}, \quad \mu_k^{(j)} = h^{(j)}(\eta_k^{(j)}), \quad \eta_k^{(j)} = \sum_i \beta_i^{(j)} x_{ki}^{(j)}$$

The proposed algorithm is easily implemented in GLIM for particular models. The method is illustrated by examples in time series analysis, analysis of survival data and probit analysis.

Keywords: Non-linear models, GLIM, successive relaxation, IRLS.

1. THE MODEL AND FITTING ALGORITHM

Consider the model where y_k has a distribution from the exponential family and $\mu_k = E(y_k)$ has the form,

$$\mu_k = \prod_{j=1}^{m} \mu_k^{(j)}, \quad \mu_k^{(j)} = h^{(j)}(\eta_k^{(j)}), \quad \eta_k^{(j)} = \sum_i \beta_i^{(j)} x_{ki}^{(j)}$$

The proposed algorithm for Maximum Likelihood estimation of $\underline{\beta}$ consists of a successive relaxation approach in which each iteration has m stages in which stage j estimates $\underline{\beta}^{(j)}$ considering all other $\underline{\beta}$ parameters fixed at their current estimates.

Let $c_k^{(j)} = \prod_{\ell \neq j} \hat{\mu}_k^{(\ell)}$

then stage j is equivalent to fitting

$$\mu_k = c_k^{(j)} \mu_k^{(j)}, \quad \mu_k^{(j)} = h^{(j)}(\eta_k^{(j)})$$

This is equivalent to a GLIM model except for the "multiplicative offset" $c_k^{(j)}$ which is treated as known. The problem is reduced to one of fitting a model $\mu_k = c_k h(\eta_k)$ at each stage.

Such models only fit immediately into the GLIM framework if h(.) is the exponential function. An example of such log-linear models is considered in section 3. However, other models can be fitted by specification of relatively simple "OWN" macros and for a class of distributions described in the next section an even simpler solution is available.

Compstat 1986

2. DISTRIBUTIONS WITH A POWER MEAN-VARIANCE RELATIONSHIP.

Consider a distribution from the exponential family for which;

$$E(y_k) = \mu_k \ , \quad V(y_k) \propto (\mu_k)^\alpha$$

This class of distributions includes the Normal, Poisson, gamma and inverse Gaussian distributions. For non-exponential family distributions for which the relationship holds the following method applies for quasi-likelihood estimation.

consider $V(y_k) = (\Phi/w_k)\mu_k{}^\alpha$ and $y_k{}^* = y_k/c_k$

then $V(y_k{}^*) = (\Phi/w_k{}^*)(\mu_k{}^*)^\alpha$

where $\mu_k{}^* = E(y_k{}^*) = \mu_k/c_k$ and $w_k{}^* = w_k(c_k)^{2-\alpha}$

Thus $y_k{}^*$ has the same distributional properties as y_k and estimation is equivalent for $y_k{}^*$, $w_k{}^*$ and $\mu_k{}^*$. When applied to the model in stage j of the above algorithm we obtain $\mu_k{}^* = g(\eta_k)$ and thus can apply standard GLIM fitting by transforming to $y_k{}^*$ and using weights $w_k{}^*$.

2.1 A Simple Example

Consider the following simple time series model for long-term forecasting, one of linear trend with multiplicative seasonality.

$$y_t = s_p(\alpha+\beta t) + \epsilon_t \ , \quad \epsilon_t \sim N(0,\sigma^2)$$

where p corresponds to the period of the year.

In the present framework the model can be written;

$$\mu_t = \eta_t{}^{(1)}\eta_t{}^{(2)} \ , \quad \eta_t{}^{(1)} = s_p \ , \quad \eta_t{}^{(2)} = \alpha+\beta t$$

Since the Normal distribution corresponds to α=0 this model can be fitted by the above algorithm with $w_t{}^*=c_t{}^2$. The algorithm is easily implemented using the following GLIM macro for each iteration:

```
$macro ITER
$cal YSTR=Y/FV2 : W=FV2**2  $fit S  $cal FV1=%fv $
$cal YSTR=Y/FV1 : W=FV1**2  $fit T  $cal FV2=%fv $
$endmac
```

3. LOG-LINEAR MODELS

The standard GLM model for contingency tables has the Poisson distribution for y_k and h(.)=exp(.). Application of the above algorithm produces the familiar Iterative Proportional Scaling method. Use of GLIM will not produce an efficient solution since the estimation at each stage will be iterative even though direct estimates are available. However, there is a very useful application of this approach for a closely allied model as described below.

3.1 The Proportional Hazards Model for Survival Data

In the Cox proportional hazards model we attempt to assess the effects of covariates, $\underline{x}$, on survival without defining a parametric form of the underlying hazard function. That such models could be formulated within the GLM framework was first proposed in Aitkin et al (1983) based on the ideas of Breslow (1972) and Kalbfleish and Prentice (1980). Briefly the approach is to divide the time axis into intervals with end-points defined by the observed death times, the hazard being assumed constant within each time interval. The ith individual's experience over the N_i intervals comprising the survival time is then considered. Defining e_{ij} to be the exposure time in the jth interval and d_{ij} a censoring indicator with $d_{ij}=1$ indicating death and 0 censoring the likelihood is;

$$L = \prod_{i=1}^{n} \prod_{j=1}^{N_i} h_{ij}^{d_{ij}} \exp(-h_{ij}e_{ij})$$

where $h_{ij}=\exp(\phi_j+\underline{\beta}'\underline{x}_i)$

and ϕ_j are the baseline log hazard constants.

This model can be fitted directly in GLIM using the Poisson error specification but suffers from the severe disadvantages that the vectors have to be expanded to length ΣN_i and the usually large number of log hazard constants make estimation very slow. For the example below with 65 observations, vectors are expanded to length 1570 and 40 baseline hazard parameters have to be estimated.

Formulated as a GMM with $\lambda_j = \exp(\phi_j)$ and $\gamma_i = \exp(\underline{\beta}'\underline{x}_i)$ then

$$\ell = \log L = \Sigma\Sigma e_{ij}\{ (d_{ij}/e_{ij})\log(\lambda_j\gamma_i) - \lambda_j\gamma_i \}$$

is a Poisson likelihood with $y_{ij}=d_{ij}/e_{ij}$, $w_{ij}=e_{ij}$ and $\mu_{ij}=\lambda_j\gamma_i$.

Applying the algorithm has two benefits. Firstly there are explicit estimates for λ_j considering $\underline{\beta}$ fixed, viz

$$\hat{\lambda}_j = \sum_i d_{ij}/\sum_i e_{ij}\gamma_i$$

Secondly when fitting β with λ_j fixed at $\hat{\lambda}_j$ we have;

$$\ell = \Sigma\Sigma d_{ij}\log\hat{\lambda}_j + \Sigma\Sigma(d_{ij}\log\gamma_i - e_{ij}\hat{\lambda}_j\gamma_i)$$

The first term is constant and the second term can be written as;

$$\Sigma u_i\{ (d_{i.}/u_i)\log\gamma_i - \gamma_i\}$$

a Poisson likelihood with modified y-variate $d_{i.}/u_i$ and weights

$$u_i = \sum_j e_{ij}\hat{\lambda}_j$$

Thus the problem is returned to one with n observations and an iterative fit for the covariate parameters only. The extra calculations and the extra level of iteration required by the algorithm do not render interactive analysis impossible for modest problems.

3.2 Analysis of the Stanford Heart Transplant Data

The algorithm is compared with the direct GLIM fitting used to analyse the Stanford heart transplant data studied in Aitkin et al (1983). One disadvantage of successive relaxation methods is that estimates for the (co)variances of parameter estimates do not make allowance for the estimation of the parameters at other stages and are therefore underestimates. Empirical observation indicates that this is commonly a small effect, as indicated in the following table.

	Direct GLIM		GMM algorithm	
parameter	estimate	s.e.	estimate	s.e.
AGE	0.0559	0.0223	0.0559	0.0218
SURG	-0.862	0.483	-0.862	0.481
MM	0.427	0.291	0.426	0.281
units	1570		65	
cpu time	155 secs		24.2 secs	

4. OTHER DISTRIBUTIONS

For distributions from the exponential family that do not possess the power mean-variance relationship the algorithm can still be implemented in GLIM by supplying the necessary GLIM macros for an OWN fit of

$$\mu_k = c_k g(\eta_k)$$

This only affects the macros to calculate the fitted values (%fv) and the derivatives $\partial\eta/\partial\mu$ (%dr), the other two macros are standard for the particular distribution as illustrated in the following application.

4.1 The Simple Independent Action Probit Model

Consider the model for the toxicity of a mixture of two poisons at doses $x^{(1)}$ and $x^{(2)}$. If they act in totally different physiological ways their effects may be taken as independent, so that,

$$\Pr(\text{survival}) = \Pr(X^{(1)}>x^{(1)}).\Pr(X^{(2)}>x^{(2)})$$

where $X^{(1)}$ and $X^{(2)}$ are the individual's tolerances to the respective poisons. Following the probit formulation for log dose:

$$E(y_k)= \mu_k = n_k p_k^{(1)} p_k^{(2)}$$

$$p_k^{(j)}=(1-\Phi(\eta_k^{(j)}) \ , \ \eta_k^{(j)}=\beta_0^{(j)}+\beta_1^{(j)}\log(x_k^{(j)})$$

where y_k is the number who survive out of n_k at doses $x_k^{(1)}$ and $x_k^{(2)}$ of poisons 1 and 2.

To fit this model we need OWN macros for the model;

$$\mu_k = n_k c_k(1-\Phi(\eta_k)) \text{ and } y_k \sim \text{Bino}(n_k, p_k)$$

which require the folowing GLIM macros:

```
$macro MFV $cal %fv=N*C*(1-%np(%lp)) $$endmac
$macro MDR $cal %dr=-%C/(N*C*%exp(-0.5*%lp**2)) $$endmac
```

With these specified in the $OWN directive the following macro can be used to perform one iteration of the GMM algorithm:

```
$macro ITER
$cal C=1-%np(LP2) : %lp=LP1  $fit LGX1 $cal LP1=%lp $
$cal C=1-%np(LP1) : %lp=LP2  $fit LGX2 $cal LP2=%lp $
$endmac
```

5. GENERAL *IRLS* FORMULATION

The GMM models can be formulated easily in the general IRLS method as described in Green (1984). All that is required is the evaluation of the "local design matrix", X_L (Jørgensen (1984)). This has a relatively simple form for these models:

Let $\underline{\eta}_k^{(j)} = X^{(j)}\underline{\beta}^{(j)}$

and partition $X_L = [\partial\mu_i/\partial\beta_j]$ as $[X_L^{(1)}|X_L^{(2)}|\ldots|X_L^{(m)}]$

then $X_L^{(j)} = \left\{ \bigotimes_{i \neq j} \underline{\hat{\mu}}^{(i)} \right\} \otimes g'^{(j)}(\underline{\hat{\eta}}^{(j)}) \otimes X^{(j)}$

where $\otimes$ is a direct product operator.

This can only be implemented in GLIM for specific models and cannot be generalized with the present facilities. However, for comparison purposes, a GLIM macro for the IRLS method was written for a particular application descibed below.

5.1 Comparison of GMM and IRLS methods

The methods were applied to data on the toxicity of Rotenone and Deguelin from Finney (1947). The following table summarises the results.

	GMM		IRLS	
parameter	estimate	s.e.	estimate	s.e.
l(1)	-2.706	0.328	-2.705	0.332
ROT	1.746	0.198	1.745	0.199
l(2)	-1.746	0.188	-1.746	0.189
DEG	0.894	0.0763	0.894	0.0764

The GMM method converged in 12 iterations to a deviance of 34.5 on 13 d.f. from simple starting values $\underline{\eta}^{(1)}=\underline{\eta}^{(2)}=\Phi^{-1}(1-(y./n.)^{½})$. The IRLS method converged quicker but diverged without good starting values. In conclusion the GMM algorithm provides a stable and versatile method of estimation for a wide variety of models outside the standard GLM framework.

References:

Aitkin M., Laird N., and Francis B. (1983), A Reanalysis of the Stanford Heart Transplant Data, J.Am.Statist.Ass., 78, 264-292

Breslow N. (1972), Contribution to the discussion on the paper of D.R. Cox, J.R.Statist.Soc B, 34, 187-220

Finney D.J. (1947,1952), Probit Analysis, University Press, Cambridge

Green P.J. (1984), Iteratively Reweighted Least Squares for Maximum Likelihood Estimation and some Robust and Resistant Alternatives, J.R.Statist.Soc. B, 46, 149-192

Jørgensen B. (1984), The Delta Algorithm and GLIM, International Statistical Review, 52, 3, 283-300

Kalbfleisch J.D., and Prentice R.L. (1980), The Statistical Analysis of Failure Time Data, Wiley, New York

A Convenient Way of Computing ML-Estimates: Use of Automatic Differentiation

By C. Kredler and W. Kowarschick, München

Summary:

A standard Pascal-program system for the maximization of log-likelihood functions envolving several regressor variables is presented. The user must supply a data matrix, e.g. the x - and y - variables of a linear model. Only FORTRAN-like characterizations for the densities must be available. Gradients and Hessians needed for quadratically convergent optimization algorithms are computed by automatic differentiation.

There is a wide area of application, e.g. generalized linear models, quadratic logistic discriminant analysis, nonlinear curve fitting and regression models with time series errors.

KEYWORDS: Automatic differentiation, Pascal-system Autolike, regression with time series errors, nonlinear regression, quadratic logistic discriminant analysis.

1. The models

Consider a sample $[y_i, x_i]$, $i = 1,\dots,N$, where $y_i \in \mathbb{R}$ denote the dependent and $x_i \in \mathbb{R}^p$ the regressor variables. Further let the structure of the density f corresponding to y_i be independent of the index i. This assumption and the restriction to univariate y_i is not really necessary for our approach. Nevertheless we use it for convenience of exposition.

Denote by $\beta \in \mathbb{R}^m$ the parameters to be estimated. In many applications the regressor variables go into f only as linear combinations of the form

$$v_i = \sum_{j=1}^{p} x_{ij}\beta_j \ , \quad w_i = \sum_{k=p+1}^{2p} x_{ik}\beta_k \tag{1}$$

a.s.o. In this case the log - likelihood can be written as

$$L(\beta) = \sum_{i=1}^{N} \ln \ f(\beta_i, y_i, v_i, w_i). \tag{2}$$

Compstat 1986

Before we describe how to maximize *like*lihoods using *auto*matic differentiation with the Pascal program Autolike we give two examples. These models from rather different areas of data analysis shall convince the reader that our approach covers a large field of applications.

Example 1.1: Nonlinear exponential family models

Assume that the densities of y_i belong to an exponential family

$$f(y_i, \theta_i) = c(y_i) \exp (y_i \theta_i - b(\theta_i)),$$

where the canonical parameter $\theta_i = \theta(v_i, w_i, \beta)$ contains all model information. For instance, the cumulant generating function $b(\theta) = \ln(1 + e^{\theta})$ characterizes the class of *binomial logistic* models. The quadratic discriminant approach of Anderson (1975) chooses $m = 2p + 2$

$$\theta_i = \beta_{m-1} + v_i + \beta_m w_i^2 , \quad \sum_{k=p+1}^{2p} \beta_k^2 = 1 \text{ (see (1))}. \tag{3}$$

For $\beta_m = \beta_{2p+2} = 0$ we obtain the well known logistic regression. From a geometric point of view in (3) the data are projected into the v-w-plane and then separated by a parabola.

Example 1.2: Regression with AR(1) errors

Assume a linear model

$$y = X\beta_x + u, \quad u_t = \beta_\rho u_{t-1} + \varepsilon_t , \quad t = 2,\dots,N , \ |\beta_\rho| < 1.$$

Let $u_1, \varepsilon_2, \dots, \varepsilon_N$ be independently normal,

$$\operatorname{var}(\varepsilon_t) = \beta_{\sigma^2}, t = 2,\dots,N , \ \operatorname{var}(u_t) = \beta_{\sigma^2}/(1-\beta_\rho^2), \ t=1,\dots,N \tag{4}$$

and

$$Q(\beta_x, \beta_\rho) = (1 - \beta_\rho^2) u_1^2 + \sum_{t=2}^{N} (u_t - \beta_\rho u_{t-1})^2 , \text{ then}$$

$$L(\beta_x, \beta_\rho, \beta_{\sigma^2}) = \frac{1}{2} (\ln (1-\beta_\rho^2) - N \ln \beta_{\sigma^2} - Q/\beta_{\sigma^2}) \tag{5}$$

is the likelihood to be maximized.

2. Optimization procedures

At this time Autolike is equipped with a robust optimizer for problems without constraints. There are 3 options: 'N' for modified Newton, 'B' for BFGS and 'BN' for BFGS at start and Newton near the solution to obtain ultimately quadratic convergence.
As in many statistical applications in example 1.2 we get rid of constraints by simple transformations.

$$\beta_\rho = \frac{2}{\pi} \arctan (\beta_{m-1}), \; \beta_{\sigma^2} = \beta_m^2 \;, \; m = p + 2 \tag{6}$$

fulfil the requirements $-1 < \beta_\rho < 1, \beta_{\sigma^2} > 0$. In example 1.1 the quadratic constraint is due to β_m. It can be avoided by solving two problems for

$$\theta_i = \beta_{m-1} + v_i \pm w_i^2 \; . \tag{7}$$

Nevertheless the quadratically convergent algorithm of Best, Bräuninger, Ritter, Robinson (1981) for the solution of constrained problems runs in a preliminary test version. In nonlinear regression applications with large residuals we are faced with indefinite Hessians. Provided the smallest eigenvalue λ of $\nabla^2 L$ is negative and additionally $\|\nabla L\|$ is small, the eigenvector corresponding to λ gives a good search direction. An approximation $\tilde{\lambda}$ for λ is computed by Lanczos' method. Then the desired eigenvector can be easily computed by inverse iteration.
In Autolike ∇L and $\nabla^2 L$ are computed by automatic differentiation, cf. ch. 4. A user has only to supply the following formal description of L itself.

3. Description of the log - likelihood

The formulae of $L(\beta)$ given here for example 1.1 and 1.2 are subsequently split into simpler parts to which the techniques of automatic differentiation, see ch. 4, can be applied. For quadratic discrimination of 1.1 we have to supply

```
      T[I]  = B[2*P+1]  +  V[I] ±  W[I] ** 2
                                                          (8)
  L = SUM (I = 1,N) ( Y[I] *  T[I]  - LN(1 + EXP( T[I] )) ,
```

whereas the AR(1) model of 1.2 is characterized by

$$R = \mathrm{ARCTAN}\ (B[M-1]) * PI/2$$

$$Q = (1 - R{**}2) * V[1]{**}2 + \mathrm{SUM}(I = 1,N)(V[I] - R*V[I-1]){**}2 \qquad (9)$$

$$L = (\ln (1-R{**}2) - N*\mathrm{LN}(B[M]{**}2) - Q/B[M]{**}2)/2 . \qquad (10)$$

Additionally a file containing

'N'	(e.g. choose Newton or 'B' BFGS)
N,p	(dimension of data matrix)
m	(m = 2p + 1 in 1.1, m = p + 2 in 1.2)
(y_i, x_i^T)	(observations rowwise)

should be available.

4. Automatic Differentiation

First references of automatic differentiation go back to 1959. In the system Autolike the procedures of Rall (1983) are implemented in portable standard-Pascal. Here we can only sketch the basic ideas. A more detailed description can be found in Rall(1983).

4.1 Characterization of f and basic functions

Suppose the density f in (2) is composed of +, -, *, /, ** (power), standard functions (exp, ln, sin, cos, tan, ...) and so called basic functions $b_1(\beta) = \beta_1, \ldots, b_m(\beta) = \beta_m$ and $c_d(\beta) \equiv d$ = const. Define for any function f the triple

$$F(\overline{\beta}) := [f(\overline{\beta}),\ f_G(\overline{\beta}),\ f_H(\overline{\beta})] \qquad (11)$$

of function, gradient and Hessian at an arbitrary but fixed $\overline{\beta} \in \mathbb{R}^m$. Denote by e_j the j-th canonical unit vector, 0_m the m-vector and $0_{m,m}$ the m x m matrix of zeros, respectively. Then we have

$$B_j(\overline{\beta}) = [\overline{\beta}_j,\ e_j,\ 0_{m,m}]\ ,\ j = 1,\ldots,m\ ,$$

$$C_d(\overline{\beta}) = [d,\ 0_m,\ 0_{m,m}]\ ,\ d \in \mathbb{R}. \qquad (12)$$

4.2 Divide and differentiate

Like in any compiler f is subsequently *devided* into simpler parts. Finally we end up with basic functions. To explain this consider a simple log-linear model with density

$$f(\beta) = f(\beta_j, y_i, x_i) = (\beta_2 - x_i\beta_1)y_i - e^{\beta_2 - x_i\beta_1} . \qquad (13)$$

We obtain

$$f(\beta) = r_1(\beta) - r_2(\beta),$$
$$r_1(\beta) = r_3(\beta) \cdot c_{y_i}(\beta)$$
$$r_3(\beta) = b_2(\beta) - r_4(\beta)$$
$$r_4(\beta) = c_{x_i}(\beta) \cdot b_1(\beta)$$
$$r_2(\beta) = \exp(r_3(\beta)).$$

For the second part *differentiate* use well known rules like given in

Table 4.1 : Rules for triples $R = [r, r_G, r_H]$

devide	*differentiate*
$r = a + b$	$r_G = a_G + b_G$ $r_H = a_H + b_H$
$r = a \cdot b$	$r_G = ba_G + ab_G$ $r_H = ba_H + ab_H + b_G a_G^T + a_G b_G^T$
$r = a / b$	$r_G = (a_G - rb_G)/b$ $r_H = (a_H - rb_H - b_G r_G^T - r_G b_G^T)/b$
$r = a^b$	$z = ba_G/a + b_G \ln a$ $r_G = rz$ $r_H = r[(ba_H + a_G b_G^T + b_G a_G^T - ba_G a_G^T/a)/a + b_H \ln a + zz^T]$
$r = s \circ a$ s stand. function	$r_G = s'(a)a_G$ $r_H = s''(a)a_G a_G^T + s'(a)a_H$

5. Numerical results

Anderson's quadratic model, see 1.1, was applied to data from steel quality control with up to N = 180 observations, p = 3 and p = 14 x-variables, respectively. The CPU-times on the CDC-Cyber 875 of the Leibniz-Rechenzentrum München are for modified Newton

N	p	Autolike	analytic gradient and Hessian
100	3	18.5 sec	0.64 sec
180	14	59.1 sec	5.15 sec

Hence, Autolike in its present version cannot compete with well working methods using analytic derivatives.
Nevertheless automatic differentiation is a powerful tool during the process of investigation. To demonstrate this consider the data $[y_i = i^2,\ x_i^T = (1,i)]$, $i = 1,20$ as an AR(1) regression model. It is not very hard to obtain the first derivatives of L in (5), e.g.

$$\partial L/\partial \beta_{\sigma^2} = -(N - Q/\beta_{\sigma^2})/\beta_{\sigma^2}, \qquad (14)$$

but the Hessian for general $p > 1$ is somewhat more complicated. Now the input (9), (10) for L of (5) is compared with a slightly altered version eliminating β_{σ^2} according to (14)

$$L = (\ln(1 - R ** 2) - N * \ln(Q/N) - N)/2. \qquad (15)$$

The following table gives CPU-times of Autolike and number of iterations for BFGS and modified Newton, starting from the initial guess $(\beta_1, \beta_x, \beta_\rho, \beta_{\sigma^2})_0 = (1,1,0,10)$.

parameters	input	mod. Newton	BFGS
$\beta_1, \beta_x, \beta_\rho, \beta_{\sigma^2}$	(9) + (10)	(8.1 sec, 14 it)	(no convergence within 100 it)
$\beta_1, \beta_x, \beta_\rho$	(9) + (15)	(5.9 sec, 10 it)	(7.5 sec, 75 it)

With the help of Autolike we conclude that we should eliminate β_{σ^2} before optimizing and that BFGS may lead to trouble in this sort of problem.
To obtain this result we have only to supply data and the formal descriptions (9), (10) and (15), respectively. Subroutines for ∇L or $\nabla^2 L$ are not requested.

References

Anderson J.A. (1975), Quadratic Logistic Discrimination, Biometrika, 62, 149-154.

Best M.J., Bräuninger J., Ritter K. and Robinson S.M. (1981), A Globally and Quadratically Convergent Algorithm for General Nonlinear Programming Problems, Computing, 26, 141-153.

Rall L.B. (1983), Differentiation and Generation of Taylor Coefficients in Pascal - SC. In: Kulisch U.W. and Miranker W.L. (eds.), A New Approach to Scientific Computation, Academic Press, New York.

On the Numerical Solutions of Bounded Influence Regression Problems

By A. Marazzi, Lausanne

Several iterative procedures to solve numerically the problem of robust multiple regression have been proposed in Huber (1972), Huber and Dutter (1974) and Dutter (1977a). According to Dutter (1977b), the H-algorithm distinguished itself for its programming simplicity and rapid computation. The H-algorithm was adapted to the problem of bounded influence regression by Marazzi (1980) and introduced in ROBETH. A similar adaptation was also described by Samarov and Welsch (1982) and introduced in TROLL. Experience shows that when high influential leverage points are present the convergence of the H-algorithm is very poor and the reliability of the result is doubtful. In this paper, alternative procedures with better convergence are discussed. These procedures have been implemented in ROBSYS, a conversational package for robust statistical computing (see Marazzi and Randriamiharisoa (1985)).

1. The problem of bounded influence linear regression

As usual the model is described by $y=X\Theta+e$ where y is an nx1 response variable, X is a full rank nxp matrix containing p explanatory variables, Θ is a p-vector of unknown parameters and e is the error n-vector such that its components $e_1,\ldots,e_n$ are i.i.d. according to some symmetric distribution $L(\cdot;\sigma)$ and σ is a scale parameter. We consider a class of estimators $(\hat{\Theta},\hat{\sigma})$ defined as the value of (Θ,σ) which minimises

$$Q(\Theta,\sigma) = \sum_{i=1}^{n} \rho\left(\frac{r_i}{\sigma w_i}\right)\sigma w_i^2 + \sigma\text{CONST} \qquad (1)$$

where $r_i=y_i-x_i^T\Theta$ denotes the i-th residual, x_i^T the i-th row of X, CONST is a given number, the w_i's are given weights and ρ a given function. Very often, $\rho(s)=\rho_c(s)$ where $\rho_c(s)=s^2/2$ for $|s|<c$, $\rho_c(s)=c|s|-c^2/2$ for $|s|>c$ and c is a user specified value.

It is usually assumed that ρ is convex, $\rho\geq 0$, $\rho(0)=0$. Thus, unless the minimum occurs on the boundary $\sigma=0$ it can be characterised by the equations

Compstat 1986

$$\sum \psi(r_i/(\sigma w_i))w_i x_{ij} = 0 \qquad j=1,\ldots,p \qquad (2)$$

$$\sum \chi(r_i/(\sigma w_i))w_i^2 = \text{CONST} \qquad (3)$$

with $\psi(s)=\rho'(s)$ and $\chi(s)=s\psi(s)-\rho(s)$.

The weights w_i are introduced in order to bound the influence of "outlying rows of X" (see for ex. Hampel et al. (1986)) on the estimates. When $w_i=1$ for all i one obtains a Huber estimate. Replacing w_i, y_i, x_i by $\tilde{w}_i:=\sqrt{w_i}$, $\tilde{y}_i:=\tilde{w}_i y_i$, $\tilde{x}_i:=\tilde{w}_i x_i$ in (1) one obtains an estimate of the Mallows class. In general (1) leads to an estimate of the Schweppe class and there is no loss of generality in considering only this case. The w_i's are computed in a preliminary step as a function of the metric distance of the point x_i from the origin. The metric is given by a robust covariance matrix (see for ex. Hampel et al. (1986)). Finally CONST is normally chosen so that $\hat{\sigma}$ is asymptotically consistent for normal errors, that is CONST$=(\frac{n-p}{n})\cdot\sum w_i^2 E(\chi(Z/w_i))$, $Z\sim N(0,1)$.

2. ALGORITHMS

Huber (1981) describes several algorithms for minimising $Q(\Theta,\sigma)$ when $w_i=1$ for all i and provides convergence proofs. Algorithms and proofs can be adapted to bounded influence regression.

Most algorithms alternate between improving trial values for Θ and σ. For example, an updated estimate of σ^2 is $\sigma_1^2:=\sum\chi(r_i/(\sigma w_i))(\sigma w_i)^2/\text{CONST}$ (which is an improved solution of (3)). However, one could also replace (3) by some other criterion and set σ_1 equal to the MAD of the current residuals. In the following, attention is restricted to the Θ-step and σ is supposed fixed. (We write $Q(\Theta)$ instead of $Q(\Theta,\sigma)$).

The Newton algorithm (NA) is a natural choice which is justified by the approximate quadratic behaviour of Q near its minimum. We have :

$$\nabla Q := \frac{\partial Q}{\partial \Theta} = -\frac{1}{\sigma} X^T r^* \qquad \text{and} \qquad H := \frac{\partial^2 Q}{\partial\Theta\partial\Theta^T} = \frac{1}{\sigma} X^T D X$$

where $r^*=(r_1^*,\dots,r_n^*)^T$, $r_i^*=\psi(r_i/(\sigma w_i))\sigma w_i$ for $i=1,\dots,n$ and $D=\mathrm{diag}(\psi'(r_i/(\sigma w_i)))$. An improvement step τ is then found by solving $X^TDX\tau=X^Tr^*$. In other words, one approximates $Q(\Theta+\tau)$ by $V(\tau)=Q(\Theta)+\tau^T\nabla Q+\frac{1}{2}\tau^TH\tau$. The well known drawbacks are : a) H and its inverse must be recalculated at each iteration; b) H can become singular (and a positive definite generalised inverse has to be used; c) in order to prevent oscillations an adaptive braking factor must be introduced.

Various tricks can be devised for overcoming these difficulties. In the following algorithms, D is replaced by some approximation $\tilde{D}$ and $V(\tau)$ by functions of the form $\sum U_i((r_i-x_i^T\tau)/(\sigma w_i))\sigma w_i^2$ where $U_i(s)=a_i+b_is+c_is^2$ and 2n coefficients are determined so that $U_i(s/(\sigma w_i))$ and $\rho(s/(\sigma w_i))$ have the same value and the same tangent at $s=r_i$ for all i. Therefore, the comparison function becomes $\tilde{V}(\tau)=Q(\Theta)+\tau^T\nabla Q+\frac{1}{2}\tau^T\tilde{H}\tau$ with $\tilde{H}=X^T\tilde{D}\,X$.

The H-algorithm (HA) sets $\tilde{D}=I$ i.e. $\tilde{H}=X^TX$ which must be inverted only once. This is equivalent to fixing $c_i=1/2$ for all i. If $0\le\rho''\le 1$, it is easy to show that $\tilde{V}(\tau)\ge Q(\Theta+\tau)$ for all τ and therefore no braking factor is needed.

The W-algorithm (WA) sets $\tilde{D}=\mathrm{diag}(g(r_i/(\sigma w_i))$ where $g(s)=\psi(s)/s$. (Noting that $\psi'(s)=g(s)+sg'(s)$ one just ignores the second term.) This amounts to fixing $b_i=0$ for all i. If ρ is convex and symmetric and $\psi(s)/s$ is bounded and monotone decreasing for $s>0$ then $\tilde{V}$ is the best quadratic approximation that lies above Q. No braking factor is therefore needed; however, $\tilde{H}$ changes at each step.

The I-algorithm (IA) (see Hampel et al. (1986) pp. 263-264 for an introduction) sets $\tilde{D}=\mathrm{diag}(E\psi'(Z/(\sigma w_i)))$, where the expectation is taken over a tentatively estimated error distribution, for example the standard Gaussian one. $\tilde{H}$ does not change at each iteration. However, it is not possible to guarantee that $\tilde{V}(\tau)\ge Q(\Theta+\tau)$ for all τ ($\tilde{H}$-H is not positive definite) but using a braking factor convergence can be obtained.

The S-algorithm (SA) described in Dutter (1977a) can also be adapted to bounded influence regression. It is essentially the Newton algorithm for the special case $\rho(s)=\rho_c(s)$ which allows the algebraic exact solution to be determined.

3. Some experiments

Table 1 gives some results (intercept, slope, number of iterations (NIT) and CP-time) obtained with HA, IA, WA and NA for the computation of 12 regression lines using Krasker-Welsch weights (see Samarov and Welsch (1982)), $\rho(s)=\rho_c(s)$ and $c=1.05*\sqrt{2}$. Each data set contains 14 "good points", namely (-3, -3.6), (-2, -1.75), (-1, -.5), (0, .2), (1, .95), (2, 1.85), (3, 2.85) repeated twice and 3 coincident "outliers" whose components x and y are indicated in the table. The three digit solutions given under NA are exact. All problems were run on a CDC Cyber 170/855.

Table 1 : comparison of different algorithms

x	y	WLAR		HA				IA				WA				NA			
		Θ_1	Θ_2	Θ_1	Θ_2	NIT	CP	Θ_1	Θ_2	NIT	CP	Θ_1	Θ_2	NIT	CP	Θ_1	Θ_2	NIT	CP
5	5	.090	.920	.099	.929	18	.14	.099	.929	4	.06	.098	.929	7 (6)	.11	.099	.929	3	.05
5	0	.050	.900	.117	.874	86	.62	.118	.873	16	.16	.118	.873	27 (13)	.43	.118	.873	4	.07
10	10	.090	.920	.090	.933	45	.32	.090	.933	8	.09	.090	.933	8 (5)	.13	.090	.933	3	.05
10	0	.200	.825	.134	.864	222	1.58	.132	.867	31	.27	.132	.867	28 (10)	.44	.132	.867	2	.03
20	20	.090	.920	.086	.934	135	.96	.085	.935	8	.09	.085	.935	8 (5)	.13	.085	.935	3	.05
20	0	.200	.825	.144	.858	566	4.00	.139	.864	51	.43	.139	.864	31 (10)	.49	.139	.864	2	.03
40	40	.090	.920	.084	.934	397	2.80	.083	.936	15	.15	.083	.936	9 (5)	.14	.083	.936	3	.05
40	0	.200	.825	.154	.850	1298	9.14	.142	.862	87	.72	.142	.862	34 (11)	.54	.142	.862	2	.03
160	160	.090	.920	.086	.928	2081	14.62	.081	.937	58	.48	.081	.937	11 (5)	.18	.081	.937	2	.04
160	0	.200	.825	.181	.825	9	.07	.145	.861	299	2.47	.145	.861	41 (13)	.65	.145	.861	2	.03
1600	1600	.090	.920	.091	.920	4	.04	.080	.937	582	4.79	.080	.937	13 (7)	.21	.080	.937	2	.03
1600	0	.200	.825	.181	.825	9	.07	.146	.860	2967	24.18	.146	.861	53 (16)	.84	.146	.861	2	.04

The scale parameter σ is set to .390 which is the ordinary LS estimate on the good points. Iterations are stopped when, for all j, the j-th absolute step component $|\tau_j|$ is smaller than 10^{-5} times the standard deviation of Θ_j. This is estimated by means of the j-th diagonal element of $\sigma(X^TDX)^{-1}(X^TCX)(X^TDX)^{-1}$ with $C=\text{diag}\,(E\psi'_c(Z/w_i))$ and $Z \sim N(0,1)$ (see Samarov and Welsch (1982)).

The starting values are given by the solution (WLAR) of $\sum w_i|r_i|=\min!$ (Despite its computational simplicity, ordinary LS or even weighted LS is generally a very poor starting value in bounded influence regression problems). In NA, when H is singular it is replaced by $X^T\mathrm{diag}(g(r_i/(\sigma w_i)))X$. Braking factors are estimated by cubic interpolation as suggested in Fletcher and Powell (1964).

4. Discussion

Dutter (1977b) compared iteration numbers and CP times of HA, WA and SA for the case of robust regression without outliers in factor space. No important difference was found. However, Table 1 clearly indicates that when high influential points are present the differences are striking. The convergence of HA becomes extremely slow. In fact the improvement steps become very small. Sometimes the convergence criterion is satisfied when the current parameter values are still "far" from the solution, which means the reliability of the result is doubtful. It is easy to see that this drawback originates in the use of $\tilde{H}=X^TX$, which means that there are a few dominating quadratic terms in $\tilde{V}(\tau)$.

The influence of outlying rows of X on the step size can be reduced in various ways. For example, IA downweights distant points regardless of the value of the corresponding residual. However, Krasker-Welsch weights are roughly proportional to $1/|x_i|$ and some quadratic terms can still become unbounded when $|x_i|\to\infty$. This partially explains the mediocre performance of IA in Table 1. The strategies of WA and NA are more efficient. When $\rho\equiv\rho_c$, NA completely rejects distant x_i's with a large residual and produces the largest steps, but the smooth rejection of WA does not generate a singular $\tilde{H}$. A compromise can be obtained by using $\tilde{D}=\mathrm{diag}(g(r_i/(\sigma w_i))^{\alpha})$ with $\alpha>1$. Notice that WA uses $\alpha=1$, NA uses $\alpha=\infty$ and HA uses $\alpha=0$. The column in parenthesis in Table 1 indicates the iteration numbers obtained with $\alpha=3$.

In conclusion, the results of this paper demonstrate the importance of downweighting outlying rows of X when approximating the Hessian matrix in Newton like-methods. The strategies discussed appear to show strong promise for providing simple but effective solutions to this problem.

Acknowledgements. Thanks are due to Frank Hampel and Werner Stahel for stimulating discussions and to Alex Randriamiharisoa for programming assistance.

REFERENCES

Dutter R. (1977a), Algorithms for the Huber Estimator in multiple regression, Computing 18, pp. 167-176.

Dutter R. (1977b), Numerical solution of robust regression problems, computational aspects, a comparison, J. Statist. Comput., Simul., 5, pp. 207-238.

Fletcher R. and Powell M.J.D. (1964), A rapidly convergent descent method for minimization, Computer Journal 6, pp. 163-168.

Hampel F.R., Ronchetti E.M., Rousseeuw P.J., Stahel W.A. (1986), Robust Statistics, Wiley, New York.

Huber P.J. (1972), Robust regression : asymptotics, conjectures and Monte Carlo, The Annals of Statistics 1, pp. 799-821.

Huber P.J. (1981), Robust Statistics, Wiley, New York.

Huber P.J. and Dutter R. (1974), Numerical solution of robust regression problems, COMPSTAT 1974, Proc. in Comput. Stat., pp. 165-172.

Marazzi A. (1980), ROBETH : a subroutine library for robust statistical procedures, COMPSTAT 1980, Proc. in Comput. Stat., pp. 577-583.

Marazzi A. and Randriamiharisoa A. (1985), ROBETH-ROBSYS : a software for robust statistical computing, Cahiers de recherche et docum., Institut Universitaire de Médecine Sociale et Préventive, Lausanne.

Samarov A. and Welsch R.E. (1982), Computational procedures for bounded-influence regression, COMPSTAT 1982, Proc. in Comput. Stat., pp. 412-418.

Testing of Statistical Algorithms and Programs with the Help of the Multivariate Sample, Having Described Values of Empirical Parameters

By E.-M. Tiit, Estonian SSR

The construction of multivariate samples having given values of sample marginal moments and given sample correlation matrix is described. Such "exact samples" are useful for testing the validity and correctness of the algorithms and programs of multivariate statistical analysis. For one family of test distributions - the family with constant correlation matrix - the exact values of parameters of classical multivariate procedures (regression, factor, component, canonical analyses) are given, depending on the given correlation and dimensionality.

Keywords: testing statistical algorithms, sample correlation matrix, factor analysis, canonical analysis, partial correlation, multiple correlation

1. The problem of testing of software in multivariate statistics

In the situation when a very great amount of statistical software is available for numerous types of computers, exploited by great numbers of users (most of them non-mathematicians), the problem of testing, comparing and estimating of algorithms and programs is of increasing interest, see Francis et al (1975), Vahle and Weber (1982).

Usually for testing of algorithms of multivariate statistics the following two methods are used:

1° The "classical" (empirical) examples, described in some earlier published papers.

2° The Monte-Carlo method.

The shortage of the first method is its randomness; besides, the correct exact values of the parameters investigated are, as a rule, unknown.

Compstat 1986

The second method gives the possibility of systematical research, but, it demands large series of trials, needing much computing time and completing software.

In the report the alternative method for testing classical procedures of multivariate statistical analysis is suggested. The method is useable for all types of computers, including the most modest personal ones.

2. The estimation of the algorithmical error

For the procedures of classical multivariate statistics the sufficient statistics are the empirical first and second moments. That means, our task is the following:

1) to describe such a family Θ of moments' vectors ϑ, $\vartheta = (\mu^T \vdots (\mathrm{vec}\,\Sigma)^T)$, that for every ϑ ($\vartheta \in \Theta$) the vector β of parameters of statistical procedure investigated can be computed exactly (analytically),

$$\beta = \beta(\vartheta) \tag{1}$$

2) to find the algorithm for generating the finite sample X^* having the empirical parameters exactly equal to ϑ.

If A is some algorithm for estimating β on the basis of any sample (X; n), then the estimation $\tilde{\beta} = \tilde{\beta}(A; X^*)$ must coincide with the true value of β, computed analytically (1). In some cases $\tilde{\beta} = \beta + b(\beta, n)$, where $b(\beta, n)$ is the bias due to the finity of the sample, $\lim_{n \to \infty} b(\beta, n) = 0$; then take instead of (1) $\beta^* = \beta - b(\beta, n)$, that is also computable analytically by ϑ and n.

In the case

$$|\tilde{\beta}(A; X^*) - \beta^*| < \eta, \tag{2}$$

where η is some given small constant, the algorithm A is correct (ϑ belongs to the admissibility region of algorithm A). Otherwise the algorithm A induces an algorithmical error

$$\varepsilon = \tilde{\beta}(A; X^*) - \beta^*$$

in the estimated parameter β of statistical procedure.

3. The families of distributions with given moments and the computation of parameters of statistical procedures

In Tiit (1984) several families of distributions, such as constant correlation matrix, block correlation matrix and others were introduced, and the formulae for exact computation of regression parameters were given. In the similar way, for these families the exact values of the parameters of discriminant, factor, component, and canonical analyses and the partial correlations may be computed analytically, so far as they are the functions of inverse matrix, eigenvalues or eigenvectors of the correlation matrix or its simple functions.

For instance, in the case of constant correlation matrix $R(\rho)$ of order k we have the following results.

1° The coefficient of partial correlation (when the influence of s variables is eliminated, $1 \leqslant s \leqslant k-2$): $\rho/(1+s\rho)$.

2° The coefficients of regression (by s terms, $1 \leqslant s \leqslant k-1$) are

$$b_i = \frac{\rho}{1+(s-1)\rho}, \quad i = 1,\dots,s; \; b_o = 0;$$

the multiple correlation

$$R = \frac{\rho}{\sqrt{1+(s-1)\rho}}$$

the estimated variance s^2 of regressand

$$s^2 = \frac{(1-\rho)(1+s\rho)n}{(1+(s-1)\rho)(n-s-1)},$$

and the values of F for testing the significance of the whole s-term model and one term correspondingly:

$$F = \frac{\rho^2(n-s-1)}{(1-\rho)(1+s\rho)},$$

$$F(b) = \frac{\rho^2(n-s-1)}{(1+(s-2)\rho)(1+s\rho)}.$$

3° All the loadings of the first principal component are equal,

$$f_{1i} = \sqrt{\frac{1+(k-1)\rho}{k}}, \quad i = 1,\dots,k.$$

All the other principal components are not uniquely defined. In the case of stopping rule $\lambda_i \leqslant 1$ only one component must be computed.

The individual factor scores are

$$l_{1i} = \frac{1}{\sqrt{(1+(k-1)\rho)k}}, \quad i = 1,\dots,k.$$

4° The <u>classical factor analysis</u> (with iterations) gives one factor only, so as the reduced correlation matrix has the eigenvalues $\lambda_1 = k\rho$, $\lambda_2 = \dots = \lambda_k = 0$,

$$f_{1i} = \sqrt{\rho},$$

$$l_{1i} = \frac{\sqrt{\rho}}{1+(k-1)\rho}, \quad i = 1,\dots,k.$$

5° The <u>canonical correlations</u> in the case when there are in the first group q and in the second group p $(p \leqslant k-q)$ variables:

$$\rho_1 = \rho\sqrt{\frac{pq}{(1+(p-1)\;)(1+(q-1)\;)}},$$

$$\rho_2 = \dots = \rho_s = 0, \quad s = \min(p, q),$$

and the coefficients of canonical variables of the first and second groups are correspondingly:

$$\frac{1}{\sqrt{q(1+(q-1)\rho)}}, \quad \frac{1}{\sqrt{p(1+(p-1)\rho)}}.$$

4. Construction of k-variate sample with given first and second moments

<u>Lemma 1</u>. There exists an exact k-variate sample with $\bar{x}_i = 0$ and $s_i^2 = 1$, $r_{ij} = 1$, $i,j = 1,\dots,k$, consisting of two points $(1,\dots,1)$ and $(-1,\dots,-1)$.

<u>Lemma 2</u>. Let k be fixed, $k \geqslant 2$. There exist a k-variate sample with $\bar{x}_i = 0$, $s_i^2 = 1$ and $r(i,j) = 0$, $i \neq j$, $i,j = 1,\dots,k$, of size $n = 2^h$,

$$h = [\log_2 k] + 1. \tag{3}$$

<u>Proof</u> gives the construction of the sample X ($k \times n$-matrix)

by its rows $x_{(i)} = (x_{i1}, \dots, x_{in})$, $i = 1, \dots, k$ that is similar to some construction of ortogonal contrasts:

$$x_{(1)} = y_{(1)};$$

when $x_{(1)}, \dots, x_{(2^l-1)}$ are already defined, then

$$x_{(2^l)} = y_{(l+1)},$$

$$x_{(2^l+i)} = y_{(l+1)} * x_{(i)}, \quad i = 1, \dots, s,$$

$$s = \begin{cases} 2^l - 1, & \text{if } l < h-1, \\ k - 2^{h-1}, & \text{if } l = h-1, \end{cases}$$

where $y_{(i)} = (y_{i1}, \dots, y_{in})$ $(i = 1, \dots, h)$ are auxiliary vectors, defined in the following way:

$$y_{ij} = \begin{cases} 1, \text{ if } j = 2l \cdot 2^{h-i} + 1, \dots, (2l+1)2^{h-i}; \\ -1, \text{ if } j = (2l+1)2^{h-i} + 1, \dots, (2l+2)2^{h-i}; \\ \quad l = 0, \dots, 2^{i-1} - 1, \; i = 1, \dots, h; \; j = 1, \dots, n. \end{cases}$$

The properties of the sample

$$X = \begin{pmatrix} x_{(1)} \\ \dots \\ x_{(k)} \end{pmatrix}$$

follow immediately from the construction.

Theorem. Let ρ be rational number, $\rho \in [0, 1]$. Then there exists a k-variate sample of size n with the sample correlation matrix $R(\rho)$ and $\bar{x}_i = 0$, $s_i = 1$, $i = 1, \dots, k$;

$$n = 2s_1 + 2^h s_2, \tag{4}$$

where integers s_1 and s_2 fulfil the condition

$$\frac{s_1}{s_2} = \frac{2^{h-1}\rho}{1-\rho}, \tag{5}$$

h is defined by (3).

Proof. Let us define from (5) the minimal integers s_1 and s_2. Then construct the sample, taking s_1 times the sample, defined in Lemma 1 and s_2 times the sample, defined in Lemma

2. Then the size of the constructed sample is (4), and so as the correlation matrices of the subsamples are R(1) and R(0) correspondingly, the correlation matrix of the sample is constant,

$$R(r) = \frac{2s_1}{n} R(1) + \frac{2^h s_2}{n} R(0)$$

with element

$$r = \frac{2s_1}{2s_1 + 2^h \frac{s_1(1-\rho)}{2^{h-1}\rho}} = \rho \text{ , q.e.d.}$$

EXAMPLE. For $k = 7$ we have $h = 3$ and the sample sizes are following:

TABLE

ρ	0.1	0.2	0.3	0.4	0.5	0.6	0.7	0.8	0.9
n	80	10	80	40	16	20	80	40	80
s_1	8	2	24	10	8	12	56	32	72

References

Francis I., Heiberger R., Velleman P. (1975), Criteria and considerations in the evaluation of statistical programs package, Amer. Statist., 29, 52-56.

Tiit E. (1986a), Definition of mixtures with given moments, Učenye Zapiski Tartu. Gosudarst. Univ., Trudy Mat. Meh., 733, 3-13.

Tiit E. (1986b), Exact samples of multivariate distributions and their exploitation in statistical algorithm's testing, Učenye Zapiski Tartu. Gosudarst. Univ., Trudy Mat. Meh., 733, 40-63.

Tiit E. (1986c), Random vectors with given arbitrary marginals and given correlation matrix, Učenye Zapiski Tartu. Gosudarst. Univ., Trudy Mat. Meh., 733, 14-39.

Tiit E.-M. (1984), Formal computation of regression parameters, COMPSTAT 1984. Proceedings in Computational Statistics, Physica-Verlag, Wien, 497-502.

Vahle H., Weber R. (1982), Mathematische-statistische Programmsammlungen, Rechentechnik Datenverarbeitung, 19, 19-21.

Three-Mode Data Matrices

Analysis of Three-Way Data Matrices Based on Pairwise Relation Measures

By R. Coppi, Roma

SUMMARY. The basic types of 3-way data matrices are described. The special case of one or more sets of qualitative characters observed on one or more groups of individuals is then examined. Several exploratory techniques of analysis are suggested, based on appropriately defined measures of relationship between individuals or between categories of the investigated characters.

KEYWORDS. Three-way data; Simultaneous principal component analysis; Contingency tables; Association measures; Multidimensional scaling.

1. INTRODUCTION

A 3-way data matrix is to be considered as a result of the statistician's choice, who is willing to study a given "information frame" according to 3 classification criteria (e.g. individuals, variables, occasions, times, categories of an "exogenous" variable, etc.). It should be underlined that the same information frame could be examined from a 2-dimensional viewpoint (e.g. individuals-variables) or as a 3-way, or even n-way, data matrix, on the grounds of a suitable choice from among several alternatives.

In this paper we focus on 3-way data matrices. The emphasis will be on exploratory techniques of analysis, aiming at the representation of the various elements of the data matrix (individuals, variables, occasions, pairs of individuals, etc.) in low-dimensional spaces. We shall not take into account the data generating process. Therefore, we shall not consider a probability model for our data. In the next Section we introduce a general notation, useful for classifying the basic types of 3-way data matrices. In the remainder we deal only with the special case of "individuals-variables-occasions" data matrices, whose different forms are presented in Section 3. In Section 4 various euclidean spaces are illustrated, which may constitute the mathematical support for the analysis of this type of data. In this framework, a particular importance is attached to representing the pairwise relationships between elements (pairs of individuals, variables or occasions) in suitable tensor spaces. In fact, in Section 5, several strategies of analysis are suggested, based on the "relation matrices" set up by means of appropriate measures of relationship defined for all pairs of elements belonging to the same "criterion" (e.g. all pairs

Compstat 1986

of individuals). These developments refer to the case of qualitative characters as observed in one or more groups of individuals, on several occasions. As a special instance the analysis of a set of multidimensional contingency tables (referred to the same characters) is considered, based on appropriately defined association measures for pairs of categories.

2. TYPES OF THREE-WAY DATA MATRICES

We denote by I,J,K, respectively, the "ways" (or "criteria") and use the same symbol for the number of elements belonging to each way (e.g. the number of individuals, or occasions). Let i,j,k ($i = 1,\dots,I$; $j = 1,\dots,J$; $k = 1,\dots,K$) denote their respective elements (or, alternatively, their indices), and d_{ijk} the elementary (numerical or categorical) datum of the resulting 3-way matrix $D = \{d_{ijk}\}$. In order to classify the different types of matrices, which can be envisaged, let us introduce the following notation:

(I) I and K are "independent": $I \text{ ind } K \Leftrightarrow \forall k: I_k = I$ (implying: $\forall i: K_i = K$).

(II) I and K are "dependent": $I \text{ dep } K \Leftrightarrow \exists k: I_k \neq I$ (implying: $\exists i: K_i \neq K$).

Analogous definitions can be provided for the remaining pairs: (I,J) and (J,K). In (I) and (II) we denoted by I_k the set of elements belonging to I, which are associated, in D, with the k-th element of K. The same notation is adopted for the other criteria. We can, then, show that there exist four types of 3-way data matrices:

1) "Cubic" (or "Parallelepiped") matrices:

   I ind J, I ind K, J ind K

2) Sequence of 2-way matrices with only one criterion having "fixed elements" with respect to the others:

   e.g. I ind J, I ind K, J dep K

3) Sequence of 2-way matrices with one criterion having "fixed elements" only with respect to the criterion which defines the sequence:

   e.g. I ind K, I dep J, J dep K

4) Sequence of 2-way matrices with no criteria having "fixed elements" with respect to any other:

   I dep J, I dep K, J dep K

In the literature, only cases 1 and 2 have been considered in detail so far (see, e.g., Kroonenberg, 1983 for type 1 matrices, which are also called 3-mode matrices; Escoufier, 1980, Escofier and Pages, 1984, for type 2 matrices). A detailed investigation on the different types of 3-way matrices is provided by Bolasco (1986). In the ensuing discussion we shall consider only matrices of the type "individuals-variables-occasions" and shall think of the "occasions" as being the elements of an unordered set K (therefore excluding the case of ordered sequences of 2-way matrices, such as those encountered in multivariate time series analysis).

3. THREE-WAY DATA MATRICES OF THE TYPE: "INDIVIDUALS-VARIABLES-OCCASIONS"

Let x_{ijk} denote the real value taken by variable j on individual i and occasion k, and $X = \{x_{ijk}\}$ the corresponding 3-way matrix. When considering qualitative characters, we shall refer to the indicator variables pertaining to each category of each character. We may distinguish the following types of data matrices:

a) "Same individuals-same variables-occasions". In this case we get a type 1 data matrix.

b) "Same individuals-different variables on different occasions". Here the set of variables is partitioned into K groups. The "occasions" correspond to the different groups of variables. Consequently, X is partitioned into K 2-way sub-matrices $X^1, X^2, \ldots, X^K$, whose columns pertain, respectively, to $J_1, J_2, \ldots, J_K$ variables ($\sum J_k = J$), and whose rows refer to the same I individuals. We obtain a type 2 matrix, which is the usual candidate for generalized canonical analysis.

c) "Same variables observed in different groups of individuals on the various occasions". In this case, the set of individuals is split into K groups of size $I_1, \ldots, I_K$, respectively, ($\sum I_k = I$), and X is partitioned into K 2-way sub-matrices $X_1, X_2, \ldots, X_K$ whose columns refer to the same set of J variables and whose rows pertain to the different groups of individuals. If all the variables are indicators of the categories of qualitative characters, the above matrix corresponds to K multidimensional contigency tables.

d) "Different variables observed in different groups of individuals on different occasions". The data matrix belongs to type 4.

4. REFERENCE SPACES AND RELATION MATRICES

The data appearing in X can be usefully represented as clouds of points in suitable euclidean spaces. In case (a) the basic spaces are R^I, R^J, R^K endowed with their respective metrics represented by the matrices: M_I, M_J, M_K. Note that the metric (or inner product) in R^J may vary according to the remaining criteria (e.g. M_J^k, k = 1,...,K). The same can happen for M_I or M_K. In case (b) the basic spaces are R^I, R^{J_k}, R^K, but also $R^J = \bigoplus_k R^{J_k}$ should be considered in order to compare the different clouds of individuals, as seen from the viewpoint of different sets of variables, in the same space (see, e.g., Escofier and Pages, 1984). Another suitable space, in this connection, is $R^{I^2} = R^I \otimes R^I$, where each occasion, k, is represented by a vector (tensor) corresponding to the "vectorization" of the matrix, say S_k, concerning the pairwise relations between individuals:

$$S_k = \{s^k_{ii'}\} \qquad (i,i' = 1,\ldots,I) \qquad (1)$$

where $s^k_{ii'}$ measures the relationship (similarity, dissimilarity, distance, inner product, etc.) between individuals i and i', computed on the basis of the k-th group of variables. The vectors: vec $S_1,\ldots,$vec S_K (where vec(•) denotes the operation which tranforms a matrix into a vector, obtained by superimposing its columns), constitute a cloud of points in R^{I^2}. In this space the single variables can be represented by means of the vectorization of the matrices: $S_j = \{s^j_{ii'}\}$, where $s^j_{ii'}$ measures the relationship between i and i', computed on the basis of variable j. A functional link might be established between S_j and S_k, such as $S_k = \sum_{j\in J_k} p_j S_j$, where the p_j's represent suitable weights for the variables. Analogously, in R^{K^2} we may represent the different individuals by means of the "relation matrices" (between occasions): $T_i = \{t^i_{kk'}\}$, where $t^i_{kk'}$ measures the relationship between occasions k and k', computed on the basis of the 2 groups of J_{k_1} and J_{k_2} variables, as observed on individual i.

In case (c), beside the spaces $R^{I_k}, R^J, R^K, R^I = \bigoplus_k R^{I_k}$, also $R^{J^2} = R^J \otimes R^J$ can be taken into account, for representing the (vectorized) matrices: $R_k = \{r^k_{jj'}\}$, where $r^k_{jj'}$ measures the relationship between variables j and j', computed on the basis of the k-th group of individuals. In the same space the single individuals can be represented by the matrices $R_i = \{r^i_{jj'}\}$, where $r^i_{jj'}$ measures the relationship between variables j and j', as observed on individual i. A link between R_i and R_k can be established, of the type already mentioned for S_j and S_k. Eventually, we might consider R^{K^2} where we can represent the different variables by the matrices: $T_j = \{t^j_{kk'}\}$, where $t^j_{kk'}$ measures the relationship between occasions k and k', computed on the basis of the values taken by variable j on the 2 groups of I_{k_1} and I_{k_2} individuals.

Introducing the above mentioned "relation matrices" is

particularly fruitful when dealing with type 2 data matrices, where there exists a "dependence" between two criteria (I dep K, or J dep K) and the rows of X_k (columns of X^k) refer to different elements. We might, then, compare the various occasions, by letting a square matrix (of the same order) correspond to each data matrix, whose structure is thereby summarized. In the technique called STATIS (Escoufier, 1980) the following inner product matrices are utilized to this purpose: $S_k = X^k M_J^k (X^k)'$, $R_k = X_k' M_I^k X_k$. D'Ambra and Marchetti (1986) take into consideration case (b) and introduce some "between individuals relation matrices" of type (1), defined in a fairly general way, both for categorical and quantitative characters. In the next Section we shall generalize this approach to case (c). We underline that the systematic exploitation of the relation matrices allows us to carry out a full analysis of X, concerning not only the comparison among different occasions (as it is the case for the above mentioned techniques) but also the representation of the various elements of X in low-dimensional spaces. In what follows we shall provide some indications about this approach.

5. SOME STRATEGIES OF ANALYSIS BASED ON RELATION MATRICES

We shall consider only cases (b) and (c) and assume that all the characters are qualitative.

5.1. Case (b).

Let us suppose, first, that each group, k, contains only one qualitative character. We may define the following measure of relationship between individuals: $s_{ii'}^k = 1$, if i and i' fall in the same category of the k-th character; $s_{ii'}^k = 0$, otherwise. $S_k = \{s_{ii'}^k\}$ is a dissimilarity matrix between individuals. The $(I^2 \times K)$-matrix: $Y = (\text{vec } S_1, \ldots, \text{vec } S_K) = (\underline{s}_1, \ldots, \underline{s}_K)$ can be considered as a new "data" matrix, whose rows and columns may be represented, respectively, in R^K and R^{I^2}. In order to obtain a low-dimensional representation of the occasions and of the pairs of individuals, D'Ambra and Marchetti (1986) suggest a principal component analysis of Y. They show that, if the (centered) columns of Y are represented in the tensor space R^{I^2}, endowed with the metric $M_I \otimes M_I$, the principal component analysis of Y leads to the diagonalization of the relation (covariance) matrix:

$$S^* = \{\text{cov}(\underline{s}_k, \underline{s}_{k'})\} \tag{2}$$

where

$$\text{cov}(\underline{s}_k, \underline{s}_{k'}) = (1/I^2) \sum_{j,\ell} (n_{j\ell}^2 - (n_j n_\ell / I)^2) \tag{3}$$

$n_{j\ell}, n_{j\cdot}, n_{\ell}$ denoting, respectively, the joint and marginal frequencies pertaining to the 2-way contingency table concerning characters k and k'. Thus, we should carry out the eigenanalysis of S^*, whose elements are particular association measures between characters. This procedure may be generalized to the case of several categorical characters for each group by defining, for instance: $s^k_{ii'} = \sum_q s^k_{ii';q}$, where q denotes an arbitrary character in group k and $s^k_{ii';q}$ measures the relation between i and i' with respect to q. In this case, the elements of S^* are constituted by the sum of the pairwise association measures between the characters in one group and those in another group. The occasions and the pairs of individuals are represented in the factor spaces $F_{I^2} \subset R^{I^2}$ and $F_K \subset R^K$, determined by means of the principal component analysis of Y. Moreover, each categorical variable (represented by its corresponding tensor, $\underline{s}_q$) can be projected onto F_{I^2} as a "supplementary" element. The foregoing procedure can be formalized in the following way:

$$X = (X^1,\ldots,X^K) \rightarrow (S_1,\ldots,S_K) \rightarrow Y = (\underline{s}_1,\ldots,\underline{s}_K) \rightarrow S^* \rightarrow (F_{I^2}, F_K)$$

An alternative strategy could be suggested, consisting of the analysis of the set of relation matrices S_k by means of multidimensional scaling techniques. For instance, using an INDSCAL approach (Carroll and Chang, 1970) the S_k's (considered as dissimilarity matrices) can be modelled as follows:

$$s^k_{ii'} = \sum_{h=1}^{H} w_{kh}(F^h_i - F^h_{i'})^2 \qquad (4)$$

where F^h_i denotes the "score" of individual i on the h-th common factor and w_{kh} the weight attached to this factor in the k-th occasion. The estimation of the parameters in (4), on the basis of a given number H of factors and of a suitable fitting criterion, leads to a full analysis of X. Shortly, when discussing case (c), we shall provide further suggestions (based on relation matrices) which apply also to case (b).

5.2. Case (c).

Let us assume that the J variables are the indicators pertaining to Q qualitative characters with, respectively, $m_1,\ldots,m_Q$ categories ($\sum_{q=1}^{Q} m_q = J$). For each occasion (group of individuals), k, the Burt's matrix concerning the k-th Q-way contingency table is:

$$B_k = X_k'X_k = \{N^k_{qq'}\} \qquad (5)$$

where $N^k_{qq'} = \{n^k_{j\ell}\}_{q,q'}$ is the 2-way contingency table pertaining to characters q,q' on the k-th occasion. Note that $N^k_{qq} = \mathrm{diag}\{n^k_j\}_q$, where the diagonal elements are the marginal frequencies of character q in the k-th group of individuals. On the basis of (5) the matrix, R_k, of the pairwise relations between categories of the Q characters, on the k-th occasion, can be set up. The essential objectives of the analysis are: (i) the comparison of the association structures (among characters) on the various occasions; (ii) the study of how the degree of association between categories varies from occasion to occasion. Therefore we need to define an appropriate pairwise association measure between categories.

5.2.1. In our first proposal, we introduce the following measure, for characters $q \neq q'$, on occasion k:

$$r^k_{j\ell} = n^k_{j\ell} / \frac{I_k}{I}\sqrt{n_j n_\ell} \qquad (6)$$

where n_j, n_ℓ denote the overall marginal frequencies pertaining to categories j and ℓ. The index (6) is null in the case of maximum "dissociation" ($n^k_{j\ell} = 0$). Recalling that a reasonable estimate of the expected value of $n^k_{j\ell}$, assuming independence between characters q and q', is $I_k n_j n_\ell / I^2$, we see that (6) takes on the value $\sqrt{n_j n_\ell}/I$ in the intermediate case of independence (neither association, nor dissociation). Finally, (6) takes on the value $I/\sqrt{n_j n_\ell}$ in the case of maximum association ($n^k_{j\ell} = I_k$). We remark that:

$$\sum_{j\in q}\sum_{\ell\in q'} (r^k_{j\ell})^2 = (I/I_k)^2(\Phi^2_k+1) \qquad (7)$$

where Φ^2_k is the well known association measure between characters q and q' (on occasion k), computed on the basis of the expected frequencies previously recalled. When categories j and ℓ refer to the same character, we let

$$r^k_{j\ell} = \begin{cases} I/\sqrt{n_j n_\ell}, & \text{if } j = \ell \\ 0, & \text{if } j \neq \ell \end{cases} \qquad (8)$$

On the basis of the relation measure defined by (6),(8) we can set up the K "association tensors": $\underline{r}_1 = \mathrm{vec}\{r^1_{j\ell}\}, \ldots, \underline{r}_K = \mathrm{vec}\{r^K_{j\ell}\}$. If R^J is endowed with the metric: $M_J = \mathrm{diag}\{n_j/I\}$, then the "natural" metric induced in R^{J^2} is: $M_{J^2} = M_J \otimes M_J = \mathrm{diag}\{n_j n_\ell / I^2\}$. The squared euclidean distance between association tensors pertaining to occasions k and k' is then provided by:

$$d^2(\underline{r}_k,\underline{r}_{k'})=(\underline{r}_k-\underline{r}_{k'})'\,M_{J^2}(\underline{r}_k-\underline{r}_{k'})=\sum_j\sum_\ell(r^k_{j\ell}-r^{k'}_{j\ell})^2\,\frac{n_j n_\ell}{I^2}=$$

$$=\sum_{q=1}^{Q}\sum_{q'\neq q}d^2(N^k_{qq'},N^{k'}_{qq'}) \qquad (9)$$

where

$$d^2(N^k_{qq'},N^{k'}_{qq'})=\sum_{j\in q}\sum_{\ell\in q'}\left(\frac{n^k_{j\ell}}{I_k}-\frac{n^{k'}_{j\ell}}{I_{k'}}\right)^2 \qquad (10)$$

If the distance (9) is assumed as the relation measure between occasions, namely $r^*_{kk'}=d^2(\underline{r}_k,\underline{r}_{k'})$, then we can set up the matrix $R^*=\{r^*_{kk'}\}$, corresponding to matrix S^* of case (b). R^* is the distance matrix between the columns of $Y=(\underline{r}_1,\ldots,\underline{r}_K)$. In the light of formulae (9),(10) it gets a particular statistical meaning. Principal coordinate analysis (Gower, 1966) of R^*,which in this case is equivalent to a principal component analysis of Y, allows us to identify a low-dimensional space where the single occasions (Q-way contingency tables) and their mutual distances can be represented, along with the pairs of categories. Let us refer, for simplicity, to the first principal component of the occasions, which we denote by $\underline{f}_K$ (with respective eigenvalue λ). The first principal component concerning the pairs of categories is then provided by:

$$\underline{g}_{J^2}=\frac{1}{\sqrt{\lambda}}\,\bar{Y}\,\underline{f}_K \qquad (11)$$

where $\bar{Y}$ is derived from Y after its columnwise centering. As to the representation of the single categories we may utilize the singular value decomposition of $\underline{g}_{J^2}$, viewed as a $(J\times J)$-matrix. In tensor notation we have:

$$\underline{g}_{J^2}=\sum_h\gamma_h\,\underline{f}^h_J\otimes\underline{f}^h_J \qquad (12)$$

where $\underline{f}^h_J$ are the scores of the various categories on the h-th factor (pertaining to the categories) and γ_h is the h-th eigenvalue of $\underline{g}_{J^2}$. This procedure is analogous to that utilized in STATIS for the so called "intra-structure" analysis and it has the same mathematical properties. However, in our present framework, it refers to a more general type of relation matrices. Finally, it should be recalled that the single individuals can be represented in R^{J^2}, by means of their respective tensors $\underline{r}_i=\text{vec}\,R_i$.

taining to occasions k and k' is then provided by:

$$d^2(\underline{r}_k,\underline{r}_{k'})=(\underline{r}_k-\underline{r}_{k'})' M_{J^2}(\underline{r}_k-\underline{r}_{k'}) = \sum_j \sum_\ell (r^k_{j\ell}-r^{k'}_{j\ell})^2 \frac{n_j n_\ell}{I^2} =$$

$$= \sum_{q=1}^{Q} \sum_{q'\neq q} d^2(N^k_{qq'},N^{k'}_{qq'}) \tag{9}$$

where

$$d^2(N^k_{qq'},N^{k'}_{qq'}) = \sum_{j\in q} \sum_{\ell\in q'} (\frac{n^k_{j\ell}}{I_k} - \frac{n^{k'}_{j\ell}}{I_{k'}})^2 \tag{10}$$

If the distance (9) is assumed as the relation measure between occasions, namely $r^*_{kk'} = d^2(\underline{r}_k,\underline{r}_{k'})$, then we can set up the matrix $R^*=\{r^*_{kk'}\}$, corresponding to matrix S^* of case (b). R^* is the distance matrix between the columns of $Y = (\underline{r}_1,\ldots,\underline{r}_K)$. In the light of formulae (9),(10) it gets a particular statistical meaning. Principal coordinate analysis (Gower, 1966) of R^*, which in this case is equivalent to a principal component analysis of Y, allows to identify a low-dimensional space where the single occasions (Q-way contingency tables) and their mutual distances can be represented, along with the pairs of categories. Let us refer, for simplicity, to the first principal component of the occasions, which we denote by $\underline{f}_K$ (with respective eigenvalue λ). The first principal component concerning the pairs of categories is then provided by:

$$\underline{g}_{J^2} = \frac{1}{\sqrt{\lambda}} \bar{Y} \underline{f}_K \tag{11}$$

where $\bar{Y}$ is derived from Y after its columnwise centering. As to the representation of the single categories we may utilize the singular value decomposition of $\underline{g}_{J^2}$, viewed as a $(J\times J)$-matrix. In tensor notation we have:

$$\underline{g}_{J^2} = \sum_h \gamma_h \underline{f}^h_J \otimes \underline{f}^h_J \tag{12}$$

where $\underline{f}^h_J$ are the scores of the various categories on the h-th factor (pertaining to the categories) and γ_h is the h-th eigenvalue of $\underline{g}_{J^2}$. This procedure is analogous to that utilized in STATIS for the so called "intra-structure" analysis and it has the same mathematical properties. However, in our present framework, it refers to a more general type of relation matrices. Finally, it should be recalled that the single individuals can be represented in R^{J^2}, by means of their respective tensors $\underline{r}_i = \text{vec } R_i$.

5.2.2. An alternative strategy, as we already underlined when discussing case (b), consists of analyzing the set of relation matrices R_k $(k=1,\dots,K)$, whose elements are the measures (6) (or suitable transforms such as those leading to dissimilarity indices), by means of multidimensional scaling. Of course, instead of the indices (6) we may use, in this framework, the association measures based on cross product ratios which have desirable mathematical properties (Edwards, 1963; Altham,1970). One such measure, which may be particularly useful to this purpose, is the dissimilarity index between categories suggested by Di Ciaccio (1985). This is a function of the ratios:

$$\alpha^k_{j\ell j'\ell';(q,q')} = \frac{n^k_{j\ell}\, n^k_{j'\ell'}}{n^k_{j\ell'}\, n^k_{j'\ell}} \tag{13}$$

related to the contingency table $N^k_{qq'}$. Using (13) we define the (absolute) association measure for the pair (j,ℓ):

$$\mu^k_{j\ell;(q,q')} = \frac{1}{m_q m_{q'}} \sum_{j'} \sum_{\ell'} \ln \alpha^k_{j\ell j'\ell';(q,q')} \tag{14}$$

It can be shown (Di Ciaccio, 1985) that the maximum value taken by (14), in the case of maximum association between j and ℓ, is the following:

$$\max \mu^k_{j\ell;(q,q')} = \frac{(m_q-1)(m_{q'}-1)}{m_q m_{q'}} \ln \left[\frac{(2I_k-m_q-m_{q'}+2)^2}{4(m_q-1)(m_{q'}-1)} \right] \tag{15}$$

If we define

$$\lambda^k_{j\ell;(q,q')} = \mu^k_{j\ell;(q,q')} / \max \mu^k_{j\ell;(q,q')} \tag{16}$$

we may construct the following dissimilarity index between categories j and ℓ:

$$r^k_{j\ell} = \begin{cases} (1-\sqrt{\lambda^k_{j\ell;(q,q')}})/2\,, & \text{if } \lambda^k_{j\ell;(q,q')} \geq 0 \\ (1+\sqrt{|\lambda^k_{j\ell;(q,q')}|})/2, & \text{if } \lambda^k_{j\ell;(q,q')} < 0 \end{cases} \tag{17}$$

Note that $-1 \leq \lambda^k_{j\ell;(q,q')} \leq 1$. When j and ℓ belong to the same character q, we may put: $\lambda^k_{j\ell;(q,q')}=1$ or $(-1/m_q)$ according to whether $j=\ell$ or $j\neq\ell$. The set of matrices $R_k = \{r^k_{j\ell}\}$, obtained by means of (17), could be analyzed using the procedure described in 5.2.1. However, some difficulties arise as to the statistical interpretation of the "between tensors relation matrix" R^*, due to the complexity of the indices (17). Instead, we suggest to analyze the set R_k $(k = 1,\dots,K)$ by an INDSCAL procedure. This may be implemented using the current fitting criterion (see,e.g., Takane, Young and de Leeuw, 1977). Alternatively, we propose to utilize the following loss function,called "Mapping Error" (Sammon, 1969):

$$E(\rho^k,r^k) = [1/\sum_{j<\ell} r^k_{j\ell}] \sum_{j<\ell} [(\rho^k_{j\ell}-r^k_{j\ell})^2/r^k_{j\ell}] \tag{18}$$

where $\rho^k_{j\ell}$ is given by a formula analogous to (4). The adoption of the Mapping Error has some advantages (Di Ciaccio, 1985).Its utilization in our present framework is still under investigation.

REFERENCES

Altham P.M.E. (1970), The measurement of association of rows and columns for an r×s contingency table, J.R.S.S. series B, 32, 63-73.

Bolasco S. (1986), Towards a theory on the construction and the analysis of three-mode matrices, "Proceedings of the 33rd Meeting of the Italian Statistical Society", vol. 1, 183-195.

Carroll, J.D.; Chang J.J. (1970), Analysis of individual differences in multidimensional scaling via an n-way generalization of "Eckart-Young" decomposition, Psychometrika, 35, 283-319.

D'Ambra L.; Marchetti G.M. (1986), The analysis of three-way data matrices: a method based on relation measures between units, "Proceedings of the 33rd Meeting of the Italian Statistical Society", vol. 1, 171-182.

Di Ciaccio A. (1985), Representation of a new association measure between categories using multidimensional scaling, 4th International Symposium on Data Analysis and Informatics",INRIA, Centre de Rocquencourt, France, 369-378.

Edwards A.W. (1963), The measure of association in a 2×2 table, J.R.S.S. series A, 126, 109-114.

Escofier B.; Pages J. (1984), L'analyse factorielle multiple, Cahiers du B.U.R.O., Série Recherche, 42, Université Pierre et Marie Curie, Paris.

Escoufier Y. (1980), Explolatory data analysis when data are matrices, in K. Matusita (ed.), "Recent developments in statistical inference and data analysis", North Holland, Amsterdam, 45-53.

Gower J.C. (1966), Some distance properties of latent root and vector methods used in multivariate analysis, Biometrika, 53, 325-338.

Kroonenberg N.Y. (1983), Three-mode principal component analysis: Theory and applications, DSWO Press, Leiden.

Sammon J.W. (1969), A nonlinear mapping for data structure analysis, IEEE Trans. Computer, vol. C-18, n. 5, 401-409.

Takane Y.; Young F.W.; de Leeuw J. (1977), Nonmetric individual differences multidimensional scaling: an alternating least squares method with optimal scaling features, Psychometrika, 42, 7-67.

Factor Analysis of Evolution and Cluster Methods on Trajectories

By A. Carlier, Toulouse

SUMMARY

In this paper, we study the evolution of objects indexed by $i(i\in I)$ and associated with the lines of a two-way contingency table, observed at several times. At each time t $(t\in T)$, the j-th column of the slice $\{n_{ijt}, i\in I, j\in J\}$ is associated with the j-th category of one or seveveral qualitative variables. The adaptation of the Factor Analysis of Evolutions (FAE) to such a cube of data, proposed in (5) and (6) produces factors, correlation circles and graphical displays of trajectories in factorial planes. Here, after a brief overview of FAE, we propose a hierarchical agglomerative clustering (HAC) method, which is complementary to FAE. Thus the results of the HAC can be represented by mean of classes of trajectories in the factorial planes found independently using FAE. This allows us to see the agreement between the results of the two methods, and enhances the interpretation of both. In this way, the difficulty of interpreting a large number of trajectories in the factorial planes is overcome by using the HAC method. We apply this classification procedure to the example already analysed in (5) and (6).

KEYWORDS - Data Analysis - Three-way contingency tables - Local principal component analysis - Factor analysis of evolutions - Clustering methods on trajectories.

I - AN OVERVIEW OF FACTOR ANALYSIS OF EVOLUTIONS

1. Introduction

The FAE introduced by Y.Le Foll (11) may be regarded as a particular case of local principal component analysis (LPCA) which was introduced by L. Lebart ((9) and (10)) and extended in (1) and (12). The classical model of principal component analysis (PCA) (4) requires as input the triple (Y,M,D), where $Y=(y_i^j)$ is the data matrix, $M=\text{diag}(m_{jj}, j\in J)$ is the matrix defining the euclidian metric in $\mathbb{R}^p$ and $D=\text{diag}(p_i, i\in I)$ defines the weights of individuals. The LPCA requires also a set U of pairs (i,i') of $I\times I$, regarded as the set of the edges of a graph $G=(U,I)$ on I. It replaces the usual concept of covariance of two variables y^j and y^k by a local concept :

$$\text{cov}_\ell(y^j,y^k) = \frac{1}{2} \sum_{(i,i')\in U} p_i p_{i'} (y_i^j - y_{i'}^j)\ (y_i^k - y_{i'}^k) \qquad (1)$$

By substituting $I\times I$ for U in equation (1), we find the usual covariance formulation $\text{cov}(y^j,y^k)$. Then two variables are locally correlated if they wary together along the paths defined by the graph. Applied to the cube of data (n_{ijt}), the set I is replaced by the product $I\times T$ of pairs of individual and time, and U is defined by a given set U_T of pairs of $T\times T$ through

Compstat 1986

the relation : $((i,t),(i',t'))\in U \Longleftrightarrow i=i'$ and $(t,t')\in U_T$. We refer to articles (5) and (6) for a detailed description of this method, a comparison of different choices of quadruplet (Y,M,D,U) and an application to the example of section III.

2. What choice of coding ?

The choice of coding depends of the goal of the study ; let $n_{i.t} = \sum_{j\in J} n_{ijt}$, $n_{i..} = \sum_{t\in T} n_{i.t}$, and $n_{.j.} = \sum_{i\in I}\sum_{t\in T} n_{ijt}$.

The data cube (n_{ijt}) allows us to define for each individual and time the measures on J :

CODING C1 : $y_{i,t} : j\in J \rightarrow y^j_{it} = n_{ijt}/n_{i.t}$

("line with relative frequencies"), or :

CODING C2 : $y_{i,t} : j\in J \rightarrow y^j_{it} = n_{ijt}/n_{i..}$

("slice with relative frequencies").

These measures sum to one for coding C1 and to $n_{i.t}/n_{i..}$ for coding C2. The study of the evolution of the structure in J of the objects i over time uses the coding C1, and does not take into account the variations of the total $n_{i.t}$ of the object i over time. If we use the coding C2, the study of the evolution of y_{it} permits us to study both the structural and quantitative evolution of the objects at the same time.

3. The properties of distributional equivalence

Originally defined in the context of Correspondence Analysis (2) for the study of two-way contingency tables, the property of distributional equivalence is defined for FAE in (5) in the following way :

"*A method for analysis of the data cube* (n_{ijt}) *possesses the property of distributional equivalence (D.E.) on I (resp. J) if the results of the method are invariant when to proportional slices i and i' of I (resp. j and j' of J) are replaced by their sum*".

It is shown that LPCA satisfy the property of D.E. on I if the weights are chosen in such a way that the contribution of slice i to the local inertia is proportional to $n_{i..}$. For codings C1 and C2, a solution is to take the weights $p_{i,t}$ proportional to $(n_{i..})^{1/2}$. In an analogous manner, let $var_\ell(y^j) = \sigma_{\ell j}{}^2$ denote the local variance of y^j. The property of D.E. is verified on J if the contribution of the j-th variable to the local inertia is proportional to $n_{.j.}$. The χ^2 metric is one of the solutions. As in (5), we prefer the semi-reduction metric $M_{1/\sigma} = \text{diag}(1/\sigma_{\ell j}, j\in J)$ for which the contribution of variable j is equal to the local standard deviation $\sigma_{\ell j}$.

II - CLUSTERING OF TRAJECTORIES

The visualization of the trajectories on the factorial planes rapidly becomes difficult to interpret, even when they are as few as a dozen. The use of clustering methods permits us to overcome this problem, because it is possible to replace each class by the trajectory, centroïd of the class. Furthermore, since each object is represented by a curve, the representation of all the trajectories of a cluster in the principal planes permits us to verify visually the agreement of the results of the two methods. In the sequel, the HAC method used is the socalled "incremental sum of squares" method (8). We shall now discuss the choice of the distance.

Distance between trajectories

We propose a choice of three distances, which are :

$$d_1(i,i')^2 = \sum_{(j,t)\in J\times T} \mu_{jt}\,(y^j_{it} - y^j_{i't})^2 \quad \text{(Mean instantaneous distance)}$$

$$d_2(i,i')^2 = \sum_{j\in J,\,(t,t')\in U_T} \mu_{jtt'}\,((y^j_{it} - y^j_{it'}) - (y^j_{i't} - y^j_{i't'}))^2$$

(Difference distances)

$$d_3(i,i')^2 = \beta_1\, d_1(i,i')^2 + \beta_2\, d_2(i,i')^2 \quad \text{(Mixed distances)}$$

In the expression of these distances, the positive coefficients μ_{jt} and $\mu_{jtt'}$ are chosen so as to give them the property of D.E. on J, and the coefficients β_1 and β_2 such that the distances d_1 and d_2 are homogenized and weighted. Note that d_2 depends only on the difference between the slopes of the curves $t \to y^j_{it}$ and $t \to y^j_{i't}$, and that these 3 distances are approximations of the ones between processes in a Hilbert space :

$$d_1(f,g)^2 = \int_a^b ||f(t) - g(t)||^2\, dt$$

$$d_2(f,g)^2 = \int_a^b ||D_f(t) - D_g(t)||^2\, dt \quad \text{with differential operator D}$$

$$d_3(f,g)^2 = d_1(f,g)^2 + d_2(f,g)^2 \quad \text{distance in the Sobolev space } H_1 .$$

Finally, note that :

$$d_1(i,i') = 0 \iff y_{it} = y_{i't} \quad \forall t\in T$$

$$d_2(i,i') = 0 \iff \{y_{it},\ t\in T\} \text{ and } \{y_{i't},\ t\in T\} \text{ correspond by translation.}$$

III - APPLICATION TO A DEMOGRAPHIC EXAMPLE

The example considered here has been analysed by different methods in (7) and by FAE in (5) and (6). The data stem from 4 censures taken in 1954, 1962, 1968 and 1975. The set J consists of 9 socio-economic categories : farmer or agricultural worker (EA and OA), self employed non professional workers (AC), self employed professional and executives (PL), middle order executives (CM), clerks (EM), laborers (OU), unskilled service workers (SE), others (CP). Finally, I is a set of 42 rural cantons of Languedoc-Roussillon. We will present only the results of LPCA using coding C2, with M and D defined as in I.3 and with $U_T = \{(1,2), (2,3), (3,4)\}$. The interpretation of the local correlation circle (fig. 1) shows that the numbers of people of the two agricultural categories vary together over time, but are not related with the evolution of the total of AC, and have a weak negative relation with the number of people of other urban categories. Factor F1 (57 % of the total inertia)

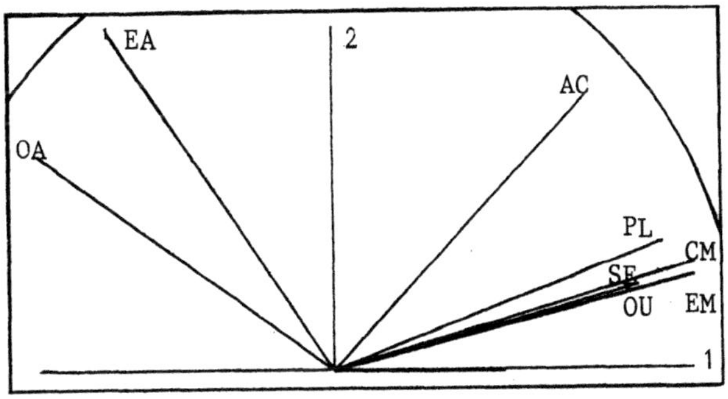

Fig.1 Local correlation circle

has a weak negative correlation with agricultural categories and shows mainly the urbanization of rural cantons (the contribution of urban categories represent 77 % of $var_\ell(F^1)$). The second factor (19 % of the inertia) is a size factor, related in particular with the evolution of rural population and with category AC. Figure 2 permits us to follow the evolution of an arbitrary set of 9 of the 42 rural cantons. The HAC methods was performed with the three distances and coding C2. For theses distances, the results were

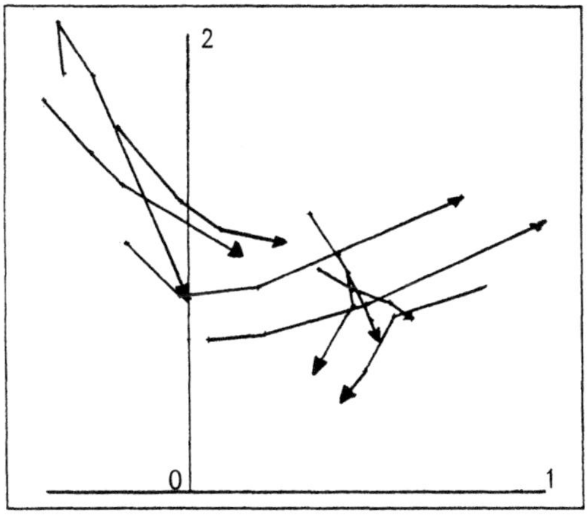

Fig. 2 An arbitrary set of 9 trajectories represented in the plane (1,2).

close, except for the cantons where the variations were small. For such cantons, d_1 groups the ones which vary in opposite directions, but having similar structures ; d_2 is more sensitive to the direction of variation, but groups cantons with diverse structures. The distance d_3 is prefered here : we give (figure 3) the dendrogram and the representation of the trajectories in 7 classes produced by the HAC. When only two classes are considered, the cantons which have a tendency to decrease (classes 3 and 4 : rural exodus ; classe 1 decrease of number of labourers of urbanized cantons ; classe 2 intermediary cantons) are separated from those were the urban categories are weakly, strongly and very strongly developping (classes 5, 6 and 7).

CONCLUSION

The methods described here can be regarded

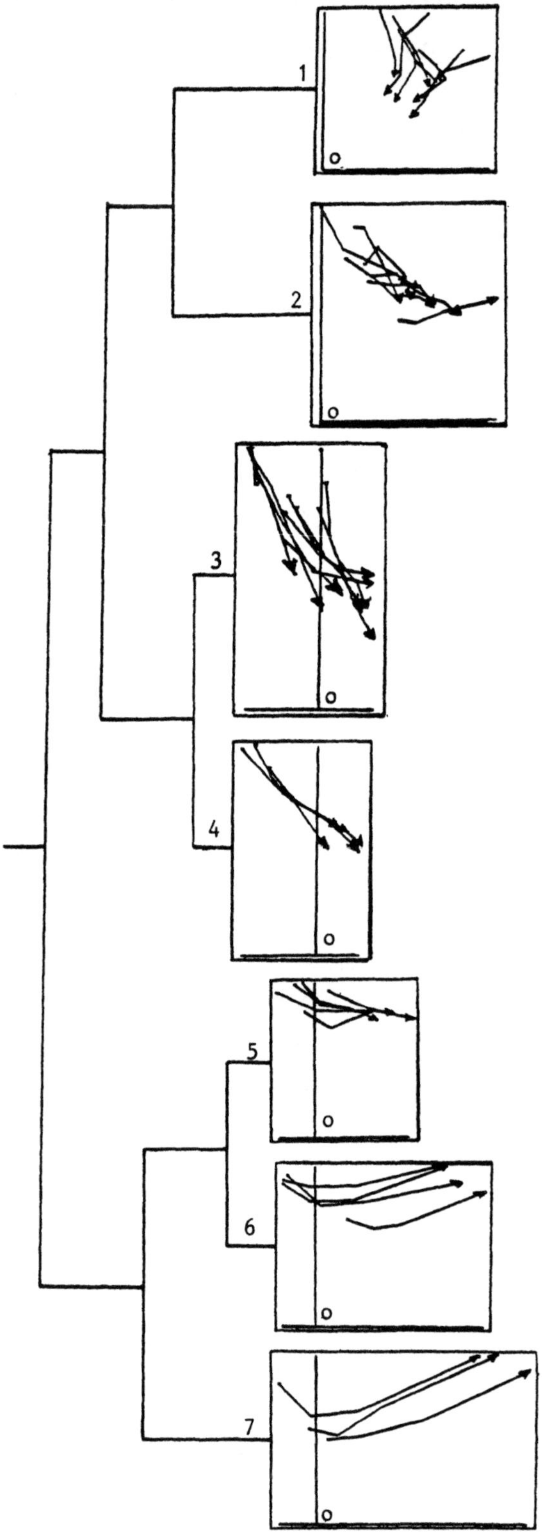

Fig. 3 Dendrogram with classes of trajectories

as good tools to perform as a first step a descriptive analysis of the variations of the objects defined by such a cube of data. The interesting aspect of these methods lies in the globality of their approach, analysing all the objects from the point of vue of all the variables. Such a mixing is allowed by natural weightings of the objects and of the variables, derived by the distributional equivalence property. The difficulties of the interpretation are not larger as for a classical PCA or HAC method, but the same precautions have to be taken in account. From these interpretations can be derived hypotheses to be confirmed or precised in a further step.

(1) Banet T.A. and Lebart L. (1974), Local and partial principal component analysis, COMPSTAT 84, 113-123.

(2) Benzecri J.P. (1973), L'analyse des données, Tome II, Dunod.

(3) Besse P. (1979), Etude descriptive d'un processus, Université Paul Sabatier, Toulouse.

(4) Caillez F. and Pages J.P. (1976), Introduction à l'analyse des données. SMASH.

(5) Carlier A. (1985), Analyse des évolutions sur table de contingence, quelques aspects opérationnels, Data Analysis and Informatics, Tome II INRIA Le Chesnay, 421-428.

(6) Carlier A. (1985), Application de l'analyse des évolutions et de l'analyse intra-périodes, Statist. et Anal. des Données, VoL. 10 n° 1, 27-53.

(7) Escoufier Y. and all (1985), Objectifs et procédures de l'analyse conjointe de plusieurs tableaux de données, Stat. et Anal. des Données Vol. 10, n° 1, 1-116.

(8) Jambu M. (1978), Classification automatique pour l'analyse des données, Tome 1, Dunod, Paris

(9) Lebart L. (1969), Analyse statistique de la contiguité, Publ. Inst. Stat. de l'Université de Paris, XVIII, 87-112.

(10) Lebart L. and Tabard N. (1971), Description statistique de certaines relations linéaires, Rapport CORDES n° 13, 132-177. CREDOC, Paris.

(11) Le Foll Y. (1972), L'analyse factorielle des évolutions, Petit sém. sur l'anal. Fact. n° 2, 23-27. Univ. Sc. Soc. de Grenoble.

(12) Le Foll Y. (1982), Pondération des distances en analyse des données. Stat. et Anal. des Données n°1, 13-31.

Cluster Analysis Methods and Algorithms

Links Between Clustering and Assignment Procedures

By A. D. Gordon, St. Andrews

Summary: In the situation in which a set of objects is summarized in terms of a smaller number of groups of similar objects, links are noted between two definitions of the word 'classification': obtaining groups of similar objects, and assigning objects to one of a set of existing groups. Some problems on which further work could be undertaken are described.

Keywords: adaptive clustering; assignment; classification; cluster analysis; discriminant analysis; hierarchical classification.

1. INTRODUCTION

In the statistical literature, the word 'classification' has been used in two different senses, to refer to

(i) the division of a set of objects into groups, with the property that objects in a group are similar to one another and dissimilar to objects in other groups;

or to

(ii) the assignment of an object to one of a set of existing groups.

Thus, Gordon (1981) presents an introduction to methodology pertaining to the first definition above, whereas Hand (1981) uses the word 'classification' in the second sense and concentrates on describing procedures applicable to that problem. In this paper, the two activities will be referred to as 'cluster analysis' and 'assignment', respectively.

The two definitions of classification given above are sometimes seen as referring to completely separate activities; the aim of this paper is to review some of the links between them.

It has been widely recognised that it is important to

Compstat 1986

specify the properties of the groups detected in a cluster analysis. The subsequent assignment of new objects to these groups would be facilitated if this specification included an explicit statement of properties required to be satisfied (or largely satisfied) by members of the group, but this is lacking in the great majority of clustering procedures; exceptions are the method of maximal predictive classification (Gower, 1974), and monothetic divisive procedures (e.g. Crawford and Wishart, 1967; Lance and Williams, 1968), these latter being a form of binary diagnostic key (Payne and Preece, 1980; Breiman *et al.*, 1984).

The two main forms in which data are summarized in a cluster analysis are:

(i) a partition into g (say) disjoint groups of objects;

(ii) a hierarchically-nested set of such partitions, commonly represented in a dendrogram, or rooted tree diagram.

The number of groups, g, and their identity are both unknown at the start of the investigation; indeed, the conclusion to be drawn from a cluster analysis may well be that any partition of the objects into groups would misrepresent the structure in the data. Sometimes, however, there will be informative representations of the data for several different values of g, when groups at one level can be related to one another at a higher level (lower value of g). The next two sections note links between clustering and assignment procedures based on, respectively, a single partition of the data set and a hierarchically-nested set of partitions.

2. LINKS BASED ON PARTITIONS

Two main examples are considered. First, in assignment procedures, the properties of the groups are usually not completely specified from theoretical considerations; instead, relevant information is obtainable from a training set of objects. For example, it may be postulated that the objects in a group can be profitably considered to be a sample from some conceptual multivariate normal population, the parameters of which are estimated from the training set. Usually, the identity of the group to which each of the objects in the training set belongs is known. However, if this information is not available, an estimate of the composition of the groups can be provided by undertaking a cluster analysis of the training set.

Secondly, there is an upper limit to the size of data set which it is computationally feasible to analyse using clustering procedures, and while one can expect computing developments to increase the upper limit, it seems likely that different strategies would still be used for the analysis of large data sets. One approach is to use clustering procedures to analyse a suitably-chosen subset of the data, and then assign other objects to the groups found in the subset.

These two problems both fit into the following general framework for the classification of a data set (Ω, say):

1. Select a sample ω of objects.
2. Use clustering procedures to obtain a classification of the objects in ω.
3. Assign the objects in $\Omega\backslash\omega$ to the classification obtained in (2).

In the latter problem, some thought has to be given to ways of selecting the subset ω of the data on which the cluster analysis will be undertaken. Little work appears to have been done on this problem. The aim is for the subset ω to contain representatives of all (or, more realistically, all major) groups present in the complete data set, Ω. On occasion, this endeavour is assisted by external information, such as the spatial location of each object in remote sensing studies (e.g. Bryant, 1979), but such information is usually lacking.

Clearly, objects with many close neighbours, few of which have already been chosen, are candidates for membership of ω. One possibly useful visual aid is a plot, for each object in $\Omega\backslash\omega$, of the average distance to (or dissimilarity with) its k nearest neighbours, for a range of small values of k. Objects whose traces lie below those of other objects over a range of small values of k are indicated as potentially valuable members of ω. A completely automated rule could be defined by specifying a single value of k, but given suitable graphics facilities the lowermost traces could be displayed to the user to enable him to select members of ω. In the analysis of large data sets, it is essential to use efficient methods of evaluating average distances; distributed-array processors (Gostick, 1979) seem potentially very valuable in this context. It would seem advisable for the plot to be sequentially updated so that the only distances considered are those between members of the set $\Omega \backslash \omega$. In the absence of such a strategy, ω will be dominated by objects from compact groups.

Further work which could usefully be done in this area includes an investigation of useful values of k, and the final size of ω; a complicating factor is that different measures of dissimilarity or distance may be appropriate in different groups.

Having obtained the subset ω, the second step is to use the methodology of cluster analysis to provide a partition into disjoint groups of the objects belonging to ω; some comments on the selection of appropriate methods of analysis are given in Gordon (1981, Chapter 6). If the clustering procedure employed does not explicitly specify an assignment rule for the objects in $\Omega \backslash \omega$, one can evaluate a measure of the distance from each group to each of the objects to be assigned.

In the third step, each object in $\Omega \backslash \omega$ is assigned to the group in the partition of ω to which it is closest, although it could be useful to regard some objects as belonging to the overlap between two or more groups, or as outliers if the distances to them from each group all exceed some threshold; such outliers would be re-examined to establish whether any of them belonged to groups that failed to contribute samples to the subset ω. It would also be informative to compare the results obtained from several different choices of the subset ω, since the membership of ω could be quite markedly affected by the choice of strategy adopted in the first step.

These ideas are illustrated under the assumption that each

group is a sample from a p-variate normal distribution, the probability density function of the ith distribution being

$$f_i(x) = \frac{1}{(2\pi)^{p/2}\,(\det \Sigma_i)^{\frac{1}{2}}} \exp\{-\tfrac{1}{2}D^2(x;\mu_i,\Sigma_i)\} \qquad (1)$$

where $D^2(x;\mu_i,\Sigma_i) = (x-\mu_i)'\Sigma_i^{-1}(x-\mu_i)$ (2)

and μ_i and Σ_i are the mean vector and covariance matrix of the distribution (Morrison, 1976).

If the prior probability of the ith distribution is π_i $(i=1,\ldots,g)$, the probability that a reading x comes from the ith distribution is

$$p(i|x) = \pi_i f_i(x) \Big/ \sum_{j=1}^{g} \pi_j f_j(x) \qquad (3)$$

By substituting into (3) sample estimates y_i and S_i of the unknown population parameters μ_i and Σ_i, one can obtain an estimate $\hat{p}(i|x)$ of $p(i|x)$ and use it for assigning objects; under certain conditions, described below, this is equivalent to a minimum-distance assignment using the criterion $D^2(x;y_i,S_i)$.

It should be noted that the probabilities $\hat{p}$ are inaccurate, for two reasons. First, instead of simply inserting estimates of the population parameters into (3), one should use a predictive approach (see Aitchison and Dunsmore, 1975, Chapter 11); in this case, this would involve replacing multivariate normal distributions by multivariate t-distributions. Secondly, because of the method of selecting the subset ω, the covariance matrices of the groups G_i $(i=1,\ldots,g)$ in ω will usually be smaller in magnitude than the corresponding Σ_i. For these reasons, the estimated probabilities

$$p^*(m|x) \equiv \max_i \hat{p}(i|x) \qquad (4)$$

will be over-estimates of the true maximum posterior probabilities. Nevertheless, the values $\{p^*(m_j|x_j) \mid x_j \in \Omega\backslash\omega\}$ can be used informally in assignment procedures.

Examples to illustrate these ideas will be given in the oral presentation; this material is not included here because of space limitations.

If it is further assumed that the determinants of the

covariance matrices $\{S_i \ (i=1,\ldots,g)\}$ are all equal, the assignment of an object x is to the group from which its Mahalanobis distance $D^2(x;y_i,S_i)$ (defined in equation (2)) is minimum; when differences within the sets $\{\det S_i \ (i=1,\ldots,g)\}$ and $\{\pi_i \ (i=1,\ldots,g)\}$ are not too great, assignment is still largely dictated by the size of the Mahalanobis distances.

This measure of distance is relevant if each group occupies an ellipsoidal region in attribute space, which may not be a realistic assumption for many data sets. With a general definition of the distance from a group G_i to an object x, the third step of the classification procedure described in this section has much in common with so-called adaptive clustering procedures, in which the algorithm adapts its strategy according to the structure encountered in the data. In particular, the agglomerative procedure described by Rohlf (1970) assigns new objects to existing groups using local distance criteria. One difference is that adaptive clustering strategies generally involve a sequential up-dating of groups and recalculation of local distances. This approach seems less feasible for the classification of large data sets, for which making assignments during a single pass through the data would seem to be indicated. Further, empirical investigations by Rohlf (1970) and Maronna and Jacovkis (1974) indicated that if groups are sequentially updated in this manner, some groups can grow disproportionately at the expense of others.

Other adaptive clustering strategies are studied by Diday and Govaert (1977), Lefkovitch (1978,1980), and Art *et al.* (1982). There would seem still to be scope for investigation of other methods of measuring the distance from a group to an object.

3. LINKS BASED ON TREE DIAGRAMS

There has been a limited amount of work concerned with adding objects to a dendrogram, but two main approaches can be described; motivations are the analysis of large and/or incomplete data sets, and the validation of classifications.

First, there exist efficient algorithms for the single link (Sibson, 1973) and complete link (Defays, 1977) clustering methods which construct a dendrogram by successively adding new objects to, and modifying, an existing tree diagram, using a pointer representation.

The second main approach, relevant for clustering procedures that aim to take a middle road between the extremes of the single link and complete link criteria, has much in common with diagnostic keys. A diagnostic key can be represented by a rooted tree diagram. An object to be identified is moved down the tree, starting at the highest internal node. At each internal node encountered, a question is asked about the properties of the object, the answer to which specifies the branch below the node down which the object proceeds. Eventually, the object reaches one of the terminal nodes, defining the class to which the object is assigned.

Similar algorithms have been developed for assigning new objects to a dendrogram (e.g. Zupan, 1982; Jambu and Lebeaux, 1983; Wishart, 1985); one difference is that a new object cannot be routed infallibly along one of the branches below an internal node, but merely resembles to a greater or lesser degree the set of objects reached via the branch. Unlike the situation with the diagnostic key, there is thus no guarantee that the correct branch has been selected at each stage, and various rules have been devised to attempt to prevent the object's being located in a inappropriate part of the tree. The final tree can be markedly influenced by the order in which objects are incorporated into it; its properties could be improved by a careful selection of the objects belonging to the original small tree, and of the order in which objects are incorporated into the tree; some of the methodology described in Section 2 may prove helpful in this respect.

Acknowledgment: An earlier version of this paper was presented at a SERC-supported research workshop on 'Statistics and Pattern Recognition', held in Edinburgh in July 1985. I am grateful to the participants in the workshop for their constructive criticism of the work.

REFERENCES

Aitchison, J. and Dunsmore, I.R. (1975), *Statistical Prediction Analysis*. Cambridge University Press, Cambridge.

Art, D., Gnanadesikan, R., and Kettenring, J.R. (1982), Data-based metrics for cluster analysis. *Utilitas Mathematica* 21A, 75-99.

Breiman, L., Friedman, J.H., Olshen, R.A., and Stone, C.J. (1984), *Classification and Regression Trees*. Wadsworth, Belmont, California.

Bryant, J. (1979), On the clustering of multidimensional pictorial data. *Pattern Recognition* 11, 115-125.

Crawford, R.M.M. and Wishart, D. (1967), A rapid multivariate method for the detection and classification of groups of ecologically related species. *J. Ecol.* 55, 505-524.

Defays, D. (1977), An efficient algorithm for a complete link method. *Computer J.* 20, 364-366.

Diday, E. and Govaert, G. (1977), Classification automatique avec distances adaptatives. *R.A.I.R.O. Informatique/Computer Sciences* 11, 329-349.

Gordon, A.D. (1981), *Classification: Methods for the Exploratory Analysis of Multivariate Data*. Chapman and Hall, London.

Gostick, R.W. (1979), Software and algorithms for the distributed - array processors. *ICL Technical Journal* 2, 116-135.

Gower, J.C. (1974), Maximal predictive classification. *Biometrics* 30, 643-654.

Hand, D.J. (1981), *Discrimination and Classification*. Wiley, Chichester.

Jambu, M. and Lebeaux, M.-O. (1983), *Cluster Analysis and Data Analysis*. North Holland, Amsterdam.

Lance, G.N. and Williams, W.T. (1968), Note on a new information- statistic classificatory program. *Computer J.* 11, 195.

Lefkovitch, L.P. (1978), Cluster generation and grouping using mathematical programming. *Math. Biosciences* 41, 91-110.

Lefkovitch, L.P. (1980), Conditional clustering. *Biometrics* 36, 43-58.

Maronna, R. and Jacovkis, P.M. (1974), Multivariate procedures with variable metrics. *Biometrics* 30, 499-505.

Morrison, D.F. (1976), *Multivariate Statistical Methods* (Second Edition). McGraw-Hill, New York.

Payne, R.W. and Preece, D.A. (1980), Identification keys and diagnostic tables: a review (with discussion). *J. R. Statist. Soc. A* 143, 253-292.

Rohlf, F.J. (1970), Adaptive hierarchical clustering schemes, *Syst. Zool.* 19, 58-82.

Sibson, R. (1973), SLINK: an optimally efficient algorithm for the single link method. *Computer J.* 16, 30-34.

Wishart, D. (1985), Estimation of missing values and diagnosis using hierarchical classifications. *Computational Statistics Quarterly* 2, 125-134.

Zupan, J. (1982), *Clustering of Large Data Sets*. Research Studies Press, Letchworth.

Identification of Linear Regression Models by a Clustering Algorithm

By B. Baldessari and A. Bellacicco, Roma

Abstract

This paper is a contribution to the problem of representation of clusters by a set of regression models.

A new divisive clustering algorithm, reffered to as CLUREG, is presented. The algorithm belongs to the class of exchange methods, is able to deal with large sets of units and is such that each cluster of units is "well represented" by a specific regression model. The quality of the representation is measured by the sum of squares of the errors of the fitted regression model.

CLUREG, by the k- median algorithm, optimally clusters the variables. Then, after an initial partition of the units, based on the clusters of the variables, reallocates the units by the optimization of the overall sum of squares of errors in the clusters.

The computational complexity is a linear function of the number of units and a non linear function of the number of variables.

Keywords: CLUSTER, K-MEDIANS, OPTIMIZATION, REGRESSION MODEL.

1- Introduction

The aim of this paper is to present a new divise clustering algorithm, able to identify k clusters, G_h , h=1,...,k, of the units S_i of a set $S=\{S_1,\dots,S_i,\dots,S_k\}$.

Compstat 1986

Each cluster is "well represented" by a linear regression model. Specifically the clustering algorithm identifies both the clusters G_h, h=1,...,k, and the corresponding models

$$Y_{(h)} = X_{(h)} b_{(h)} + e_{(h)} , \qquad h = 1,\dots,k, \tag{1}$$

where:

(i) $X_{(h)}$ is the $n_h \times p_h$ design matrix;

(ii) $b_{(h)}$ is the p_h regression coefficients vector;

(iii) $E(e_h) = 0$, $Cov(e_{(h)}, e_{(h)}) = v_h^2 I_h$, $I_h = diag(1,1,\dots,1)$

(iv) $n_1 + n_2 + \dots + n_k = n$, $p_1 + p_2 + \dots + p_k = p$.

The optimization criterion is

$$Q = \Sigma_h \; e^T_{(h)} \; e_{(h)} = \min. \tag{2}$$

Usually, clustering algorithms have exploratory purposes and no model is assumed. The representation of clusters by a model is proposed in this paper and some previous work following this approach is found in Bellacicco and Morgana (1978) and Bellacicco and Corbo' (1982).

The main features of CLUREG are that

(i) it belongs to the class of exchange methods, Spath (1980);

(ii) it uses the k- median problem solution, Hakimi (1965);

(iii) it provides a simultaneous identification of clusters of units of S and of the relevant variables in the regression models;

(iv) it is able to handle large sets of data ((n>p>k>1 ; k predetermined), mainly because it focuses on the variables and uses OLS routines.

2. Clustering algorithm

CLUREG is based on two main steps. First, by a clustring of the variables, an initial partition of the units of S, G_h^o, h=1,...,k, is identified. The units S_i - th are assigned in such a way that

$$Q = \Sigma_h \; e_h^T \; e_h = \min. \tag{3}$$

Second, the full regression model

$$Y = X b + e \tag{4}$$

is fitted to the clustes G_h^o, h = 1,....,k, and the units are reassigned in such a way that Q is decreased monotonically.

The stopping rule is: stop at the r-th reassignment if

$$Q_r - Q_{r-1} < w , \tag{5}$$

where w is a given treshold.

The main steps of CLUREG are summarized as follows.

INPUT : Data are arranged in a unit- variable matrix

X	X_1 ,.......,X_j ,.......,X_p	Y
S_1	.	.
.	.	.
.	.	.
.	.	.
S_i	.... x_{ij}	y_i
.		.
.		.
.		.
S_n		

where X= (X_1 ,..., X_p) is the n x p matrix of the predictors and Y = (y_1 ,....,y_n) is the n x 1 vector of the responces.

Step 1. Compute the correlation matrix R from the data X.

Step 2. Compute the euclidean p x p distance matrix D, whose element are

$$d_{jz}^2 = 2\,n\,(1 - r_{jz}) \quad , \quad j,z= 1,\dots,P \tag{6}$$

Step 3. For fixed k > 1, by the k-median algorithm, Hakimi (1965), identify k- median variables $X_{(h)}$, h = 1,...,k such that

$$\sum_{h=1}^{k} \sum_{j\neq p}^{p} d_{hj}^2 = \min.$$

Step. 4. Associate each of the remaining p - k variables X_j to the k- median variable $X_{(h)}$ is such a way that

$$d_{hj} = \min, \qquad h=1,\dots,k \;\; ; \;\; j = 1,\dots,P-k. \tag{7}$$

<u>Remark</u>. k clusters of variables $X_{(h)}$, h = 1,...,k , are obtained by Step 4. These clusters are used for identifying a first partition of S.

Step 5. Fit k linear regression models (1) to the units of S.

Step 6. Construct the n x p matrix E^{o} with elements

$$e_{ih,o}^2 = (y_i - \hat{y}_{i(h)})^2, \; i = 1,\dots,n \;;\; h = 1,\dots,k, \tag{8}$$

and $\hat{y}_{i(h)}$ is the value of Y for unit S_i predicted by the regressors in $X_{(h)}$.

Step 7. Assign unit S_i , i = 1,...,n , to the cluster G_h^{o} such that

$$e_{ih,o}^2 = \min \;, \;\; h = 1,\dots,k. \tag{9}$$

Remark. A first partition of S , G_h° , $h=1,\dots,k$ is obtained.

Step 8. Fit the full model (4) in each class G_h° , $h=1,\dots,k$ and evaluate Q° by (2).

Step 9. Contruct the n x p matrix E^1 with elements

$$e^2_{ih,1} = (y_i - \hat{y}_{i,(h)})^2 , \quad i(h) = 1,\dots,n , \quad h=1,\dots,k, \tag{10}$$

where $\hat{y}_{i(h)}$ is the predicted value of Y for unit S_i in the cluster G_h° .

Step 10. In each cluster G_h° , $h = 1,\dots,k$, identify

$$e^2_{h,1} = \max_{i(h)} e^2_{ih,1} , \quad i(h)= 1,\dots,n_h , \quad h = 1,\dots,k, \tag{11}$$

where $i(h)$ identify the units of S belonging to G_h° and from a list of units and of the corresponding maximum squared errors

$$\left\{ \begin{matrix} S_{q(1)} , & S_{q(2)} , & \dots, & S_{q(k)} \\ e^2_{1,1} , & e^2_{2,1} , & \dots, & e^2_{k,1} \end{matrix} \right\} \tag{12}$$

Step 11. Identify the units $S_{q(h)}$ with max $e^2_{h,1}$, and assign them to the cluster $G^{\circ}_{h'}$, , $h' \neq h$, for $h,h' = 1,\dots,k$, such that

$$Q_1 = \sum_{h=1}^{k} e^T_{h,1} \, e_{h,1} = \min , \tag{13}$$

where $G^1_h = G^{\circ}_h - \{S_q\}$ and $G^{\circ}_{h'} = \cup \{S_q\}$.

Step 12. If $Q_o - Q_1 < w$, then STOP, otherwise go to Step 9.

Remark The stopping rule at Step 12 can be conditioned at Step 11, when we delete S_q . If n^1_h is less then $p+1$, then a second term, which satisfies the second best requirement, in E^1 must be found.

3. Concluding remarks

CLUREG is able to deal with a large set of units S , because S is explored a only few times: at the first step, when we consider the clusters of predictor variables, and at the last step ,when we sequentially reassign the units to the clusters. It is easy to see that some units will never be removed either at the beginning or at the end of the run.

The algorithm follows the general approach of the exchange methods in which the objective function is maximised, or minimised, by the exchange of units among the clusters, Späth (1980). In the case of GLUREG, the improvement in the optimization algorithm is given by the choice of a good initial partition obtained by an optimal clustering of the variables.

Acknowledgements

This paper is supported by the MPI grant "Strutture di dipendenza dei dati e loro applicazioni".

REFERENCES

Bellacicco A. and and Morgana A. (1980). On a Fuzzy Clustering Procedure on Graphs: GRAFU. COMPSTAT, 1980, Proceeding in Computational Statistics, Physica Verlag, Wein, 81 - 88.

Bellacicco A. and Corbo' P. (1982). RANDEX. A MonteCarlo clustering algorithm. METRON, XL, 1-2, 80-91

Hakimi S.L., (1965). Optimal distribution of switching centers in a communication network and some related graph theoretic problems. OPERATIONS RESEARCH, 13, 462-475.

Späth h. , (1980). Cluster analysis algorithms. J. Wiley, New York.

Validity Tests in Cluster Analysis Using a Probabilistic Teacher Algorithm

By G. Celeux, Le Chesnay Cedex

Summary

It is well known that widely used clustering algorithms are related with the Gaussian mixture problem. Now, we have developed a probabilistic teacher algorithm for the mixture problem. In this paper, we develop validity tests in cluster analysis using the properties of this algorithm and we present some applications of these tests.

Keywords : Cluster analysis, validity tests, Gaussian mixture

1. Introduction

A natural framework to investigate the relevance and the validity of a classification is the Gaussian mixture model. Effectively, the most widely used clustering algorithms (e.g. the k-means algorithm) are closely related with the Gaussian mixture problem (see section 2).

Recently we developed a probabilistic teacher algorithm, the SEM algorithm (described in section 3), for the mixture problem. This algorithm presents noteworthy features which we use, in the mixture framework, to define validity tests for clustering algorithms using an inertia criterion. More precisely, the defined tests are asymptotic tests and are purposed to answer the following questions :

- Is the solution obtained by the clustering algorithm stable ? (In particular, is the solution a local optimum and not a saddle point ?)
- Is the number of clusters correct ?

Compstat 1986

2. Clustering algorithms and Gaussian mixtures

Consider a mixture of probability density functions (p.d.f.'s)

$$f(x) = \sum_{k=1}^{K} p_k \, f(x, m_k, \Gamma_k) \text{ with for } k=1,K, \; 0 < p_k < 1 \text{ and } \sum_{k=1}^{K} p_k = 1$$

and where $f(x,m,\Gamma)$ is the Gaussian p.d.f. with mean m and covariance matrix Γ.

Let $E = (x_1,\ldots,x_N)$ represent sample values of a $\mathbb{R}^d$- valued random variable (r.v. with p.d.f. $f(x)$. The Gaussian mixture problem is concerned with estimating the unknown parameters $q = (q_k=(p_k,m_k,\Gamma_k),\ k=1,K)$.

Mainly, there are two different approaches for this problem :

- The maximum likelihood approach for which $q = (q_k,\ k=1,K)$ is estimated by those values which maximise $\prod_{i=1}^{N} \sum_{k=1}^{K} p_k \, f(x_i, m_k, \Gamma_k)$. The SEM algorithm comes under this approach.
- The classification maximum likelihood method (CML) for which $(m_k,\Gamma_k,k=1,K)$ and $(k(i)$: label of the mixture's component from which x_i arises, $i=1,N)$ are estimated by those values which maximise $\prod_{i=1}^{N} f(x_i, m_{k(i)}, \Gamma_{k(i)})$. By this way, the CML method defines a partitions $P=(P_1,\ldots,P_K)$ so that each cluster P_k can be considered as a sample from a r.v. with p.d.f. $f(x,m_k,\Gamma_k)$. Then p_k is estimated by card P_k/N, for $k=1,K$.

In this paper, we use the following result : Usual clustering algorithms can be considered as resolving the Gaussian mixture problem by the CML method (Scott, Symons (1971), Celeux Diebolt (1984)). More precisely, we have :

- If, for $k=1,K$ $\Gamma_k=\Gamma$ and Γ is known, then the CML method is exactly the k-means algorithm.
- If, for $k=1,K$ $\Gamma_k=\Gamma$ and Γ is unknown, then the CML method is exactly the "adaptative distance" algorithm (Friedman, Rubin (1967)), which can be thought of as minimizing the determinant of the within-cluster matrix.
- If the matrices Γ_k are different and unknown, then the CML method is quite analogous to the "k-adaptative distances" algorithm (Diday, Govaert(1977)) which can be interpreted as minimizing the euclidian volumes of the clusters.

On the other hand, we have to mention the following results about the asymptotic behaviour of CML method (Bryant, Williamson (1978),(1985)) : Under regularity assumptions, the CML estimate $\hat{q}_N$ of q converges a.s. to q_o with, in general, $q_o \neq q$.

Thus, the CML method gives asymptotically biased results. The CML bias is tolerable only when the mixture components are well separated and the p_k's are not too for from 1/K.

3. The SEM algorithm (Celeux, Diebolt ((1985))

The SEM algorithm consists in a modified EM algorithm (see Redner, Walker (1984)) in which a probabilistic teacher step (Stochastic step) has been incorporated. It proceeds as follows :

Define K, an upper bound of the unknown number of components.

Iteration 0 : For i=1,N choose at random the initial affectation probabilities

$$(t_k^o(x_i),k=1,K) \text{ with } 0 \leq t_k^o(x_i) \leq 1 \text{ and } \sum_{k=1}^{K} t_k^o(x_i)=1$$

Iteration n(n>0) :

S-step (Stochastic) : For each sample value x_i, draw the multinomial r.v. $e^n(x_i)=(e_k^n(x_i),k=1,K)$ of order one and with parameter $(t_k^n(x_i);k=1,K)$. The realizations $e^n(x_i)$ define a partition $P^n=(P_1^n,\ldots,P_k^n)$ where :

$$P_k^n = \{x_i/e_k^n(x_i)=1\}$$

M-step : For k=1,K compute maximum likelihood estimates q_k^{n+1} of the parameters using the P_k^n as sub-samples.

E-step : For k=1,K;i=1,N compute the affectation probabilities

$$t_k^{n+1}(x_i) = p_k^{n+1} f(x_i,m_k^{n+1},\Gamma_k^{n+1}) / \sum_{k=1}^{K} p_k^{n+1} f(x_i,m_k^{n+1},\Gamma_k^{n+1})$$

The SEM algorithm has the following properties :

- The probabilistic teacher step prevents the algorithm from giving an unstable solution and the results do not depend on the starting point.
- The algorithm needs only an upper bound of the correct number of the mixture components.
- The sequence q^n is an ergodic Markov chain. Hence this sequence converges weakly to the unique stationary probability ψ_N.

The validity tests described in the next section are based on the following theorem.

Theorem : Under regularity assumptions, the SEM stationary probability ψ_N is asymptotically a normal distribution with mean q_N and covariance matrix $V/(N)^{1/2}$ where q_N is the unique strongly consistent solution of the likelihood equations and V can be expressed in terms of the exact mixture p.d.f.

Precise statement of assumptions and the demonstration of this theorem can be found in Celeux, Diebolt (1984) when q is $\mathbb{R}$-valued and in Celeux, Diebolt (1986) in the general case.

4. The statistical scheme for validity test

We present the validity test for the k-means algorithm (the principles for other clustering algorithms, mentionned in section 2, would be the same).
Let N points in R^d be classified by the k-means algorithm ; let $P=(P_1,\dots,P_K)$ be the obtained partition with centers $(g_1,\dots,g_K)$. As noted in section 2, the underlying Gaussian mixture model is the following : The components have the same covariance matrix, their means are well separated and the mixing proportions are not too different. In this model, each cluster is associated with one of the components and the unknown parameter $q = ((p_k,m_k),\ k=1,K)$ is estimated by
$\hat{q}_N = ((\frac{\text{card } Pk}{N},g_k),k=1,K)$.
In this framework, we run the SEM algorithm r times with :
- the Γ_k's are constrainded to remain constant and equal to the within-cluster matrix of P.
- the initial value of the parameter q is $q° = ((\frac{\text{card } Pk}{N},g_k),\ k=1,K)$.
Denote q^ℓ the estimate of q at the ℓth iteration of the SEM algorithm.

Set $\bar{q} = \frac{1}{r} \sum_{\ell=1}^{r} q^\ell$ and $S = \frac{1}{r} \sum_{\ell=1}^{r} (q^\ell-\bar{q})\ (q^\ell-\bar{q})'$. $\bar{q}$ ans S are the estimates of the mean and covariance matrix of the stationary probability ψ_N of this version of the SEM algorithm.
If K is the correct number of clusters and if the k-means estimate $\hat{q}_N$ of q is approximatively equal to the strongly consistent solution q_N of m.ℓ. equations in the mixture model, then $\hat{q}_N$ is a realization from ψ_N. Thus, under this hypothesis, it is easily seen from the theorem of the section 3, that asymptotically the distribution of $T^2 = \frac{r}{r+1}\ (\hat{q}_N-\bar{q})'\ S^{-1}\ (\hat{q}_N-\bar{q})$ is an Hotelling distribution with (r-1) degrees of freedom and the distribution of $F = \frac{T^2}{r-1}\ \frac{r-K-Kp+1}{K+Kp-1}$ is a Fisher distribution with (K+Kp-1) and (r-K-Kp+1) degrees of freedom. (Note that we have to choose r>K+Kp-1).

5. Applications

In the applications r=200 and all the tested partitions have been obtained by the k-means algorithm.
First application : This application has been done on the Ruspini data (see Celeux, Diebolt (1984) for a description of these data) which consist in 75 points in $\mathbb{R}^2$ well structured into 4 clusters.

1st case : We test the partition into 5 clusters which givés the best R^2 criterion value. By the SEM algorithm one cluster vanishes. So the correct K must be less than 5.

2d case : We test the partition into 4 clusters giving the best R^2 value. This partition is exactly the natural partition. By the SEM algorithm we have $\bar{q} = \hat{q}_N$ and the variance of each marginal of q is 0. So this partition appears perfectly stable.

3d case : We test the partition into 3 clusters giving the best R^2 value (R^2=0.79). By the SEM algorithm we have $\bar{q} \neq \hat{q}_N$; unfortunetaly S is not regular and we cannot compute T^2. But the difference between $\bar{q}$ and $\hat{q}_N$ shows that this partition is less stable than the previous one.

4d case : We test a partition into 3 clusters (R^2=0.66) which is not related with the natural partition. By the validity test we get T^2=2168 and F=252. So the test shows clearly that this partition is a saddle point of the k-means algorithm.

Second application : This application has been done on simulated data. We have generated 200 points in R^2 from a Gaussian mixture with parameters : K=2, $p_1=p_2=0{,}5$, $m_1=(0,0)$, $m_2 = (3,1)$, $\Gamma_1 = \Gamma_2 = \begin{bmatrix} 1 & 0 \\ 0 & 4 \end{bmatrix}$. Let us test two partitions into 2 clusters.

The first one P^1 is not related with the mixture. The second one P^2 is closely related with the mixture. The results are summarized in the following tables where m_k^x(resp. m_k^y) denotes the abscissa (resp. ordinate) of m_k. The first column gives the estimates of q by the k-means algorithm, the second one gives $\bar{q}$ and the third one gives the S.D. of each marginal of the SEM estimates (q^ℓ, ℓ=1,r).

Test on P^1 (R^2=0.39)

	VALUE	SEM-MEAN	SEM-S.D.
p_1	0.530	0.501	0.023
p_2	0.470	0.499	0.023
m_1^x	0.481	0.390	0.062
m_1^y	-0.884	-0.829	0.082
m_2^x	2.554	2.525	0.074
m_2^y	1.644	1.440	0.063

T^2 = 35.2, F = 6.7

Test on P^2 (R^2=0.36)

	VALUE	SEM-MEAN	SEM-S.D.
p_1	0.505	0.512	0.018
p_2	0.495	0.488	0.018
m_1^x	0.027	0.066	0.051
m_1^y	-0.393	-0.302	0.071
m_2^x	2.912	2.914	0.068
m_2^y	1.015	0.939	0.064

T^2= 8.2, F = 1.6

So, although P^1 seems better than P^2 using the R^2 criterion, the validity test shows that P^2 is more stable than P^1 and must be preferred.

6. Discussion

On account of the bias encountered in the CML method, the statement of the validity test is somewhat imprecise. Unfortunately, it seems difficult to determine in simple form when the bias is severe and hence the test is ineffectual. Roughly speaking, the validity test is efficient in order to find the correct number of clusters or to detect unstable partitions except if the used clustering algorithm is not designed to discover the underlying classification structure. In particular, the assumption that the clusters have quasi-equal size whenever they are not very well separated, is important when testing a partition obtained by the k-means algorithm. Some simulations show that this requirement can be somewhat relaxed when using one of the adaptative distance(s) algorithms instead of the k-means algorithm.

References

Bryant P., Williamson J. (1978), The Asymptotic Behaviour of Classification Maximum Likelihood, Biometrika Vol 65.

Bryant P., Williamson J. (1985), Maximum Likelihood and Classification : A Comparaison of Three Approaches, UCD-CBA Working paper 5.

Celeux G., Diebolt J. (1984), Reconnaissance de mélanges et classification : l'algorithme SEM, Rapport de Recherche INRIA, N° 349.

Celeux G., Diebolt J. (1985), The SEM Algorithm : A Probabilistic Teacher Algorithm derived from the EM for the mixture problem, CSQ, Vol 2 Issue 1.

Celeux G., Diebolt J. (1986), Comportement asymptotique de l'algorithme SEM, Rapport de recherche INRIA, to appear.

Diday E., Govaert G. (1977), Classification automatique avec distances adaptatives, RAIRO Inform. 11.

Friedman H., Rubin J. (1967), On Some Invariant Criteria for Grouping Data, Biometrics 33.

Redner R., Walker H. (1984), Mixture Densities, Maximum Likelihood and the EM Algorithm, SIAM, Vol 26, N°2.

Scott A., Symons M. (1971), Clustering Methods Based on Likelihood Ratio Criteria, Biometrics 27.

'ew Kinds of Graphical Representation in Clustering

By E. Diday, Le Chesnay Cedex

ummary

aving chosen a dissimilarity index, hierarchical clustering is one of the most popular echniques used to represent clusters ; in fact, the fit between the graphical epresentation of hierarchies (i.e. dendrograms) and the data is very good only when he two following conditions are satisfied.

) The dissimilarity index satisfies the "ultrametric inequality".

) there exists an order which is "compatible" with the dissimilarity index.

his paper is devoted to new kinds of graphical representation when those conditions re not satisfied. When the first condition is not satisfied at all, we introduce a omplementary notion of ultrametric called "ultramine". When the second condition is ot satisfied at all, a new dissimilarity index called "anti-pyramidal" which is an xtension of ultramine is introduced. Graphical representations of those two notions re given. We give a measure which indicate how much the data are ultrametric or ltramine and order compatible or not for the chosen dissimilarity index. Each one of he new graphical representations sheds new and interesting light on the data.

. Introduction

ccording to the easy way to represent graphically hierarchies by dendograms, ierarchical clustering is a very popular technic in clustering and has been widely tudied (see for instance Sneath and Sokhal (1973), Anderberg (1973), Diday, Lemaire, 'ouget, Testu (1982)). Given a dissimilarity index s the hierarchical approach usually akes it possible to obtain homogeneous clusters of similar objects according to s and ive generally a good fit with s for the small distances ; if two clusters overlap one s included in the other. In this paper we give three graphical extensions of lendrograms :

) anti-indexed dendrograms where classes contain "opposed" elements and have a good 'it with the large distances ; ii) pyramidal dendrograms which allows the graphical epresentation of overlapping classes ; iii) dendrograms built up on a curve instead of straight line, this curve expresses for instance an external constraint. All the esults of this paper are proved and widely developped in Diday (1986), for more letails on pyramids see Diday (1984) and Diday, Bertrand (1985).

Compstat 1986

2. Définition and properties of ultramines

2.1. Définition

An ultramine is a mapping $u : \Omega x \Omega \rightarrow \mathbb{R}^+$ which satisfies the two following conditions
i) $\forall w,w' \in \Omega \quad u(w,w') = u(w',w)$ (symetry)
ii) $\forall w,w',w'' \in \Omega \quad u(w,w'') \geq \mathrm{Min}\,\{u(w,w'), u(w',w'')\}$ (ultramine inequality)

It follows from this definition that an ultramine u is not a metric because t triangular inequality : $u(w,w'') \leq u(w,w') + u(w',w'')$ is not necessarily satisfied. also follows that an ultramine is a similarity index since from the ultrami inequality we can deduce that $\forall w,w' \in \Omega \quad u(w,w) \geq \mathrm{Min}\,\{u(w,w'), u(w',w)\} = u(w,w')$. Another important remark is that it is possible to represent a dissimilarity index s an ultramine even if it is a similarity as we will do it troughout this paper becau the fit can be very good except for the identical elements which are the le interesting couple. We have the following properties, where the length of an edges w of any triangle w w' w'' is u (w,w') :

Proposition 1

. A similarity index u is an ultramine iff any triangle of Ω is isosceles with the ba larger than the edges according to u.

Let $X = (X_1,...X_n) \in \mathbb{R}^n$ and M(X) be the set of ultramine u such that $\forall u \in M(X$ $u(w_i,w_i) = X_i$ for $i = 1,...,n$; then we have the following proposition where U is t set of ultrametrics.

Proposition 2

The map $g : U \rightarrow M(X)$ such that $g(\delta)=u$ with $u(w_i,w_j) = (\delta(w_i,w_j))^{-1}$ and $u(w_i,w_i) = X_i$ is bijection.

2.2. Computing of the ultramines

2.2.1. The upper-dominant ultramine

By definition the upper-dominant ultramine of a given dissimilarity index d is u_d su that : $u_d = \inf\,\{u/u \geq d,\ u \in M(X)\}$ where $u \geq d$ means $u(x,y) \geq d(x,y)\ \forall x,\ y \in \Omega$; it may shown that u_d is an ultramine and thus if δ_d is the subdominant ultrametric of d (i. $\delta_d = \sup\,\{\delta/\delta \leq d,\ \delta \in U\}$) we have the following inequality : $\delta_d \leq d \leq u_d$.

Let $\mathbb{D}^{-1}(X)$ be the set of the "inverse" of the dissimilarity indices s which are su that $s^{-1}(w_i,w_i) = X_i$ for $i=1,...,n$, and takes the value $(s(w_i,w_j))^{-1}$ if $i \neq j$; we ha then the following result :

Proposition 3

The upper-dominant u_d of a dissimilarity d belonging to a set of ultramine M(X) is th inverse of the subdominant of $d^{-1} \in \mathbb{D}^{-1}(X)$.

Exemple 1 : Let $\Omega = \{w_1,w_2,w_3,w_4\}$ and d be defined by the dissimilarity matrix M(d,θ given in figure 2 where θ is the order $w_1\ w_2\ w_3\ w_4$ on the elements of Ω.

$d^{-1} \in \mathbf{D}^{-1}(X)$ we obtain the dissimilarity matrix $M(d^{-1},\theta)$ of figure 3 ; then we duce the subdominant ultrametric δ_d of d^{-1} given in figure 4 (from the single link erarchy for instance) and finally we obtain from proposition 4 the upperdominant tramine of s in figure 5.

	w_2	w_3	w_4	
)	1	4	3	w_1
	0	2	5	w_2
		0	6	w_3
			0	w_4

igure 2.$M(d,\theta)$

w_1	w_2	w_3	w_4	
X_1	1	1/4	1/3	w_1
	X_2	1/2	1/5	w_2
		X_3	1/6	w_3
			X_4	w_4

Figure 3.$(M(d^{-1},\theta)$

w_1	w_2	w_3	w_4
X_1	1/4	1/4	1/4
	X_2	1/5	1/5
		X_3	1/6
			X_4

Figure 4.$M(\delta_{d^{-1}},\theta)$

w_1	w_2	w_3	w_4
X_1	4	4	4
	X_2	5	5
		X_3	6
			X_4

Figure 5.$M(u_d,\theta)$

.2. Computing of other ultramines

e have seen in proposition 2 that the inverse $\delta^{-1} \in \mathbf{D}^{-1}(X)$ of an ultrametric δ is an ltramime of M(X) ; thus to obtain an ultramine it is possible to compute an ltrametric by using the usual algorithm of hierarchical clustering and the classical nter-class agregation indices (single link, complete link, average etc...) and then to ake its inverse which is an ultramine.

y analogy with the result given in proposition 4 another way to compute an ultramine onsists of firstly computing the inverse $s^{-1} \in \mathbf{D}^{-1}(X)$ of the initial dissimilarity ndex s, secondly computing an ultrametric on s^{-1} as described just before and taking ts inverse to obtain an ultramine $u \in M(X)$.

hese two ways for the computing of an ultramine generally give different solutions ; n pratice the second is more relevant because its fit with the initial dissimilarity ndex is generally better particularily for large distances.

. Graphical representation of ultramine and anti-pyramidal index

.1. By anti-indexed hierarchies and pyramids

e say that an order θ is anticompatible with s iff $s(w,w'') \leq \text{Min}\{s(w,w'), (w',w'')\}$. A dissimilarity matrix $M(s,\theta)$ is anti-robinson iff the terms of the rows nd columns never increase when moving away from the main diagonal in either irection. From these definition we have the following result :

roposition 4

f u is an ultramine then there exists an anticompatible order θ with u, and $M(u,\theta)$ is ntirobinson.

f there exists an order which is anticompatible with s we say that s is an ntipyramidal index. From proposition 4 it is easy to see that the ultramines are ontained in the antipyramidal indices. From proposition 2 we know that there is a ijection between ultramines and ultrametrics ; as there is also a bijection between indexed hierarchies and ultrametrics it results that any ultramine may be represente by a hierarchical dendrogram. In practice, starting from a dissimilarity index s w compute a) its inverse $s^{-1} \in \mathbf{D}^{-1}(X)$; b) an ultrametric δ (single link or complet link etc...) on s^{-1} ; c) the hierarchical dendrogram indexed by f : f(h) = Ma $\{\delta(x,y)/x,y\in h\in H\}$ associated to the hierarchy H induced by δ ; d) the ultramine $u = \delta^{-}$ $\in \mathbf{D}^{-1}(X)$ and the index F : $F(h) = \text{Min}\,\{u(x,y)/x,y\in h\in H\}$.

From the definition of F it follows that it satisfies the following property :
h h' => $F(h) \geq F(h')$; in this case we say that (H,F) is an anti-indexed hierarchy ; th dendogram associated to (H,F) has an unusual interpretation : the level associated t the smallest clusters contains the most opposed elements according to the initia dissimilarity index s. The representation of a dissimilarity index s by a pyramid ma be made in an analogous way : we obtain an anti-indexed pyramid (P,F) and th associated pyramidal dendrogram has the same kind of interpretation. All these result may be sumerized in figure 6.

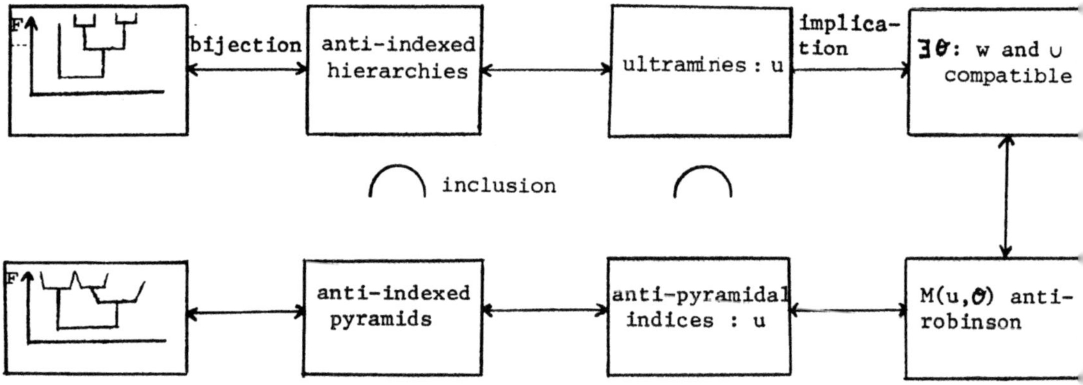

Figure 6

Example 2

Starting from the dissimilarity index d defined in figure 2 and choosing X_i = Max $\{u_d(w_\ell,w_k)/\ell,k\}$ = 6, the anti-indexed hierarchy, associated to the ultramine u_d and defined in figure 5 is given in figure 8, whereas the single link hierarchy associated to d is given figure 7. This is also, the case with pyramids as shown in figure 9 and 10.

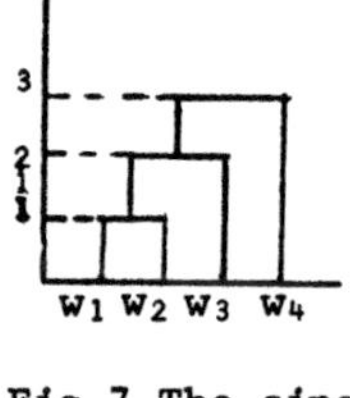

Fig.7 The single hierarchy (H,f) associated to s

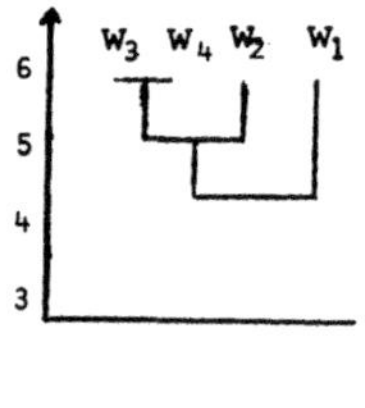

Fig.8 The anti-indexed hierarchy (H,F)

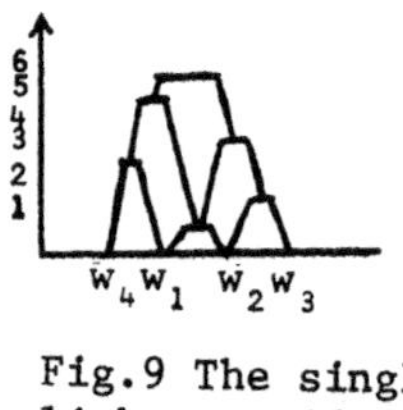

Fig.9 The single link pyramid (P,f)

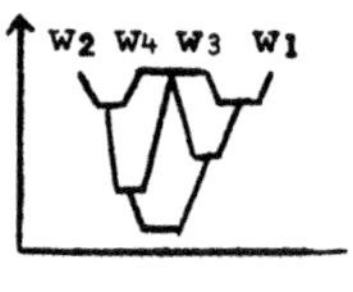

Fig.10 **The anti-index pyramid (P,F)**

.2. Representation by "compatible" curves

We say that a curve is compatible if any ordered triple $w_i\ w_j\ w_k$ set on this curve satisfies the following property : $d(w_i,w_k) \geq \text{Max}\ (d(w_i,w_j),\ d(w_j,w_k))$ (1) where d is the euclidien distance. A straight line is a simple example of such a curve.

In figure 9 we give several examples of compatible curves and non-compatible curves. As we know, to any ultrametric and to any pyramidal index it is possible to associate an order which satisfies the inequality (1), therefore a compatible curve may be another possible way of representing a dissimilarity index.

Analyticaly, the problem may be settled for example as follows : find the curve C of the plane represented by a polynomial map h : $\mathbb{R}^2 \to \mathbb{R}^2$: $h(y) = z$ such that if the elements w_i of Ω represented by $z_i = h'(w_i)$ are ordered according to a given order θ on the curve, then the following criterion : $W(h,h') = \sum_{i,j} (s(w_i,w_j) - d(z_i,z_j))^2$ (where s is the initial dissimilarity index and d the euclidean distance) is minimized among all the compatible polynomial curves of the plane and all the possible ordered positions according to θ, of the z_i on this curve (s and d are supposed normalized for instance by dividing them by their greatest value).

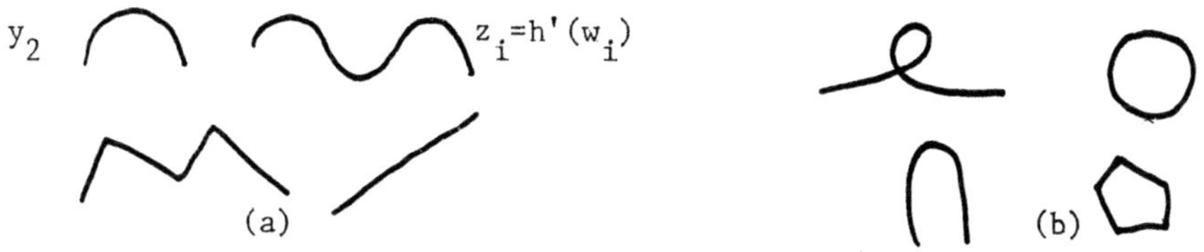

Figure 9.(a)Compatible curve (b) Non-compatible curve

They are many ways of simplifing this general problem for instance by fixing the desired curve, semi-circle in the example 3) and to give euristical approaches (see Diday (1986)).

If we substitute the inequality (1) by the following inequality : $d(w_i,w_k) \leq \text{Min}\ (d(w_i,w_j),\ d(w_j,w_k))$ in the definition of a compatible curve, we obtain an anti-compatible curve ; unfortunatly it seems that this kind of curve has not a visual representation ; however it is possible to represent an ultramine or an antipyramidal index by looking first for the associated ultrametric or pyramidal index obtained by inversion and its associated compatible curve and afterwards by indexing the distances with the values given by the ultramine or the antipyramidal index.

Example

$$\text{Let } M(s,\theta)\times 100 = \begin{pmatrix} & w_1 & w_2 & w_3 & w_4 & \\ & X_1 & 25 & 16 & 14 & w_1 \\ & & X_2 & 40 & 25 & w_2 \\ & & & X_3 & 58 & w_3 \\ & & & & X_4 & w_4 \end{pmatrix} \text{ and } M(s^{-1},\theta) = \begin{pmatrix} w_1 & w_2 & w_3 & w_4 \\ X_1 & 4 & 6 & 7 \\ & X_2 & 2.5 & 4 \\ & & X_3 & 1.7 \\ & & & X_4 \end{pmatrix}$$

where s is an anti-pyramidal index. If we choose a semi-circle for the robinson curv an ordered position of the elements w_1 w_2 w_3 w_4 on this curve is given in figure 10.

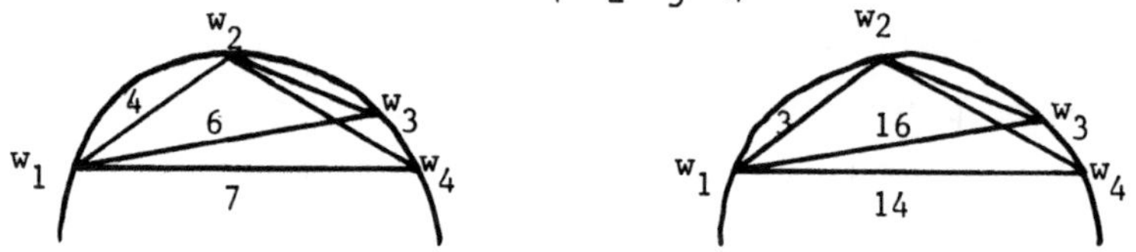

(a) (b)

Figure 10.(a) Compatible curve representation of $M(s^{-1},0)$; (b) representation of t anti-pyramidal index s.

This approach may be useful for instance, when we wish to represent a hierarchy und constraint expressed by a dissimilarity index (a geographical distance for instance There are several ways to obtain a consensus order, between the given hierarchy and t constraint (see for instance DIDAY (1982)) ; it is thus possible to represent hierarchy on a compatible curve given by the constraint instead of a straight line usual. In figure 11.b we give the representation of the hierarchy of figure 11.a on t compatible curve of figure 10.b defined by the constraint given by the anti-pyramid index s. Another kind of such a representation is given in (c) and (d) of the sa figure.

They are four ways of representing a hierarchy on a compatible curve depending on if represents an ultrametric or an ultramine and if the order is compatible anticompatible with the dissimilarity index defined by the constraint. All the representations may be made in an analogous way in the case of a pyramid under external constraint.

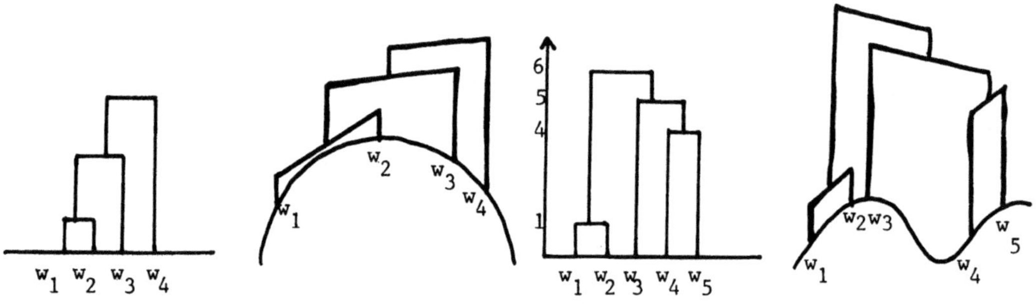

(a) (b) (c) (d)

Figure 11. The hierarchy of figure (a) (resp(c)) is represented under constraint on (b (resp.(d)).

4. Criterion to measure if the data are more or less ultrametric or ultramine for dissimilarity index

They are many ways of defining such criterion ; for instance let s be a dissimilarit index and T the set of all the triples of Ω ; each $t \in T$ defines a triangle whose lengt of a side ww' is s(w,w') ; let a (t) $\geq$ b (t) $\geq$ c (t) the length of the three sides o he triangle t and $E(t) = (a(t)-b(t))-(b(t)-c(t)) = a(t)-2b(t)+c(t)$; thus sign (E(t)) qual 0 if t is equilateral, +1 if it is isosceles with the base smaller then the other ides and -1 if it is isosceles with the base greater then other sides. If N is the umber of non-equilateral triangles we have for instance the following criterion :

$$(\Omega,s) = \frac{1}{N} \sum_{t \in T} \text{sign } E(t).$$

t is easy to see that $W(\Omega,s) = 0$ if $\forall i,j,k,l$ $s(w_i,w_j) = s(w_k,w_l)$ (i.e. all the riangle are equilateral) ; $W(\Omega,s) = -1$ if s is an ultrametric ; $W(\Omega,s) = 1$ if it is n ultramine and $W(\Omega,s) \in [-1,1]$.

onclusion

f a dissimilarity index s is such that there exists an order θ : $M(s,\theta)$ is nti-robinson (i.e. a "chaotic" order) an ultramine and its graphical representation by n anti-indexed dendrogram will have a better fit with s then the usual dendrograms epresenting an ultrametric. The criterion W may give an idea of the degree of chaotisation" (i.e. existence of interesting oppositions) or "coherence" (existence of nteresting clusters) of the data for a given dissimilarity index. In practice M(s,O) s not completely robinson or anti-robinson and the two approaches will give omplementary light on the data.

eferencies

nderberg M.R. (1973), "Cluster analysis for application", Academic Press New-York.

iday E. (1982), "Crossing orders and ultrametrics : application to visualization of onsensus for comparing classifications", Compstat 1982, Physica-Verlag.

iday E. (1982), "Croisements, ordres et ultramétriques", Math.Sc.Hum. 21ème année n°83, 31-54.

Diday E., Lemaire J., Pouget J., Testu F. (1982), "Elements d'analyse des données" Dunod.

Diday E. (1984) "Une représentation visuelle des classes empiétantes", Rapport n° 291, Rocquencourt 78150 Rocquencourt, France.

Diday E. (1986), "Nouvelles représentations graphiques en classification automatique", Rapport INRIA, 78150 Rocquencourt, France.

Diday E., Bertrand P. (1986), "An extension of hierarchical clustering : the pyramidal representation", Pattern Recognition in practice II, North Holland.

Hubert L. (1974), "Some applications of graph theory to clustering", Psychometrica 39, 283-309.

neath PHA, Sokhal RR. (1973), "Numerical Taxonomy", Freeman.

Projection on an Acute Symmetrical Simplicial Closed Convex Cone and its Application to Star Graphs

By B. Fichet, Marseille

SUMMARY

Additive trees extend hierarchical representations, and a one-to-one correspondence has been set up between these trees and quadripolar semi-distances, i.e. semi-distances satisfying the four-point condition. Among these dissimilarities, star semi-distances are noteworthy. They verify : $d(i,j) = a_i + a_j$, for all distinct units i and j, and yield a special representation, called a star graph.

After exploring some geometrical properties of star semi-distances, the least squares approximation of a dissimilarity by such a semi-distance is examined. It corresponds to a least squares problem with positive constraints. This problem is shown to be related to the projection of a vector x on a very special cone, the closed convex hull of independent vectors which have the same acute angle.

It is proved that a coordinate of the solution is null when the associated coordinate of x is non-positive. Then a descent algorithm follows, and at each step, a very simple analytic expression yields the projection on the corresponding face. Moreover, when the coordinates of x are ranked by increasing order, the suitable face for the solution is directly obtained. This gives a second algorithm.

Finally, some specific properties of the least squares star semi-distance are discussed.

KEY WORDS

Dissimilarity, star semi-distance, additive tree, least squares problem, cone.

1. QUADRIPOLAR SEMI-DISTANCES AND STAR SEMI-DISTANCES

Let I be a finite set, with cardinality n. We denote by $\mathcal{D}$ the set of real functions $d : I^2 \mapsto \mathbb{R}$ satisfying : $\forall i \in I$, $d(i,i) = 0$; $\forall (i,j) \in I^2$, $d(i,j) = d(j,i)$. $\mathcal{D}$ is a vector space with dimensionality $n(n-1)/2$ and a *dissimilarity* is an element of the orthant $\mathcal{D}^+$. In the following, $\mathcal{D}$ will be furnished with the scalar product defined by $< d,d' > = (1/2) \sum_{i,j} d(i,j)d'(i,j)$, and the associated norm will be denoted by $\|.\|$.

A *quadripolar semi-distance* is a semi-distance satisfying the following four-point condition :

$\forall (i,j,k,\ell) \in I^4$, $d(i,j)+d(k,\ell) \leqslant \max[d(i,k)+d(j,\ell), d(i,\ell)+d(j,k)]$.

If $\mathcal{D}^q$ denotes the set of quadripolar semi-distances, it is very easy to show that $\mathcal{D}^q$ is a closed cone which generates $\mathcal{D}$. Moreover, to every d in $\mathcal{D}^q$ corresponds an *additive tree* (see for instance Buneman [2], Sattah and Tversky [7], Dobson [5]), which extends hierarchical representation.

In the set $\mathcal{D}^q$, *star semi-distances* play a special role. Let us define the subset $\mathcal{D}^s$ of $\mathcal{D}$ whose elements d are such that there exists a family of real numbers $\{a_i, i \in I\}$ satisfying : $\forall (i,j) \in I^2$, $i \neq j$, $d(i,j) = a_i + a_j$. When the coefficients a_i are non-negative, d is said to be a star semi-distance, and the set of these semi-distances will be denoted by $\mathcal{D}^{s+}$.

For every i in I, let us introduce the dissimilarity δ_i such that : $\forall j \in I$, $j \neq i$, $\delta(i,j) = 1$. Then, it is obvious that $\{\delta_i, i \in I\}$ is an independent system in $\mathcal{D}$ which generates $\mathcal{D}^s$, and $\mathcal{D}^{s+}$ is the closed convex hull of vectors $\delta_i, i \in I$. Consequently, $\mathcal{D}^{s+}$ is a closed convex simplicial cone. Moreover, the edges $\delta_i, i \in I$ have the same acute angle. We have : $<\delta_i, \delta_j> = 1$ for $i \neq j$, or $(n-1)$ for $i = j$. Some other geometrical properties may be established. For instance, the orthogonal subspace of $\mathcal{D}^s$ is given by the elements of $\mathcal{D}$ satisfying : $\forall i \in I$, $\sum_j d(i,j) = d(i,.) = 0$.

Star semi-distances yield special additive trees, called *star graphs*. They have been investigated by several authors such as Carroll [3], Le Calvé [6], and occur in the decomposition of quadripolar semi-distances : every dissimilarity in $\mathcal{D}^q$ can always be decomposed into the sum of an ultrametric plus an element in $\mathcal{D}^s$ (see Brossier [1]).

2. PROJECTION ON AN ACUTE SYMMETRICAL SIMPLICIAL CLOSED CONVEX CONE

Let E be an n-dimensional Euclidean vector space and $\{e_1,\ldots,e_n\}$ a basis of E satisfying : $\forall i,j = 1,\ldots,n,\ i \neq j,\ \langle e_i, e_j\rangle = \cos\theta$ $(0 < \theta \leqslant \pi/2)$; $\forall i = 1,\ldots,n,\ \|e_i\| = 1$. The closed convex hull C of the basis vectors has the same properties as $\mathcal{D}^{s+}$. Given $x = \Sigma x_i e_i$ in E, let us consider the least squares problem :

$$(1)\ \min \{\|x - z\| \mid z \in C\}\ .$$

General results of optimization could be used. However, here we are going to establish two short algorithms which take into account the special geometrical nature of the cone C.

First, we give a lemma ; its proof is very easy.

Lemma 1

For $p < n$, the projection e'_{p+1} of e_{p+1} on the space spanned by $e_1,\ldots,e_p$ satisfies :

i) $e_{p+1} = \dfrac{\cos\theta}{1+(p-1)\cos\theta}\,(e_1+\ldots+e_p)$

ii) $\|e_{p+1} - e'_{p+1}\|^2 = \dfrac{(1-\cos\theta)(1+p\cos\theta)}{1+(p-1)\cos\theta}$

Let us denote by $y = \Sigma y_i e_i$ the solution of problem (1). If all the coordinates of x are non-negative, then $y = x$. So, we assume the existence of an index i such that : $x_i < 0$. Note that a coordinate of y is necessarily null.

Proposition 1

If i is such that $x_i = 0$, then $y_i = 0$.

Proof : let E' be the subspace generated by the vectors e_j such that : $x_j \neq 0$. Denote by $\hat{y}$ the projection of y on E'. By lemma 1, the coordinates of $\hat{y}$ are non-negative. Since $\| x-\hat{y}\| \leq \| x-y\|$, it follows $y = \hat{y}$. □

Thus, without loss of generality, we may suppose that the coordinates of x are non-null.

Proposition 2

For every i such that $x_i < 0$, the coordinate y_i is null.

For the proof, we need the following lemmas.

Lemma 2

If y has only one coordinate equalling 0, then the corresponding coordinate of x is strictly negative.

Proof : without loss of generality, we may assume that $x_1, \dots, x_p$ are the negative coordinates of x. Suppose that y_{p+1} is the unique null coordinate of y and define $r = n - (p+1)$, $(r \geq 0)$. Then, y is the projection of x on the space spanned by $e_1, \dots, e_p$, $e_{p+2}, \dots, e_n$. Thus, by lemma (1) :

$$(2) \quad \| x - y\|^2 = \frac{x_{p+1}^2 \; (1 - \cos\theta)[1 + (p + r)\cos\theta]}{1 + (p + r - 1)\cos\theta}$$

Since $y_i > 0$ for every i equalling 1,...,p, the same lemma yields :

$$(3) \quad \forall i = 1, \dots, p \; ; \; -x_i < \frac{x_{p+1} \cos\theta}{1 + (p+r-1)\cos\theta}$$

Define $z = [x_{p+1} + (x_1 + \dots + x_p)\cos\theta]e_{p+1} + x_{p+2}e_{p+2} + \dots + x_n e_n$.

By the previous inequalities : $z \in C$.

After some easy calculation, we obtain :

$$\| x - z\|^2 = (1 - \cos\theta)[(x_1^2 + \dots + x_p^2) + (x_1 + \dots + x_p)^2 \cos\theta]$$

Using (3), we deduce : $\| x - z\|^2 < \dfrac{p(1-\cos\theta)\cos^2\theta(1+p\cos\theta)}{[1 + (p+r-1)\cos\theta]^2} \, x_{p+1}^2$

Then, according to equality (2), it may be proved that $\|x - z\| < \| x - y\|$, which contradicts the definition of y. □

Lemma 3

There exists an index i such that : $x_i < 0$, $y_i = 0$.

Using lemmas 1 and 2, the proof is inductive with respect to the dimensionality n.

Hence, by a similar inductive reasoning, proposition 2 holds.

The previous two propositions allow us to state a descent algorithm. At each step the negative component of the vector is cancelled and lemma 1 gives the projection on the corresponding face of the cone.

Algorithm 1

Step k. The vector is : $x^k = \Sigma x_i^k e_i$

Denote : $I_k^+ = \{i \mid x_i^k > 0\}$; $I_k^- = \{i \mid x_i^k < 0\}$; $I_k^0 = \{i \mid x_i^k = 0\}$.

- *If* $I_k^- = \emptyset$, *then* x^k *is solution*
- *If* $I_k^+ = \emptyset$, *then* 0 *is solution*
- *If* $I_k^- \neq \emptyset$, *and* $I_k^+ \neq \emptyset$, *then* x^{k+1} *is defined by*

$$\forall i \in I_k^- \cup I_k^+ ,\ x_i^{k+1} = 0 \ ; \ \forall i \in I_k^+ ,\ x_i^{k+1} = x_i^k + \frac{\cos\theta}{1+(|I_k^+|-1)\cos\theta} \sum_{j \in I_k^-} x_j^k$$

Note that the formulae are homogeneous according to the coordinates. Hence, the algorithm stays valid when the length of basis vectors has any given constant value.

When the coordinates of the given vector $x = \Sigma x_i e_i$ are ranked by increasing order, the solution may be directly obtained. Indeed, some easy geometrical considerations justify :

Algorithm 2

Supposing $x_1 \leqslant \ldots \leqslant x_n$, putting $x_0 = 0$ and denoting:

$$\forall i = 1,\ldots,n \quad t_i = \frac{\cos\theta}{1+(n-i)\cos\theta} \sum_{j=0}^{i-1} x_j$$

- *If for every* i *(or for every* i *such that* $x_i > x_{i-1}$ *and* $x_i > 0$*) we have* $x_i + t_i \leqslant 0$, *then* 0 *is solution.*
- *If not : let* i_0 *be the smallest index such that* $x_{i_0} + t_{i_0} > 0$ (4) *(or (4),* $x_{i_0} > x_{i_0-1}$ *and* $x_{i_0} > 0$*)*

  Then : $y_1 = \ldots = y_{i_0-1} = 0$ *(if* $i_0 > 1$*)*

  $\forall i = i_0,\ldots,n \quad y_i = x_i + t_{i_0}$.

THE STAR SEMI-DISTANCE BEST FITTING A DISSIMILARITY

In order to obtain a star semi-distance fitting a given element d of $\mathcal{D}$, the least squares problem is the following :

$$\min\{\sum_{\substack{i,j\\ i\neq j}} [d(i,j) - (a_i + a_j)]^2 \mid \forall i = 1,\ldots,n,\ a_i \geqslant 0\}$$

It corresponds to the projection of d on the cone $\mathcal{D}^{s+}$. It is obvious that the solution is the projection on $\mathcal{D}^{s+}$ of the projection of d on the space spanned by $\mathcal{D}^{s+}$, i.e. $\mathcal{D}^{s}$. Le Calvé [6] has obtained the analytic expression of this last projection; the corresponding coefficients verify :

$\forall i \in I,\ a_i = [n/(n-2)][d(i,.) - nd(.,.)/2(n-2)]$

(where : $\forall i \in I,\ d(i,.) = (1/n) \sum_j d(i,j)$; $d(.,.) = (1/n) \sum_i d(i,.)$).

Taking into account the geometrical nature of the cone $\mathcal{D}^{s+}$, the projection on this cone may be obtained with our algorithms. Note that : $\cos\theta = 1/(n-1)$.

Finally we give this last proposition.

Proposition 3

If d is a non-null dissimilarity, then its projection on $\mathcal{D}^{s+}$ is also non-null.

For the proof we may use either geometrical considerations associated with algorithm 1, or the polar cone of $\mathcal{D}^{s+}$; F. Critchley has established the equations of this cone.

4. REFERENCES

[1] G. BROSSIER (1985) *Approximation des dissimilarités par des arbres additifs*. Math. Sci. Hum. n°91, 5-21.

[2] P. BUNEMAN (1974) *A note on the metric properties of trees*. Journal of Combinatorial Theory (B). 17, 48-50.

[3] J.D. CARROLL (1976) *Spatial, non-spatial and hybrid models for scaling*. Psychometrika, vol.41-4, 439-463.

[4] G. DE SOETE (1983) *A least squares algorithm for fitting additive trees to proximity data*. Psychometrika, vol. 48-4, 621-626.

[5] A. DOBSON (1974) *Unrooted trees for numerical taxonomy*. J. Appl. Prob. 11, 32-42.

[6] G. LE CALVE (1985) *Distance à centre*. Statistique et Analyse des Données. Vol. 10, n°2, 29-44.

[7] S. SATTAH and A. TVERSKY (1977) *Additive similarity trees*. Psychometrika. Vol. 42, n°3, 319-345.

Design of a Command Language for Interactive or Batch Use in a Statistical Package

By J. Henderson and D. Wishart, Edinburgh

SUMMARY

The paper describes a new command language for the statistical package Clustan which is especially suitable for interactive use from a computer terminal or PC workstation. It features immediate error correction, flexible help (and more help) facilities, defaults for unspecified parameters, verification before execution, simulation of execution and simultaneous storage of commands as a program. The interactive method of operation is fully compatible with batch operation by a Clustan program.

KEYWORDS: interactive, command language, statistical package, help system, cluster analysis, user interface, micro-computer, error correction.

Introduction

A cluster analysis is specified by a sequence of commands which call Clustan procedures. Each command has option keywords and parameter keywords which specify the form and parameters to be applied in the analysis. The available keywords are unique to the command, and thus have limited options or values which are valid in the context of that command or the user's data. This validity is checked and errors are generated at the terminal for any invalid specifications. In addition, suitable default options or parameter values are invoked for keywords which are omitted from the command.

The above description of the command language could apply to various statistical packages capable of batch operation. Where Clustan's command language differs is in its additional support for the interactive user. When errors occur, they are reported to the terminal user and only the invalid keywords or values need to be corrected; commands or keywords which have been accepted as valid do not have to be re-entered. If the keyword has been omitted from the specifications, or a parameter is incorrect, the user need only supply the missing keyword or corrected parameter.

The user can request help (or more help) from a documented help file to assist with the correction of an error. When the specifications are completed the command can be reviewed before execution to amend further, reject or submit it, as appropriate. The current status of the user's internal data file can also be interrogated to display variables, user settings, graphical options, current files in use and intermediate results.

Three methods of operation

The command language provides three methods of operating Clustan:

Batch mode	- by a Clustan program, typed in or read from a file
Interactive mode	- keyword specifications, error correction and help file
Workstation mode	- menu specifications, error correction and help file

Compstat 1986

A prime design consideration was that a single, unified command language should be provided for all three methods of use. The only differences are in the treatment of errors and prompts, which in interactive or workstation mode provide for the immediate correction of errors.

In interactive and workstation mode, commands and their parameters can be entered and immediately executed, with results directed to the user's terminal or to a file, line printer or graphics device (eg high resolution plotter). At the same time, or instead of immediate execution, Clustan can create and store a program from an interactive session, which can be re-run later in batch mode. The various possibilities are illustrated in figure 1.

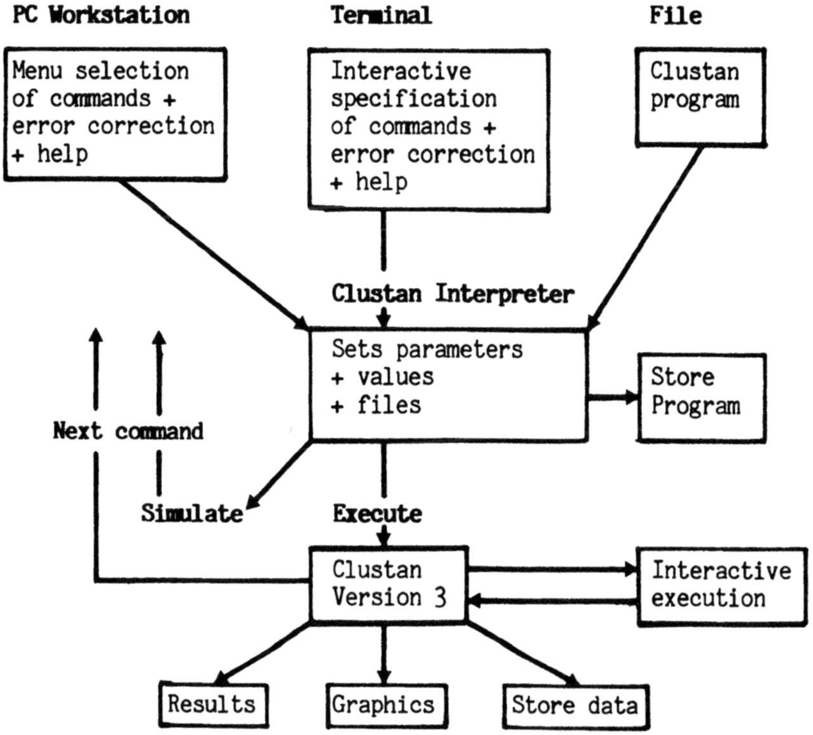

Figure 1. The Clustan command language has 3 modes of operation, with error correction and help assistance in interactive or workstation mode, program storage and optional simulation of execution.

Design of the Clustan command language

The emphasis in the design of the Clustan command language has been on simplicity, to assist the novice user, with ample error correction and help facilities when in interactive mode. Once a user has gained confidence and experience, however, there are shortcuts available to reduce the effort needed to specify a Clustan analysis.

An analysis is specified as a sequence of **commands**, each of which invokes a Clustan **procedure**. Associated with these are **command specifications** which give values to the mandatory or optional parameters of each command.

```
Read data, cases 200, variables multistate 1, continuous 2-5,
     binary 6, missing 6*-1
          [ data inserted here, or specified from a file ]
Assign file=trainer, spec=cluster.res
Cluster, method Wards, measure Seuclid, save clusters min 2 max 6,
     results outfile=trainer, transform ranges, print fusions,
     icicle, centres min 2 max 6, members min 2 max 6
Rules
Plink
Stop
```

Figure 2. Clustan program for a Ward's cluster analysis with mixed data, missing values, range transformations, save clusters and results file, various print options, Rules significance test for the "best" number of clusters and a high-resolution Plink graph plot of the tree.

The example in figure 2 illustrates various important features of the command language. Firstly, a Clustan program is **readable**, in the sense that English commands and sub-commands are self-explanatory and provide documentation of the specified analysis. Of course it is necessary to understand the principles of the underlying statistical method, and these are set out in an accompanying user manual, but given this knowledge the purpose of the program should be clear.

Secondly, syntactic rules are kept to a minimum. Each command must start on a new line, but the command specifications which follow can appear in any order and are fully supported by **default values** for any non-essential keywords that have been omitted. They are continued by a comma at end-of-line. Keywords are unique **in context** (for example, the various uses of min and max to qualify parameters of Cluster in figure 2) and can be abbreviated (to the first 4 letters, or fewer if unique). Sub-commands determine the context. For instance, figure 3 illustrates a 4-level contextual nesting of the Cluster specifications in figure 2, in which the same keyword names (eg min, max) recur but are interpreted properly in the context of the preceding superior keyword.

```
Cluster
        Method Wards
        Measure Seuclid
        Save   clusters min 2 max 6
               results outfile=trainer
        Transform ranges
        Print  fusions
               icicle
               centres min 2 max 6
               members min 2 max 6
```

Figure 3. The simplicity of the Cluster command in figure 2 conceals 4 levels of contextual nesting of sub-command keywords. Thus, for example, the keywords min and max recur 3 times but are interpreted correctly in each context.

Thirdly, the parameter values can be flexibly specified. They take the form of integer or real numbers, lists of integers or reals, text strings or file names. Parameters can be separated from their values by blanks or an "=" sign. A sub-command is terminated when Clustan encounters a keyword introducing another sub-command. For example, in the "Cluster" command illustrated in figure 3, the "Save" keyword is terminated by the "Transform" keyword, even though other parameters of "Save" could have been specified.

Interactive error correction
When operating in interactive mode, Clustan checks for 4 types of errors:

- Unsatisfied pre-conditions
- Syntax errors in command specifications
- Missing, invalid or unrealisable keywords or parameter values
- File specification errors

It can be irritating to discover, having typed a command followed by lengthy specifications, that the command is out of sequence or contains an error in the first of several lines. For this reason, if Clustan detects such an error it is immediately reported to reduce further wasted effort, and the user is told what pre-condition has not been satisfied. For instance, if the command "Hierarchy" was typed without a distance or similarity matrix having previously been read or computed (an essential pre-condition for Hierarchy), Clustan would report the error after the first command line had been entered, even though the command keywords and parameters had not been fully specified.

Errors are reported when an invalid keyword or inconsistent parameter value is typed, or an attempt is made to execute a command without specifying all the **essential** keywords. In each of these cases, the interpreter enters a restricted command mode in which the user has the option to amend or negate the keywords which have been rejected. Only the keywords which are incorrect or missing in the specifications need to be re-specified. Error checking proceeds in this way until a command and its associated keyword specifications are acceptable to the interpreter. To assist in the process of interactive error correction, all the current command's keyword values can be displayed using a "DISPLAY" function before executing the command.

In addition, the user can set a global "verify" mode to display and verify the parameter settings, or simulate execution to obtain interactive error checking before, say, submitting a long production run to a batch queue.

In batch mode, execution of the program halts when an error is detected. However, Clustan continues to scan the remaining command specifications for further possible errors, in its "simulation" mode, to check for further possible errors which can be corrected before the program is resubmitted.

Saving a Clustan program during interactive operation
Most interactive packages only operate as interactive packages: to run them you always have to work through all the menus, or answer all the questions every time you use the package. With Clustan's integrated approach, by contrast, the user can opt to store the commands developed during an interactive session - all the commands, sub-commands, keywords and parameters specified interactively are then simultaneously copied to a file, which can subsequently be submitted to Clustan as a batch program. So an interactive session can be used to **develop** a Clustan program which can be re-run without repeating the terminal dialogue. Moreover, such a program can also be edited using a conventional text editor, to make minor changes, where this would be quicker than repeating the full interactive dialogue.

Hierarchical HELP system
The HELP system is a relatively sophisticated hierarchical directory of information available to the interactive Clustan user. At the level of immediate assistance, when the user types "?" in answer to a prompt the system responds with information on what reply is expected.

It is also possible to enter the HELP system by typing "HELP <command>", to obtain more general information about a specific command, or "HELP <topic>" and "HELP <topic> <sub-topic>" to get further levels of detailed information about particular keywords or keyword parameters within a command. A "wildcard" * can be used to obtain information about all the available sub-topics in the current context: the system remembers the context of the last "HELP" and allows further inquiries within that context. This "hierarchical" structuring of the HELP system follows the hierarchical nesting of sub-command keywords, as illustrated in the example of figure 3. It enables the user to refine successive inquiries, to select the specific information he requires, on his screen, without paging sequentially through the HELP directory.

The HELP directory for Clustan is distributed as a sequential text file, with indexing parameters separating the entries, and a separate program is provided with Clustan which creates the pages of HELP text on a direct access file, for rapid retrieval during use. This design has two special advantages. Firstly, it is possible to translate the file into other languages (an earlier version of this system was translated into German and distributed to German-speaking installations, and is supported by a German edition of the user manual - Wishart (1984)).

Secondly, the HELP directory can be modified or extended to include local information. This is particularly useful to academic users offering Clustan case studies in undergraduate courses, because it allows course-specific guidance and local operating details (eg file names for exercises) to be entered in the HELP directory.

Internal file management
The Clustan interpreter manages several pre-defined temporary files for the user. They have the internal names $INPUT, $OUTPUT, $DATA, $CLASS and $DENDRO, which appear in the default specifications of many commands. Alternatively, the user can assign permanent files - for example, to store classifications or dendrograms permanently for subsequent re-use, or where the Clustan data file is to be saved for a later re-start. In addition, other optional output files can be assigned names and referred to by name in the specification of optional output requests from particular commands. An example is the assignment of a Cluster results file called "Trainer" to the permanent file "cluster.res" in figure 3, and the use of this file as input to Classify in figure 4. All file maintenance is handled by the command language interpreter, and is generally transparent to the user.

Remote workstation operation
Clustan 3 is designed to run on an IBM PC, XT, AT or RT micro-computer (or compatible), either on a stand-alone basis or as a workstation attached to a mainframe computer. In local operation it functions in much the same way as a mainframe, except that advantage has been taken of the functional specifications of the IBM PC Monitor to design a menu approach to specifying command parameters. This additional facility is provided for PC users who

are comfortable with the menu method of specifying applications. Of course, once such users gain confidence in their use of Clustan they can switch to specifying commands in the normal way, thus saving time.

The micro-computer version of Clustan will also allow the user to transmit Clustan specifications to a remote mainframe. This provides a greater degree of flexibility, since Clustan programs can be specified interactively at the workstation and, if appropriate, tested locally on a sample of data, before being submitted for production running on a mainframe. We envisage that this will be particularly attractive at installations using PC workstations as intelligent terminals, since it should help to reduce the use of expensive mainframe time currently occupied servicing online terminals and running test jobs.

```
Assign file=trainer, spec=cluster.res
Classify, infile=trainer, search tree, clusters min 2 max 4,
          print coefficients, neighbours, update tree
Stop
```

Figure 4. In this program, the output file created by Cluster in the example of figure 2 is read in as a training set to Classify, which identifies new cases interactively by reference to the hierarchical classification previously computed and stored.

Interactive execution

Finally, certain recent Clustan commands now provide interactive execution as an option (Wishart, 1985). This allows the user to change values and make decisions during the actual execution of a Clustan analysis. Obviously it could only happen if Clustan is already being run in interactive mode, in which case the transfer from the command interpreter to the package results in Clustan run-time messages and input options being directed to the terminal. On completion of each command, control again returns to the interpreter for the next command specification.

In the example illustrated in figure 4, the hierarchical classification created by the program in figure 2 is assigned and read as input to Classify. This procedure then reads new cases interactively at the terminal or PC workstation, finds the cluster of closest fit at the 2, 3 and 4 levels in the tree, identifies the nearest neighbour, prints coefficients and updates the tree dynamically by the addition of each new case into its cluster or clusters of best fit. Further details of this procedure and sample outputs are given in Wishart (1985).

References

Wishart, D (1983), Clustan User Manual: Fourth Edition, Program Library Unit, University of Edinburgh, Edinburgh, Scotland (under revision).

Wishart, D (1984), Clustan User Manual: German Edition, Gustav Fischer Verlag, Stuttgart.

Wishart D (1985), Estimation of Missing Values and Diagnosis Using Hierarchical Classifications, Computational Statistics Quarterly, 2(1), 125-134.

Cross Association Measures and Optimal Clustering

By F. Marcotorchino, Paris

Introduction

The purpose of this short paper is to show how the introduction of the concept of 'Paired Comparisons' allows us to linearize most of the cross association criteria on Contingency Tables in Statistics. These statistical cross association measures derived from contingency tables amount to 'at least' quadratic mathematical formulations in the classical statistical approach of cross association for categorical variables (qualitative nominal variables). This possibility of linearizing some of these criteria allows us to transform 'classification problems' into Linear Programming ones without any type of hypothesis of cluster size fixation. The 'paired comparisons principle' plays a prominent part in this paper. Let us mention here that the first mathematician who spoke about 'paired comparisons' was A. Condorcet in 1785, in order to solve the difficult problem of 'Collective Choice' in Preferences Aggregation. We shall not speak, once more, about the axiomatic of the Condorcet's criterion, widely discussed in [15], nor about the solving of the associated aggregation problems, for which we shall find details in [14]. We shall only focus on the 'central' part of the paired comparisons principle' in a 'broad sense', thus enlarging the possibility of this concept in a context not reserved to Similarity aggregation but to a very large theory, called ***"Maximal Association"***.[1] In the paired comparisons approach, we compare the partitions induced by two modality variables noted : Y and C, pairs of objects with pairs of objects two by two. y_{ij} and c_{ij}, that needs to consider, starting from n objects, data tableaux of size n^2. On the contrary, in the statistical approach, for two categorical variables, one builds a contingency table, crossing both the variables Y and C. The distribution of values in this table n_{uv} represents the number of objects having simultaneously the modality u of C and the modality v of Y. Then, these tables are of size $p \times q$, where p is the number of modalities of C and q the one of Y.

This opposition between the two representation spaces will be the basis of this work.

Cross association criteria for categorized data

As mentioned by L.A. Goodman and W. Kruskal in their excellent reference book [4], the problem in this context is not the existence of criteria but their too large number. We shall give here briefly a survey of some of these cross association measures. Readers interested in a much more complete survey of both classical and new cross association measures, can read [12].

The usual cross association criteria

We shall only present here the most important classification criteria, or those to them the recent scientific developments refer. These criteria or cross association measures will be presented without mention of all their properties. For further details see [12]. A list of association measures can be found in [8] but some fundamental criteria are missing in the book; such as the Rand's index (1971) or the 'related to Jordan' criterion (1983) etc.. For all the indexes or cross association measures presented here, we start from the following situation:

C represents a categorical variable with p modalities (partition with p classes of a set with N objects) and Y represents a partition with q classes of this same set. This involves the following notations:

n_{uv} = Number of objects having the u modality of C and the v modality of Y.

---

1 A particular case of this problem being precisely the Similarity Aggregation.

Compstat 1986

$n_{u.}$ = Number of objects having the modality u of $C = \sum_v n_{uv}$, $n_{.v}$ = Number of objects having the modality v of $Y = \sum_u n_{uv}$. The grand total commonly written $n_{..} = \sum_u \sum_v n_{uv} = N$

Let us now define the classical criteria:

The Chi-squared criterion (1904)

It is defined by the following formula:

(1) $$\chi^2 = \sum_{u=1}^{p} \sum_{v=1}^{q} \frac{N\left(n_{uv} - \frac{n_{u.}n_{.v}}{N}\right)^2}{n_{u.}n_{.v}}$$

Although commonly used under this expression referring to the statistical test of the same name, there exists index' versions normalized between 0 and 1, such as the Cramer's or Tchuprow's ones. Although very used by practitioners, the χ^2 is not the only measure of association, and for some authors not the best; as stated by Goodman and Kruskal themselves (see [4]).

The Belson's criterion (1959)

We shall afterwards call Belson's criterion the following measure of association:

(2) $$B(C,Y) = \sum_{u=1}^{p} \sum_{v=1}^{q} \left(n_{uv} - \frac{n_{u.}n_{.v}}{N}\right)^2$$

Some historical backgrounds and some properties concerning this criterion are given in [12] and in the book of M. Hughes [7].

The Rand's cross association measure (1971)

This measure is given here slightly different from the one introduced in 1971 by W. Rand [15] but completely ***equivalent***.

(3) $$R(C,Y) = \frac{2\sum_{u=1}^{p} \sum_{v=1}^{q} n_{uv}^2 - \sum_{u=1}^{p} n_{u.}^2 - \sum_{v=1}^{q} n_{.v}^2 + N^2}{N^2}$$

If W. Rand was the first to propose this measure in statistical contingency notations, see for example Anderberg ph.1], and also a recent paper of L. Hubert and P. Arabie [6] etc..; it is also true that its real meaning in paired comparisons notations (i.e the 'symmetric difference analogue') was found before 1971, for instance E.T. Zahn (1964) [19], S. Regnier [17] (1965), Green and Rao [5] (1969). In its new version the Rand's measure varies from $\frac{1}{N}$ to 1. $R(C,Y) = 1$ when C and Y are identical; $R(C,Y) = 1/N$ if C has an unique class and Y has N classes, (or the reverse). In that case the traditional Rand's index is equal to 0.

Unusual association measures

These following criteria, although characterized by interesting properties are poorly used by statisticians in practice.

Mean-interaction measure of association (derived from an idea of C.Jordan (1927))

This criterion first proposed in [12] (1983) is derived from a basic idea of C. Jordan [8] (1927).
It is given by:

(5) $$J(C,Y) = \sum_{u=1}^{p} \sum_{v=1}^{q} \frac{n_{uv}}{N}\left(n_{uv} - \frac{n_{u.}n_{.v}}{N}\right)$$

Properties of J

1. J(C,Y) = 0 in case of 'independence'

2. J(C,Y) = 0 also in case of 'indetermination' $n_{uv} = \frac{n_{u.}}{q} + \frac{n_{.v}}{p} - \frac{N}{pq}$ $\forall u$ and v.
3. J(C,Y) is maximum in case of 'complete association'
4. J(C,Y) is minimum for fixed margins in case of 'ambiguousness'

Although poorly known comparing its interesting structural properties, this measure has been recently rediscovered as a measure of association for 'object triples' in contingency fields. Thus S. Chah in [3] (1985) showed that the 'derived from Jordan $J(C, Y)$ measure occurs when one deals with 'triplets or triples' in categorical variables analysis. On the same way, L. Hubert and P. Arabie in their recent paper [6] (1985) ended up at the same conclusion with their ***Con - Dis*** measure on triples which yields to a formulation analog to $J(C, Y)$.

Goodman-Kruskal Light-Margolin criterion

In 1954, L. Goodman and W. Kruskal in [4], have proposed a very interesting index, called :'proportional prediction index' which consists of dissymmetrize the association between C and Y, assuming that C is known. This index, also called τ_b is given by:

$$(6) \qquad \tau_b = \frac{\sum_{u=1}^{p}\sum_{v=1}^{q}\frac{n_{uv}^2}{N n_{u.}} - \sum_{v=1}^{q}\frac{n_{.v}^2}{N^2}}{1 - \sum_{v=1}^{q}\frac{n_{.v}^2}{N^2}}$$

$\tau_b = 0$ if $n_{uv} = \frac{n_{u.}\,n_{.v}}{N}$ $\forall u, v$ 'independence' and $\tau_b = 1$ in case of 'complete association' This index redefined in terms of 'variance interpretation' by Light and Margolin [11], is from now on rediscovered and used in factor analysis, see for example the recent works of N. Lauro and L. D'Ambra [10].

Other criteria

There exist other criteria relevant of linearization process further information about them can be found in [12].

Paired comparisons and cross association measures

The concept of paired comparisons appears first in the statistical literature in the Kendall's paper of 1939 [9]; followed by many authors, however, the first reference is due to A. de Condorcet in (1785) as mentioned by P. Michaud [15].

General presentation of paired comparisons

To define in a first step the notion of paired comparisons, suppose we have a categorized variable C with p modalities. (The process presented here is also suitable for other types of relational variables: linear order, semi-order, tournaments etc..). see [14].

Paired comparisons tableau associated to a partition

All the partitions as C will be represented by a 'paired comparisons' matrix (also written C); this matrix will be a squared binary matrix of size $n \times n$, of general term c_{ij} defined as follows.

$c_{ij} = 1$ if i is in the same cluster as j

$c_{ij} = 0$ otherwise

Some properties of the C tableau

An equivalence relation is a binary relation: ***reflexive, symmetric and transitive***. These three properties can be turned into the following linear expressions:

• ***Reflexivity***:

(7) $c_{ii} = 1 \quad \forall i$

• ***Symmetry***:

(8) $c_{ij} = c_{ji} \quad \forall i \neq j$

• ***Transitivity***:

(9) $c_{ij} + c_{jk} - c_{ik} \leq 1 \quad \forall\ i,j,k$

To these previous properties, we have to add the following ones:

(10) $c_{i.} = \sum_{j=1}^{n} c_{ij}$ = size of the cluster i

(11) $c_{..} = \sum_{i=1}^{n} \sum_{j=1}^{n} c_{ij}$

Additivity of the paired comparisons tableaux

Let us assume that we have at our disposal a set of m categorized variables: $\{C^1, C^2 ... C^k, ... C^m\}$ of paired comparisons tableaux associated to m partitions. One of the advantages of the 'paired comparisons' lies in the possibility of ***adding*** these tableaux to get a ***collective*** tableau having a very important meaning. This 'collective' tableau (noted K), is given by:

(12) $K = \sum_{k=1}^{m} C^k$

where:

(13) $\kappa_{ij} = \sum_{k=1}^{m} c_{ij}^k$

The meaning of κ_{ij} = Number of variables for which i and j have the same modality.

Linearization in paired comparisons

We shall prove in this paragraph that the contingency cross association measures lead to linear formulations in terms of y_{ij} since we shall consider C as a constant known variable.

Linearization formulas

We shall give here all these linearization formulas without giving their proof; these proofs are detailed in [13].

(14) $\sum_{u=1}^{p} \sum_{v=1}^{q} n_{uv}^2 = \sum_{i=1}^{n} \sum_{j=1}^{n} c_{ij}\, y_{ij}$

With the convention $c_{ii} = 1$.

(15) $\sum_{u=1}^{p} n_{u.}^2 = \sum_{i=1}^{n} \sum_{j=1}^{n} c_{ij} = c_{..}$

(16) $\sum_{v=1}^{p} n_{.v}^2 = \sum_{i=1}^{n} \sum_{j=1}^{n} y_{ij} = y_{..}$

(17) $$\sum_{u=1}^{p}\sum_{v=1}^{q} n_{uv}\, n_{u.}\, n_{.v} = \sum_{i=1}^{n}\sum_{j=1}^{n} \frac{(c_{i.} + c_{.j})}{2}\, y_{ij}$$

(18) $$\sum_{u=1}^{p}\sum_{v=1}^{q} n_{u.}\, n_{uv}^2 = \sum_{i=1}^{n}\sum_{j=1}^{n} \frac{(c_{i.} + c_{.j})}{2}\, c_{ij}\, y_{ij}$$

(19) $$\sum_{u=1}^{p}\sum_{v=1}^{q} \frac{n_{uv}^2}{n_{u.}} = \sum_{i=1}^{n}\sum_{j=1}^{n} \frac{2\, c_{ij}}{(c_{i.} + c_{.j})}\, y_{ij}$$

(20) $$\sum_{v=1}^{q}[\sum_{u=1}^{p} n_{u.}\, n_{uv}]^2 = \sum_{i=1}^{n}\sum_{j=1}^{n} c_{i.}\, c_{.j}\, y_{ij}$$

(21) $$\sum_{v=1}^{q}[\sum_{u=1}^{p} \frac{n_{uv}}{n_{u.}}]^2 = \sum_{i=1}^{n}\sum_{j=1}^{n} \frac{y_{ij}}{c_{i.}\, c_{.j}}$$

(22) $$\sum_{u=1}^{p}\sum_{v=1}^{q} n_{u.}^2 n_{.v}^2 = \sum_{i=1}^{n}\sum_{j=1}^{n} c_{..}\, y_{ij}$$

Linearization of some cross association measures

Linearization of the Belson's criterion

The Belson's measure can be developed and under its new expression, we can directly use the formulas (14), (15), and (17) independently and thus obtain the paired formulation of the Belson's criterion written: $P[B(C, Y)]$:

(23) $$P[B(C, Y)] = \sum_{i=1}^{n}\sum_{j=1}^{n} \left(c_{ij} - \frac{c_{i.} + c_{.j}}{N} + \frac{c_{..}}{N^2}\right) y_{ij}$$

Linearization of the Rand's criterion

This result comes trivially from (14) and (15):

(24) $$P[R(C, Y)] = \frac{\sum_{i=1}^{n}\sum_{j=1}^{n} (2c_{ij}\, y_{ij} - c_{ij} - y_{ij} + 1)}{N^2}$$

This formula can be simplified in:

(25) $$P[R(C, Y)] = \frac{\sum_{i=1}^{n}\sum_{j=1}^{n} (c_{ij}\, y_{ij} + \bar{c}_{ij}\, \bar{y}_{ij})}{N^2}$$

Where $\bar{c}_{ij} = 1 - c_{ij}$. It is nothing but the expression of the Condorcet's criterion for one voter.

Linearization of the Jordan's like criterion

Under its developed form the Jordan's like criterion allows to use formulas (14) and (17), we get the following paired expression:

(26) $$P[J(C, Y)] = \frac{1}{N}\sum_{i=1}^{n}\sum_{j=1}^{n} (c_{ij} - \frac{c_{i.} + c_{.j}}{2N})\, y_{ij}$$

Linearization of the Light-Margolin Haldane's criterion

We called Light-Margolin-Haldane's criterion the numerator of the Goodman-Kruskal τ_b. The reasons for the choice of this criterion instead of the Goodman-Kruskal's are given in [12].

$$L(C,Y)=\sum_{u=1}^{p}\sum_{v=1}^{q}\frac{n_{uv}^2}{n_{u.}}-\frac{1}{N}\sum_{v=1}^{q}n_{.v}^2$$

The linearization in y_{ij} of this measure needs the formulas (19) and (16), thence we get:

$$\text{(27)}\qquad P[L(C,Y)]=\sum_{i=1}^{n}\sum_{j=1}^{n}\left(\frac{2c_{ij}}{c_{i.}+c_{.j}}-\frac{1}{N}\right)y_{ij}$$

This new expression is then linear in y_{ij}.

The same process can be applied to the other criteria presented beforehand, by using the different linearization formulas.

Maximal Association

The general model

The Maximal Association is an extension of the Williams-Lambert principle, the general Association Problem amounts to start with m given categorized variables C^k with respectively p_k modalities, and to look for an ***unknown*** Y equivalence relation (partition or categorized variable) ***maximizing*** the global association with each of the m given C^k variables. Whence, writing $\Lambda(C^k, Y)$ for the generic expression of a given ***linearized*** (in y_{ij}) cross association measure, chosen among the list presented in the previous chapter, we are looking for Y maximizing the following L.P. problems:

$$\underset{Y}{Max}\sum_{k=1}^{m}\Lambda(C^k,Y)$$

$$y_{ij}-y_{ji}=0 \qquad \text{(symmetry)}$$

$$y_{ij}+y_{jk}-y_{ik}\leq 1 \qquad \text{(transitivity)}$$

$$y_{ij}=0 \text{ or } 1 \qquad \text{(binarity)}$$

Where the constraints added to the model represent the fact that Y must be a ***partition*** (as shown before). The solutions of these problem are ***equivalence relations***, in other words ***partitions***. Then, we get a ***classification*** of the population of the N objects without fixing ***in advance*** the number of requested clusters, as usually done by other classification techniques. The general solving algorithms are given in [14].

One example of association problems (Rand's Criterion)

We shall just give here an example to illustrate the previous approach. If we replace in the previous model $\Lambda(C^k, Y)$ by $R(C^k, Y)$, the general objective function can be turned into:

$$\underset{Y}{Max}\sum_{k=1}^{m}\Lambda(C^k,Y)=\underset{Y}{Max}\sum_{k=1}^{m}R(C^k,Y)$$

By reformulating the $R(C^k, Y)$ values in terms of paired notations, we get the following linear function of Y:

$$N^2\sum_{k=1}^{m}R(C^k,Y)=\sum_{k=1}^{m}\sum_{i=1}^{n}\sum_{j=1}^{n}\left[c_{ij}^k y_{ij}+\bar{c}_{ij}^k(1-y_{ij})\right]$$

Inverting the summation signs and using the "relational additivity principle" with:

$$\kappa_{ij} = \sum_{k=1}^{m} c_{ij}^{k} \text{ and } \bar{\kappa}_{ij} = \sum_{k=1}^{m} \bar{c}_{ij}$$

we get the aggregated expression of the Rand's Maximal association function multiplied by N^2:

$$N^2 \sum_{k=1}^{m} R(C^{k,Y}) = \sum_{i=1}^{n} \sum_{j=1}^{n} \left[\kappa_{ij} y_{ij} + \bar{\kappa}_{ij} (1 - y_{ij})\right] \tag{28}$$

Hence, we can recognize here in this maximimization problem, the classical ***Central Partition Problem*** introduced by S. Regnier in (1966) [17], and moreover if we suppose that the C^k are no longer partitions but similarity relations we deal with the ***similarity aggregation problem*** solved in [14]. It is interesting to remark on this occasion that the problem formulated in contingency notations ends up at an ***impossibility*** condition.

References

[1] M. ANDERBERG: *Cluster Analysis for Applications*, Academic Press, New-York, (1973).

[2] W. BELSON: *Matching and Prediction on the Principle of Biological Classification*, Applied Statistics, Vol VII, (1959).

[3] S. CHAH: *Comparaisons par Tiplets en Classification Automatique*, IBM Scientific Center Technical Report No F086, Paris, (1985).

[4] L. A. GOODMAN and W. H. KRUSKAL: *Measures of Association For Cross Classification*, Springer Verlag, Berlin,New-York, (1979).

[5] P.E. GREEN and V. R. RAO: *A Note on Proximity Measures and Cluster Analysis*, Journal of Marketing Research No 6, (1969).

[6] L. HUBERT and P. ARABIE: *Comparing Partitions*, paper to be published and presented at the 9th international Meeting on Classification, Karlsruhe, (1985).

[7] M. HUGUES: *Segmentation et Typologie*, Bordas Management, Paris, (1970).

[8] C. JORDAN: *Les Coefficients d'Intensité Relative de Körösy*, Revue de la Société Hongroise de Statistique, Vol 5, Budapest, (1927).

[9] M. G. KENDALL: *A new Measure of Rank Correlation*, Biometrika No 30, (1939).

[10] N. LAURO and L. D'AMBRA: *Analyse non Symétrique des Correspondances*, Proceedings of the Third International Symposium on Data Analysis and Informatics, North Holland, Amsterdam, (1984).

[11] R. J. LIGHT and B. H. MARGOLIN: *An Analysis of Variance for Categorical Data*, Journal of the American Statistical Association, Vol 66, No331, (1971).

[12] F. MARCOTORCHINO: *Utilisation des Comparaisons par Paires en Statistique des Contingences: Partie I*, IBM Scientific Center Technical Report No F 069, Paris, (1983).

[13] F. MARCOTORCHINO: *Utilisation des Comparaisons par Paires en Statistique des Contingences: Partie II*, IBM Scientific Center Technical Report No F 071, Paris, (1984).

[14] F. MARCOTORCHINO and P. MICHAUD: *Agrégation des Similarités en Classification Automatique*, Revue de Statistique Appliquée, Vol 30, No2, (1981).

[15] P. MICHAUD: *Agrégation à la Majorité I : Hommage à Condorcet*, IBM Scientific Center Technical Report No F 051, Paris, (1982).

[16] W. H. RAND: *Objective Criteria for the Evaluation of Clustering Methods*, Journal of the American Statistical Association, Vol 66, (1971).

[17] S. REGNIER: *Sur Quelques Aspects Mathématiques des Problèmes de Classification Automatique*, ICC Bulletin, Rome, (1965).

[18] W. T. WILLIAMS et J. LAMBERT: *Multivariate Methods in Plant Ecology, I: Association Analysis in Plant Communities*, Journal of Ecology, Vol 47, (1959).

[19] C. T. ZAHN: *Approximating Symmetric Relations by Equivalence Relations*, SIAM Journal of Applied Mathematics No12, (1964).

Robustness in Multivariate Analysis

Multivariate Data Analysis, Contributions and Shortcomings of Robustness in Practice

By W. J. J. Rey, Eindhoven

ABSTRACT. We advocate the use of weighted-mean vectors and weighted-covariance matrices in order to perform analyses with the ordinary linear algebra techniques. The weights are allocated in placing the emphasis either on the essential data part or on the outlying observations.

We are routinely analysing large multivariate data sets and we try to make profit of the experience gained with various robust methods. Unfortunately, what we do is a long way from the clean theoretical robust approaches and this is because we cannot afford to be clean and theoretical. Indeed, it is essential to get trustworthy results in a routine way, that is to say in a way which does not require a great deal of statistician time nor a great deal of computer time.

By opposition with clean-and-theoretical robust approaches, we must consider efficient methods. But what are the shortcomings of robustness? They are essentially two :

- Robust methods require much computer time due to the manipulation of non-linear equations.
- Robust methods are up to now very limited in scope. It is known how to evaluate location parameters and less clear how to deal with scatters. The multivariate domain is hardly covered by robustness and the nice powerful techniques we are used to are all non-robust. They are based on linear algebra derivations.

Throughout this text, we will consider a genuine data set. We have selected a "small" problem to keep the complexity of the analysis within bounds.

A DATA SET

The technologies applied in the fabrication of integrated circuits are rapidly changing and in view of the development of a new product, experimental chips are designed and manufactured; in the present case they are 16K SRAM MOS memories, i.e. each chip is a memory with 16384 bit cells. These test products have been measured in many respects and, here, the question is addressed as to whether there is any relation between the number of functional bit cells and some electrical parameters; here we will only deal with the correlation matrix estimation.

As is shown in Table 1, the original data consists in 2557 records derived from measurements on as many chips. The number of faulty cells is Entry 1, varying from 12 to 16384; the other seven entries

Compstat 1986

are the results of some functional tests, e.g. Entry 2 is a leakage current.

DATA SCREENING

In this context following Fisher's views (1922) are of interest: "The object of statistical methods is the reduction of data. A quantity of data, which usually by its mere bulk is incapable of entering the mind, is to be replaced by relatively few quantities which shall adequately represent the whole, or which, in other words, shall contain as much as possible, ideally the whole, of the relevant information contained in the original data."

TABLE 1
The original data consists in 8 measurements on 2557 chips

Entry	No	MEAN	STDV	MIN	MAX	
1	2557	2145.515	4857.764	12	16384	Error sum
2	2557	-3.03E-07	4.80E-06	-2.05E-05	2.05E-05	(Ampere)
3	2557	4.04E-06	3.05E-05	-2.05E-04	2.05E-04	(Ampere)
4	2557	0.006460	0.032742	-1.00E-04	0.205000	(Ampere)
5	2557	0.008796	0.034883	-1.00E-04	0.205000	(Ampere)
6	2557	0.019985	0.038457	0.001470	0.205000	(Ampere)
7	2557	-1.003566	1.304172	-3.220000	0.510000	(Volt)
8	2557	-6.74E-06	2.38E-05	-2.05E-04	0	(Ampere)

In order to give an idea of the type of data we meet with, let us provide a schematic description of Entry 2. First note that the measurements have been performed by equipment with automatic scale adjustment giving three significant digits; the equipment resolution is 0.01 microampere and the maximum current was limited to 20.5 microamps. It can be seen in Table 1 that the positive and negative extremes have been observed although they are at 4.2 standard deviations from the mean; this is somewhat exceptional seeing the sample size 2557. We may enter a little more into the details.

TABLE 2
Condensed frequency table of Entry 2 in Table 1

Freq.	Cumul. freq.	Relative freq.	value or range (microampere)
72	72	0.02816	-20.5
67	139	0.02620	(-20.5 ====> 0)
22	161	0.00860	[0 ====> 0.05]
600	761	0.23465	0.06
1746	2507	0.68283	[0.07 ====> 0.33]
1	2508	0.00039	0.57
1	2509	0.00039	0.74
1	2510	0.00039	6.36
47	2557	0.01838	20.5

Table 2 gives more information; we find 119 observations at the extreme values and almost all other measurements in the range from 0 to 0.33 microamps. The mean value, -0.3 microamps, corresponds with the quantile at probability level 0.05 and 93 percent of the observations in the range [-1, 1] standard deviation. The distribution is definitely not normal, nor uniform, nor standard in any way. Such a non-normality is more the rule than the exception.

Clearly some action must be taken in order to clean such "dirty" data. What action? First let us note that the data set is not that dirty, it just belongs to the real world rather than to the textbooks and, discussing the matter with the technologists, it appears that its only pecularity is that the meter exhibits a zero offset of some 0.06 microamps; all other features (even if not intended) were representative of the measured devices. The offset has no consequence for the statistical analysis.

What should we do or, rather, what are we able to do? An action which rejects part of the data is not justified when based on arbitrary limits; a data reject seems to be grounded only in case of transcription error and inasmuch as the interest is not related to these errors. In this context we could decide for non- or semi-parametric approaches, but the action we present here consists in weighting all data entries in order to obtain mean vectors and covariance matrices such that we can apply the ordinary linear algebra methods. One convenience of the weighting is that the missing data are very easily dealt with: you allocate zero weights.

DATA WEIGHTING

The methods used for weight allocation as well as for further dealing with the weights stem from robustness. Briefly:

- In order to deal with the essential data part we allocate weights which are one (or little inferior) to the entry-components which are close to the entry central values. Outliers are severely down-weighted.
- The weight allocation is performed according to the robust "reweighted" least squares technique often applied while evaluating M-estimators.
- Conversely, when interested in the data surrounding the essential part, we apply a complementary weight. We allocate weights which are one (or little inferior) to the entry-components which are far from their entry central values. Central values are severely down-weighted.

Before giving further precisions, let us insist on the fact that the weight allocation is performed in a component-wise fashion separately for each entry - For our illustration, separately for each of the 8 entries and on each set of 2557 components - This is sensible in most contexts, although not always founded.

The M-estimators used in the one-dimensional location-scatter problem are well investigated. Hence we will only present the main equations without dealing with the underlying issues. The relevant theory can be found for instance in Huber (1981) although, for sake of uniformity and for further developments, we will only refer to Rey (1983, sec. 6.4, pages 146-147, 110-111, 180-181 and Eq. 4.36).

Let {x1, x2, ..., xi, ..., xn} be a set of scalar data - In our illustration, we consider a single coordinate at a time and the sample is {x1, x2, ..., x2557} for n=2557 - Then the location m minimizes the next Eq. (1) where the scatter s is derived by Eq. (2).

$$M = \sum_{i=1...n} f(u_i), \quad (1)$$

$$u_i = (x_i - m) / s$$

$$s^2 = k2\, s^2\, M / (n-1) \quad (2)$$

The set (1)-(2) defines a "quasi-robust" estimator when the function f is robust. For sake of simplicity, whenever u is small we impose the validity of the approximation

$$f(u) = u^2 . \quad (3)$$

The function "Fair" is recommendable and it can be found in Table 3 as well as a few constants such as k2 in Eq. (2). It is convenient to transform the set (1)-(2) into

$$w_i = f'(u_i) / (2\, u_i) \quad (4)$$

$$u_i = (x_i - m) / s$$

$$f'(u) = (d/du)\, f(u)$$

$$m = \sum_{i=1...n} w_i x_i \Big/ \sum_{k=1...n} w_k \quad (5)$$

$$W2 = \sum_{i=1...n} w_i^2$$

$$s^2 = \frac{k3\ n}{(n-1)\ W2} \sum_{i=1...n} [w_i (x_i - m)]^2 \quad (6)$$

The coefficients wi are to be used as weights whenever the interest lies in the data essential part,

$$\text{weight}(x_i) = w_i; \quad (7a)$$

while interested by the outliers, they become

$$\text{weight}(x_i) = 1 - w_i. \quad (7b)$$

The two weight definitions (7a) and (7b) naturally follow from Eqs. (3) and (5). They prohibit the use of the so-called redescending functions f. Starting with approximate values of m and s, the set (4)-(6) of equations must be evaluated a few times to yield the coefficients wi; in practice, we start with the arithmetic mean and the standard deviation and, then, run 5 iterations with function "Fair" at relative efficiency 0.95. Note that Eq. (6) is not standard; it has been derived as indicated in the Appendix for its multivariate analogue, Eq. (9).

Before proceeding further, it might be of interest to see what is the effect of the weighting scheme on the data we used as an illustration. Table 4 corresponds with Table 2; some 5 percent of the data lie at the extremes of the covered range and they receive less than one thousandth of the total weight.

TABLE 3
Function "Fair"

$$f'(u) = (d/du)\, f(u) = 2u \,/\, (1 + |u|/k1)$$

Efficiency	0.950	0.900	0.850	0.800
k1	1.399503	0.635016	0.333313	0.176009
k2, in Eq. (2)	1.662693	2.359038	3.430700	5.335323
k3, in Eq. (6)	1.932691	2.869564	4.286763	6.776039

"Efficiency" is the power relative to the arithmetic mean estimator if the underlying distribution is the standard normal distribution.

$$\text{Efficiency} = [\text{integral}(f''(u))]^2 \,/\, \text{integral}((f'(u))^2)$$

$$k2 = 1 \,/\, \text{integral}(f(u))$$

$$k3 = \text{integral}((f'(u)/u)^2) \,/\, [\text{integral}(f'(u)^2)]$$

"integral" stands for integration on u distributed according to the standard normal distribution.

TABLE 4
Relative frequencies and weights for Entry 2
Weights derived by "Fair": k1 =1.399503, k3 := 1.932691

Freq.	Relative freq.	Relative weight	value or range (microampere)
72	0.02816	0.00052	-20.5
67	0.02620	0.00327	(-20.5 ====> 0)
22	0.00860	0.00852	[0 ====> 0.05]
600	0.23465	0.24233	0.06
1746	0.68283	0.74462	[0.07 ====> 0.33]
1	0.00039	0.00021	0.57
1	0.00039	0.00017	0.74
1	0.00039	0.00002	6.36
47	0.01838	0.00035	20.5

MULTIVARIATE DATA ANALYSIS

As already stated we have weighted all data entries in order to obtain mean vectors and covariance matrices such that we can apply the ordinary linear algebra methods. A weight wij is allocated according to the weight definitions (7a) or (7b) to each of the components xij of the observation

$$xi = (xi1, \ldots, xij, \ldots, xip), \quad i = 1 \ldots n.$$

For our illustration we have as before n=2557 and, treating the data of Table 1, p=8.

The mean vectors M and the covariance matrices COV are evaluated by formulae similar to (5) and (6), namely by

$$M_j = \sum_{i=1\ldots n} w_{ij}\, x_{ij} \Big/ \sum_{k=1\ldots n} w_{kj} \qquad (8)$$

$$W2_j = \sum_{i=1\ldots n} w_{ij}^2$$

$$COV_{kl} = \frac{k3\; n}{n-1} (W2_k\, W2_l)^{-1/2} \sum_{i=1\ldots n} [w_{ik}(x_{ik} - M_k)(x_{il} - M_l) w_{il}] \qquad (9)$$

We remark that the selection of one or the other weight definitions (7) gives very different meanings to M and COV; they are either representative of the data essential part or representative of the extreme values. This dichotomy copes with the moderate non-linearities always present in the relations between variables. With the data of our example, we have derived correlation matrices; they are reported in Tables 5a and 5b, they can be compared with the ordinary correlation matrix to be found in Table 6.

TABLE 5a
Correlation matrix (percent) for Table 1 data
Treatment of essential data part by Rule (7a)

	1	2	3	4	5	6	7	8
1	100	8	33	72	56	73	10	-6
2	8	100	22	5	5	3	31	-38
3	33	22	100	37	31	45	20	-28
4	72	5	37	100	62	78	15	-9
5	56	5	31	62	100	67	8	-5
6	73	3	45	78	67	100	20	-13
7	10	31	20	15	8	20	100	-59
8	-6	-38	-28	-9	-5	-13	-59	100

TABLE 5b
Correlation matrix (percent) for Table 1 data
Treatment of extreme values by Rule (7b)

	1	2	3	4	5	6	7	8
1	100	18	22	50	56	59	14	-25
2	18	100	8	16	20	14	2	-10
3	22	8	100	26	23	17	6	-8
4	50	16	26	100	92	77	16	-18
5	56	20	23	92	100	79	18	-18
6	59	14	17	77	79	100	18	-21
7	14	2	6	16	18	18	100	-19
8	-25	-10	-8	-18	-18	-21	-19	100

In general the entries of Table 6 are intermediate between those of Table 5a and Table 5b. But there are a few interesting values. For instance, the correlation between Entry 2 and Entry 7, r(2,7), is high only in Table 5a; the correlation existing in the main part of the data, 31 %, is concealed when the extreme values are considered. It is obvious that such features can have the utmost importance in data modeling.

TABLE 6
Ordinary correlation matrix (percent) for Table 1 data
Observations are not weighted (wij=1)

	1	2	3	4	5	6	7	8
1	100	15	27	54	57	65	13	-25
2	15	100	9	14	17	12	3	-9
3	27	9	100	33	30	26	7	-11
4	54	14	33	100	92	79	16	-20
5	57	17	30	92	100	80	17	-19
6	65	12	26	79	80	100	17	-23
7	13	3	7	16	17	17	100	-18
8	-25	-9	-11	-20	-19	-23	-18	100

The complexity of the method described by the set (8)-(9) is so great that it seems difficult to precisely give any confidence statement; this is a theoretical drawback of the approach which can easily be circumvented by randomization. By random permutation of the 2557 values of all 8 entries, we constructed a series of zero-correlated samples with the same marginal distributions; the correlation matrices of these samples have been evaluated and, then, used to assess the significance of the original correlation matrix. Such a zero-correlated matrix is displayed in Table 7, it corresponds with Table 5a. It is clear that most correlation coefficients were very significant.

TABLE 7
Zero-correlation matrix (percent) for Table 5a
Data = randomized data set used in Table 5a

	1	2	3	4	5	6	7	8
1	100	1	0	-1	3	-1	0	1
2	1	100	-1	1	3	1	0	-3
3	0	-1	100	-1	1	1	-6	-1
4	-1	1	-1	100	1	1	-1	-2
5	3	3	1	1	100	-4	0	2
6	-1	1	1	1	-4	100	-3	2
7	0	0	-6	-1	0	-3	100	2
8	1	-3	-1	-2	2	2	2	100

REFERENCES

Fisher R. A., On the mathematical foundations of theoretical statistics, Philos. Trans. Roy. Soc. London, A-222 (1922) 309-368.

Huber P.J., Robust Statistics, Wiley (1981).

Rey W.J.J., Introduction to Robust and Quasi-robust Statistical Methods, Universitext Series, Springer-Verlag (1983).

APPENDIX

We briefly sketch the derivation leading to Eq. (9) and to its one-dimensional equivalent, Eq. (6).

Let a weighted sample {xi1, ..., xip; wi1, ..., wip}, i = 1...n be given. Then the mean vector M has components Mj given by Eq. (8)

$$M_j = \sum_{i=1\ldots n} w_{ij} x_{ij} \Big/ \sum_{k=1\ldots n} w_{kj}$$

In order to derive the covariance, COV, of the above weighted-sample, we first evaluate the covariance, Cov(M), of the mean vector. Then, we pre- and post-multiply Cov(M) by the diagonal matrix of the square roots of the effective sample sizes.

The infinitesimal jackknife estimate of Cov(M) is now obtained. First, one weights each of the n observations by a factor {v1, ..., vn} and the mean vector components become

$$M_j = \sum_{i=1\ldots n} v_i w_{ij} x_{ij} \Big/ \sum_{k=1\ldots n} v_k w_{kj}.$$

Then, the elements of Cov(M) are given by

$$\mathrm{Cov}(M)_{kl} = \frac{n}{n-1} \sum_{i=1\ldots n} [(d/d\,v_i)\, M_k]\, [(d/d\,v_i)\, M_l],$$

the partial derivatives being estimated in vi=1.

By Eq. (5), the mean weight of Entry j is

$$W_j = \sum_{i=1\ldots n} w_{ij} w_{ij} \Big/ \sum_{k=1\ldots n} w_{kj}$$

and the corresponding effective sample size comes out as

$$N_j = (1 / W_j) \sum_{i=1\ldots n} w_{ij}.$$

Finally, Eq. (9) is obtained as

$$\mathrm{COV}_{kl} = k_3\, N_k^{1/2}\, \mathrm{Cov}(M)_{kl}\, N_l^{1/2}$$

where k3 is a factor which scales COV in order to equate the ordinary covariance matrix if the sample is drawn from a multivariate normal distribution.

On the Use of Bootstrap and Jackknife in Covariance Structure Analysis

By A. Boomsma, Groningen

SUMMARY

Applications of bootstrap and jackknife for covariance structure analysis under violation of standard maximum likelihood assumptions (small sample size, analysis of correlation matrices, and nonnormality) are discussed. Procedures are illustrated for a factor analysis model, using LISREL. Facts and failures of the resampling methods are covered. For bootstrap and jackknife a computer program with graphical facilities is introduced.

Key words: covariance structure analysis, bootstrap, jackknife, robustness, interval estimation, cross-validation, influence analysis.

INTRODUCTION

In a study on the robustness of LISREL (Boomsma, 1983) two findings are that in covariance structure analysis maximum likelihood (ML) estimators are not robust against (i) analysis of correlation matrices, irrespective of the sample size n, and (ii) skewness of discrete observed variables. When correlation matrices are analysed, or when distribution functions are rather skew, estimated confidence intervals for model parameters can be very unreliable; it was shown that the variance of the standardized parameter estimates was frequently too small or too large.

Given the results of this research it is of interest to investigate whether ML-estimation can be improved under such conditions. The potential merits of jackknife and bootstrap (Efron, 1981, 1982) then primarily raised the question whether they will give improved estimates for standard errors. In short, the challenge to investigate is whether bootstrap and jackknife lead to more accurate interval estimation than standard ML-procedures in covariance structure analysis when model assumptions are violated.

This paper explores the practical feasibility of applying both methods when the LISREL program (Jöreskog & Sörbom, 1984) is used, and it gives results on a number of trials for an artificial factor analysis model. The practical relevance of this exploration is that (1) very often, traditionally and for reasons of interpretation, researchers analyze correlation structures instead of covariance structures; (2) in many applications severe deviations from normality occur; (3) frequently the sample size is not large enough to allow for cross-validation.

Compstat 1986

BOOTSTRAP AND JACKKNIFE PROCEDURES

General purposes of bootstrap and jackknife are (i) bias reduction (not important here); (ii) estimation of standard errors, often in situations where they are otherwise difficult to obtain; (iii) cross-validation, stability analysis, influential data analysis. Both techniques have been studied for a variety of statistical models. However, rather little is known about their usefulness in covariance structure analysis. Beran & Srivastava (1985) examined the validity of the bootstrap for functions of the covariance matrix. They found that 'under natural assumptions, the nonparametric bootstrap distribution for the sample covariance matrix converges to the actual distribution, as sample size increases' (p. 96).

The bootstrap. The statistic $\hat{\theta}(X_1, X_2, \ldots, X_n)$ is a function of n i.i.d. observations. If N independent random samples of size n are drawn from the empirical distribution function $\hat{F}$ this results in N bootstrap replications $\hat{\theta}^*(j)$; see Efron (1982). The bootstrap estimate for θ, and the estimated standard error are defined as follows:

$$\hat{\theta}_B = \Sigma\ \hat{\theta}^*(j)/N, \quad \hat{\sigma}_B(\hat{\theta}) = [\Sigma\ \{\hat{\theta}^*(j) - \hat{\theta}_B\}^2/(N-1)]^{1/2}. \qquad (1)$$

Under approximate normality assumptions the statistic $T_B=(\hat{\theta}_B-\theta)/\hat{\sigma}_B(\hat{\theta})$ has roughly a Student t-distribution with N-1 degrees of freedom. In constructing confidence intervals for θ normal theory approximations are used: $\hat{\theta}_B \pm t_\alpha \hat{\sigma}_B(\hat{\theta})$, where t_α is the α-point of the t_{N-1}-distribution. This parametric approach is simple and straightforward. Nonparametric methods (e.g. Efron, 1981) primarily attempt to capture the right degree of asymmetry. In our study there was no first need to apply them for calculating bootstrap limits, because most empirical distributions were rather symmetric.

The jackknife. Let $\hat{\theta}$ be the estimate when the total sample of size n is analyzed, and let $\hat{\theta}_{-i} = \hat{\theta}(X_1, \ldots, X_{i-1}, X_{i+1}, \ldots, X_n)$, i.e. deletion of one observation case. The i-th pseudo-value is defined as $\tilde{\theta}_i = n\hat{\theta}-(n-1)\hat{\theta}_{-i}$. The jackknife estimate for θ, and its estimated standard error are given by

$$\hat{\theta}_J = \Sigma\ \tilde{\theta}_i/n, \text{ and } \hat{\sigma}_J(\hat{\theta}) = [\Sigma\ \{\tilde{\theta}_i - \hat{\theta}_J\}^2/\{n(n-1)\}]^{1/2}. \qquad (2)$$

If pseudo-values could be treated roughly as an independent, normal sample the statistic $T_J=(\hat{\theta}_J-\theta)/\hat{\sigma}_J(\hat{\theta})$ would be approximately distributed as Student's t_{n-1}. Confidence limits for θ would then be given by $\hat{\theta}_J \pm t_\alpha \hat{\sigma}_J(\hat{\theta})$.

A COVARIANCE MODEL

As an example, consider an oblique two-factor model of simple structure (Boomsma, 1983, 1985; Model 4CM). The model is defined by $\mathbf{x} = \mathbf{\Lambda\xi} + \mathbf{\delta}$, where $\mathbf{x}$(8x1) is a vector of observed variables, $\mathbf{\Lambda}$(8x2) a matrix of loadings on two common factors ξ_1 and ξ_2, and $\mathbf{\delta}$(8x1) a vector of measurement errors. Under usual assumptions the population covariance matrix of $\mathbf{x}$ has the form $\Sigma = \mathbf{\Lambda\Phi\Lambda}' + \mathbf{\Psi}$, where $\mathbf{\Phi} = E(\mathbf{\xi\xi}')$, and $\mathbf{\Psi} = E(\mathbf{\delta\delta}')$ is diagonal with the variances of the unique scores as elements. The first four variables have a positive loading on ξ_1 [in single subscript notation: $\lambda_1=\lambda_2=0.6$, $\lambda_3=\lambda_4=0.8$] and a zero loading on ξ_2, the other four variables have a positive loading on ξ_2 [$\lambda_5=\lambda_6=0.6$, $\lambda_7=\lambda_8=0.8$] and a zero loading on ξ_1. Variances of x_i are normalized: $\mathbf{\Psi} = I - \text{diag}(\mathbf{\Lambda\Phi\Lambda}')$; i.e. uniquenesses ψ_i equal $1-\lambda_i^2$, here. The correlation ϕ between ξ_1 and ξ_2 equals 0.3.

Given this model, and a sample covariance matrix S based on n observations, the object of the covariance structure analysis is to estimate the k=17 elements of $\mathbf{\theta}'=[\lambda_1,\lambda_2,\lambda_3,\lambda_4,\lambda_5,\lambda_6,\lambda_7,\lambda_8,\phi,\psi_1,\psi_2,\psi_3,\psi_4,\psi_5,\psi_6,\psi_7,\psi_8]$, and their corresponding standard errors, and to examine the fit of the model (19 degrees of freedom). This was accomplished by using iterative maximum likelihood estimation procedures in the LISREL-VI program.

The model was studied for a sample size n=25,50,100, using two random samples from the multivariate normal population $N_8(0,\Sigma)$ each. For these cases, denoted as 25.1, 25.2, 50.1, 50.2, 100.1, and 100.2, correlation matrices **R** were analyzed. Boomsma (1983) found that by analyzing **R** instead of S, confidence intervals for θ were often too large; the sampling distributions of standardized parameters being too narrow compared with an expected N(0,1) distribution. Hopefully, bootstrap and jackknife lead to smaller estimates of standard errors, and thus to smaller confidence intervals.

In addition, two random samples (n=100) were taken from a discrete, skew population distribution with covariance structure Σ (see Boomsma 1983; Model 4CMD, variation S2). Among the 8 variables the range in number of categories was [2,5], the range of skewnesses [0.75,2]. For these two cases, denoted as D100.1 and D100.2, covariance matrices S were analyzed. For such nonnormality Boomsma (1983) found that confidence intervals were often too small; sampling distributions of standardized parameters having too large a variance. Here, it is hoped that the resampling techniques give larger estimates of standard errors, hence larger confidence intervals.

The number of bootstrap replications was fixed at N=300 for all cases.

For both techniques a general purpose program BOJA was written in FORTRAN (Boomsma, 1986), enabling (i) sampling procedures for a given total sample X(nxp), resulting in (re)sample covariance matrices **S**, which were used as input for LISREL analyses; (ii) for the bootstrap, analysis of N vectors $\hat{\boldsymbol{\theta}}^*(j)$, and for the jackknife, analysis of the vector $\hat{\boldsymbol{\theta}}$ and n vectors $\hat{\boldsymbol{\theta}}_{-i}$, all of length k, resulting in estimates defined by (1) and (2), and additional statistical information, e.g. confidence intervals; (iii) graphical displays of empirical sampling distributions, like histograms and QQ-plots.

RESULTS

1. Non-convergence. Problems occurred far more often than expected on the basis of results from Boomsma (1983, 1985). For Cases 25.1 and 25.2 the bootstrap percentage of nonconvergence was 18% and 39%, respectively; for the other cases hardly any problems occurred. For the jackknife, in Case 25.2 two subsamples did not converge. Therefore, the resampling methods are unattractive for n=25, and details of Cases 25.1 and 25.2 are deleted.

Instead of ML, it might be tempting to use instrumental variable (IV) methods, which turned out to be rather promising for factor analysis models (Hägglund, 1986). For our purposes they would give an enormous reduction of computing costs (of at least a factor 10), and no problems of convergence. However, using the IV-method of the LISREL program did not work out well (more bias, large confidence intervals, unsuitable for analyses of **R**).

2. Improper solutions. More often than expected, negative estimates of variances ψ_i occurred. For Cases 50.1, 50.2, 100.1, 100.2, D100.1, and D100.2, bootstrap percentages of improper subsamples were 11%, 33%, 0%, 0%, 17%, and 0%, respectively, while the jackknife gave no improper solutions. For n=25 results for both methods were influenced substantially by extreme outliers due to improper solutions. To some degree, this also holds for the bootstrap in Case 50.2. In such circumstances trimming and Winsorizing could be considered for constructing robust alternatives. In this study, estimates outside their admissible range were always included.

3. Bias. When the total sample is analyzed the ML-estimate is denoted as $\hat{\theta}$, and the corresponding standard error based on Fisher's information matrix as $\hat{\sigma}(\hat{\theta})$. Detailed results on the bias of individual parameters give evidence of close agreement among $\hat{\theta}$, $\hat{\theta}_B$, and $\hat{\theta}_J$. A summary of results is presented in Table 1. No systematic improvements of $\hat{\theta}_B$ and $\hat{\theta}_J$ over $\hat{\theta}$ were detected.

Table 1. Mean absolute bias, and mean differences in estimated standard errors by comparing total sample, bootstrap, and jackknife ML-estimates.

	50.1	50.2	100.1	100.2	D100.1	D100.2
$\Sigma\ \lvert\hat{\theta} - \theta\rvert/k$	.075	.084	.067	.099	.114	.075
$\Sigma\ \lvert\hat{\theta}_B - \theta\rvert/k$	.072	.089	.067	.098	.120	.071
$\Sigma\ \lvert\hat{\theta}_J - \theta\rvert/k$	.077	.085	.068	.100	.111	.079
$\Sigma\ [\hat{\sigma}_B(\hat{\theta}) - \hat{\sigma}(\hat{\theta})]/k$	-.013	.028	-.014	-.011	.023	.024
$\Sigma\ [\hat{\sigma}_J(\hat{\theta}) - \hat{\sigma}(\hat{\theta})]/k$	-.011	.002	-.014	-.009	.023	.029

4. Standard errors. Summary comparisons between $\hat{\sigma}_B(\hat{\theta})$, $\hat{\sigma}_J(\hat{\theta})$, and the estimated standard error $\hat{\sigma}(\hat{\theta})$ from the total sample, are given in Table 1. For Cases 50.1, 100.1, and 100.2, both $\hat{\sigma}_B(\hat{\theta})$ and $\hat{\sigma}_J(\hat{\theta})$ are smaller than $\hat{\sigma}(\hat{\theta})$, as was hoped. This holds especially for the loadings. The reverse result for Case 50.2 is due to some outlying estimates. For the nonnormal cases $\hat{\sigma}_B(\hat{\theta})$ and $\hat{\sigma}_J(\hat{\theta})$ are almost uniformly larger than $\hat{\sigma}(\hat{\theta})$, as was hoped. Here too, bootstrap and jackknife give virtually identical results. Therefore, it seems more attractive to apply jackknife than bootstrap, because it is less costly (50 LISREL analyses against 300 for n=50). It may also imply that our choice for N was too large (cf. Efron, 1981, p. 591).

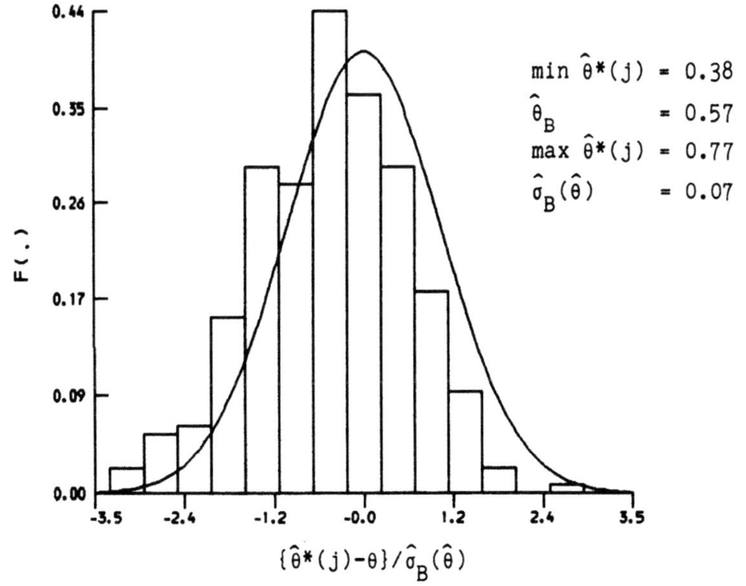

Figure 1. Standard normal fit of the empirical sampling distribution of standardized bootstrap replications for $\theta = \lambda_2 = 0.60$; Case 100.2, N=300.

5. Confidence intervals. The results on bias and standard errors indicate that our expectations with respect to confidence intervals are confirmed. When analyzing correlation matrices two-sided 95% confidence intervals for most parameters are smaller, especially for the loadings. Under nonnormality these intervals for θ are systematically larger for both methods.

6. Distributional behavior. In general, the empirical distributions of bootstrap replications $\hat{\theta}^*(j)$ and pseudo-values $\tilde{\theta}_i$ did not deviate substantially from theoretical expectations. Figure 1, which is produced by the BOJA program, shows that, except for some bias, the empirical distribution of $\{\hat{\theta}^*(j)-\theta\}/\hat{\sigma}_B(\hat{\theta})$ closely follows the standard normal distribution, as expected. Notice, that the statistics in the plot refer to bootstrap replications $\hat{\theta}^*(j)$. With graphical options from BOJA the user may effectively illustrate (in)stabilities and influential data in a one sample result.

CONCLUSION

The detailed results indicate that bootstrap and jackknife are potential means for correcting the usual maximum likelihood estimate of standard error, both when correlation matrices R are analyzed, and when apparent nonnormality occurs. Further research, studying many more cases, will be needed to assess the stability of these results, and to show whether and to what extent such findings can be generalized, e.g. to other models.

REFERENCES

Beran, R., & Srivastava, M.S. (1985). Bootstrap tests and confidence regions for functions of a covariance matrix. The Annals of Statistics, 13, 95-115.

Boomsma, A. (1983). On the robustness of LISREL (maximum likelihood estimation) against small sample size and non-normality. Amsterdam: Sociometric Research Foundation. (doctoral dissertation)

Boomsma, A. (1985). Nonconvergence, improper solutions, and starting values in LISREL maximum likelihood estimation. Psychometrika, 50, 229-242.

Boomsma, A. (1986). BOJA: A program for bootstrap and jackknife analysis: User's guide. Groningen: Vakgroep Statistiek & Meettheorie.

Efron, B. (1981). Nonparametric estimates of standard error: the jackknife, the bootstrap and other methods. Biometrika, 68, 589-599.

Efron, B. (1982). The jackknife, the bootstrap, and other resampling plans. Philadelphia: Society for Industrial and Applied Mathematics.

Hägglund, G. (1986). Factor analysis by instrumental variables methods. University of Uppsala, Department of Statistics. (doctoral dissertation)

Jöreskog, K.G., & Sörbom, D. (1984). LISREL VI: Analysis of linear structural relationships by maximum likelihood, instrumental variables, and least squares methods: User's guide (3rd. ed.). Mooresville, IN: Scientific Software.

On a Class of Robust Methods for Multivariate Data Analysis

By V. Dobrić, Zagreb

A class of robust methods for multivariate data analysis was proposed to enable processing of data that do not satisfy assumptions necessary for the application of classical data analysis methods. Canonical covariance analysis (Momirović, Dobrić and Karaman, 1983) was developed as the general method to analyse relationships between two sets of variables. A model of robust redundancy analysis (Prot, Bosnar and Momirović, 1983), as well as models of robust regression (Štalec and Momirović, 1983) and discriminant (Dobrić and Momirović, 1984) analysis, were derived as special cases from the canonical covariance model. In order to clarify the interpretation of the obtained results, relationships between robust methods and corresponding classical methods were described in our previous papers. A set of programs for all proposed methods was written in GENSTAT and SS languages, and the behaviour of the methods was examined on real data.

KEYWORDS: Data analysis. Robust methods. Canonical covariance analysis. Robust redundancy analysis. Robust regression. Robust discriminant analysis.

1. INTRODUCTION

Development of some robust methods for multivariate data analysis has been initiated few years ago at University Computing Centre in Zagreb, in order to satisfy users demands for processing of data that are not appropriate for the application of classical data analysis methods. The first proposed method was canonical covariance model (QCR) for the analysis of relationships between two sets of quantitative variables, followed by the robust model for redundancy analysis, regression and discriminant analysis, derived as special cases from QCR.

The advantages of the proposed methods are in their robustness both to outliers and to the artificial associations generated by the few pairs of variables from two different sets. Furthermore, this variables need neither be normally distributed nor their covariance matrices need be regular.

Although some heuristic criterion has been suggested for the number of retained latent dimensions, until 1985 there were no statistical tests for parameters obtained, so that all proposed methods were considered more as data analysis than statistical methods. However, some test statistics for robust parameters has been recently found to be asimptotically χ^2 distributed for

Compstat 1986

the general class of elliptically distributed variables (Lužar, 1985; 1986). Canonical covariance model is not matric invariant, so that all proposed methods could be defined in the spaces with special metrics; Guttman's and Harris spaces are, of course, of special theoretical and practical interest (Momirović, Dobrić and Gredelj, 1985).

In order to clarify the interpretation of obtained results, all the relationships between robust methods and corresponding classical methods were described in our previous papers. Proposed algorithms were programmed in GENSTAT and SS languages and their behaviour on real data was also inspected.

2. METHOD

Let $Z_1(n,m_1)$ and $Z_2(n,m_2)$ be two centered data matrices, obtained as a description of set E of n entities over two sets V_1 and V_2 of quantitative, elliptically distributed variables. Maximization of covariances between linear composites of variables belonging to the sets V_1 and V_2 is defined, under some constraints, as

$$\begin{array}{l|ll} Z_1X_p = L_p & \rho_p = L_p^T K_p \frac{1}{n} \rightarrow \max & \\ Z_2Y_p = K_p & \rho_p > \rho_{p+1} & p=1,\ldots,r-1;\ r=\min(m_1,m_2) \\ & X_p^T X_q = Y_p^T Y_q = \delta_{pq} & \end{array} \quad (1)$$

and can be reduced to the singular value decomposition of the matrix $C_{12}=Z_1^T Z_2 \frac{1}{n}$.

2.1. CANONICAL COVARIANCE ANALYSIS (QCR)

In general case when $m_1>1$ and $m_2>1$, and symmetrical relations between V_1 and V_2 can be assumed, QCR model is obtained. QCR algorithm seeks for such linear composites L_p and K_p of variables from V_1 and V_2 to maximize successively their covariances ρ_p under the constraints of orthonormality of transformation vectors $\{X_p\}$ and $\{Y_p\}$ respectively. Solution of the problem could be obtained from the characteristic equation

$$C_{12}C_{12}^T X_p = \rho_p X_p \qquad p=1,\ldots,r \qquad (2)$$

with covariances ρ_p equal to $\rho_p = X_p^T C_{12} Y_p$.

As was recently found (Lužar, 1985), the test statistic

$$n\ (M\rho)^T (M\Psi M^T)^{-1} (M\rho) \qquad (3)$$

is asimptotically χ^2 distributed with r-1 degrees of freedom; here, ρ is vector of covariances ρ_p, M is hipotetic matrix and Ψ is a matrix with elements estimated from the elements of crosscovariance matrix of quasicanonical variates $\{L_p,K_p\}$, $p=1,\ldots,r$.

Obtained quasicanonical dimensions $\{L_p\}=L$ and $\{K_p\}=K$ are not orthogonal; if $C_i=Z_i^TZ_i\frac{1}{n}$ are covariance matrices of variables from v_i, $i=1,2$ respectively, then

$$W_1=L^TL\frac{1}{n}=X^TC_1X \neq \text{diag} \quad \text{and} \quad W_2=K^TK\frac{1}{n}=Y^TC_2Y \neq \text{diag}.$$

Denoting by D_i^2 =diagW_i, $i=1,2$ variances of linear composites $\{L_p\}$ and $\{K_p\}$ respectively, the structure matrices F_i, $i=1,2$ of the first and second set of quasicanonical factors could be obtained as

$$F_1 = C_1XD_1^{-1} \quad \text{and} \quad F_2 = C_2YD_2^{-1}$$

together with their pattern matrices

$$A_i = F_i(D_i^{-1}W_iD_i^{-1})^{-1}, \qquad i=1,2.$$

Quasicanonical correlations Γ, for diagonal matrix ρ, are obtained as

$$\Gamma = D_1^{-1}\rho D_2^{-1}$$

and if η_1 is the first canonical correlation, then $\gamma_1<\eta_1$. This and other relations between QCR and canonical correlation analysis can be found in Momirović and Dobrić, 1985.

2.2. ROBUST REDUNDANCY MODEL

When $m_1>1$ and $m_2>1$, and when sets v_1 and v_2 have a logical status of predictors and criteria, redundancy analysis is more appropriate than canonical analysis.

Robust redundancy model is very simple and is defined as maximization of redundancy

$$\rho_{Ap}^{*2} = A_p^TA_p \rightarrow \max; \qquad A_p = Z_1^TK_p^*\frac{1}{n}$$

and $\qquad p=1,\ldots,r$

$$\rho_{Bp}^{*2} = B_p^TB_p \rightarrow \max; \qquad B_p = Z_2^TL_p^*\frac{1}{n}$$

of composites $L_p^*=Z_1X_p^*$ and $K_p^*=Z_2Y_p^*$ respectively; here A_p and B_p are crosscovariance vectors of one set of variables and redundancy variates of another set of variables, under the constraints of orthonormality of transformation vectors X_p^* and Y_p^* respectively. It can be easily shown that $\rho_{Ap}^{*2}=\rho_{Bp}^{*2}=\rho_p^2$ and that $X_p^*=X_p$ and $Y_p^*=Y_p$, $p=1,\ldots,r$ so that maximization of covariances of linear composites is equivalent to the maximization of redundancy of each composite.

2.3. SRA MODEL OF ROBUST REGRESSION ANALYSIS

When $m_1>1$ and $m_2=1$, SRA model of robust regression analysis is obtained.

Let Z_1 be predictors and $Z_2=K$ the criterion variable. If $Z_1\tilde{X}=L$ is linear composite of predictors, SRA model is defined as the maximization of covariance $\tilde{\rho}$ between L and criterion K

$$K^T L \frac{1}{n} = \tilde{\rho} \rightarrow \max \tag{4}$$

under the contraint

$$\tilde{X}^T\tilde{X} = 1.$$

If we denote $Z_1^T K = C_{1K}$, solution of the problem could be obtained from the characteristic equation

$$C_{1K}C_{1K}^T\tilde{X} = \tilde{\rho}^2\tilde{X}$$

that is equivalent to (2), which makes it obvious that SRA model is a special case of QRC. Analogously as before, SRA multiple correlation $\tilde{\gamma}=\tilde{\rho}\tilde{\sigma}^{-1}$ is less or equal to the least squares regression (LSR) multiple correlation η, $\tilde{\sigma}^2$ being the variance of L. Furthermore,

- covariance between LSR and SRA estimates of K is equal to $\tilde{\rho}$
- correlation between LSR and SRA estimates of K is equal to $\tilde{\gamma}|\eta$
- scalar product of LSR and SRA weights vectors is equal to $\eta^2|\tilde{\rho}$
- covariance between residuals obtained under LSR and SRA models is equal to $1-\eta^2$
- correlation between residuals obtained under LSR and SRA models is equal to the ratiobetween standard deviations of residuals obtained under LSR and SRA models.

All that relations between SRA and least squares regression could be found in Dobrić, Štalec and Momirović, 1984.

2.4. SDA MODEL OF ROBUST DISCRIMINANT ANALYSIS

SDA model is defined by the successive maximization of variances

$$\hat{\rho}_p^2 = \hat{K}_p^T\hat{K}_p\frac{1}{n} \rightarrow \max \qquad p=1,\dots,r$$

of linear composites $\hat{K}_p=S(S^TS)^{-1}S^TZ_1\hat{\gamma}_p$, under the constraints of orthonormality of transformation vectors $\hat{\gamma}_p^T\hat{\gamma}_p=\delta_{pq}$. Here $S(S^TS)^{-1}S^TZ_1$ is matrix of standardized group means of variables from v_1 centered to common zero mean, S is a selector matrix denoting the membership of each entity to one and only one of the g groups and $r = \min(m_1, g-1)$.

The obtained SDA variables $Z_1\hat{Y}_p$ are not orthogonal; if $\{\hat{Y}_p\}=\hat{Y}$, covariance matrix $\hat{W}=\hat{Y}^T C_1 \hat{Y} \neq \text{diag}$. Denoting the variances of SDA variates with $\hat{D}^2=\text{diag}\hat{W}$, the structure matrix $\hat{F}$ of SDA factors is equal to $C_1\hat{Y}\hat{D}^{-1}$ and their pattern matrix $\hat{A}$ is equal to $C_1\hat{Y}(\hat{Y}^T C_1 \hat{Y})^{-1}\hat{D}$.

If we define $Z_2=S(S^TS)^{-1}S^TZ_1$, then SDA also maximizes covariances $L_p^{*T}K_p^*\frac{1}{n}$ between linear composites of variables $Z_1X_p^*=L_p^*$ and linear composites $Z_2Y_p^*=K_p^*$ of these variables projected into the space spanned by vectors from S. These covariances are maximized on the same vectors $X_p^*=Y_p^*=\hat{Y}_p$ and are equal to the $\hat{\rho}_p^2$. So, SDA is a special case of QRC model too. This fact was recently utilized to define the test statistic for the parameters $\hat{\rho}_p^2$ (Lužar, 1986). Obtained expression, analogous to (3), is again asimptotically χ^2 distributed, with specially defined matrices M and Ψ.

Furthermore, SDA also maximizes redundancy $R_p^TR_p$ of SDA factors defined over the same $\hat{Y}_p$ as $R_p=Z_1^TS(S^TS)^{-1}S^TZ_1\hat{Y}_p$, with their redundancy equal to $\hat{\rho}_p^4$, $p=1,\dots,r$ (Momirović and Dobrić, 1985).

3. PROGRAMS

Five basic programs and a number of modifications were written in GENSTAT and SS languages to inspect the behaviour of the proposed algorithms on real data. Programs are implemented on an UNIVAC 1100 machine; they may handle up to 250 variables in each set and up to 10000 entities, and are stored in public libraries SRCE*GENS-MACRO and SRCE*SS-MACRO of the University Computing Centre. All of them are available to all the users from the University of Zagreb.

4. CONCLUSION

Proposed robust methods could be introduced as a general class of data analysis methods. They could be considered as the principal component analysis of the matrix C_{12} (Gredelj, Momirović and Dobrić, 1986). Contrary to the classical methods that are based on the centered data matrices left eigenvectors, our methods could be derived directly from the centered data matrices. This fact enables the derivation of the strict relations between classical methods and corresponding robust methods for data analysis.

REFERENCES

Dobrić V. and K. Momirović (1984), An algorithm and program for robust discriminant analysis, Proceedings of 8th Bosnian symposium in informatics "Jahorina 84", Sarajevo, 213 1-5.

Dobrić V., J. Štalec and K. Momirović (1984), Note on some relationships between least squares and robust regression analysis, Proceedings of 6th International symposium "Computer at the University", Dubrovnik, 507 1-7.

Dobrić V., K. Momirović and M. Gredelj (1985), Quasicanonical relationships of variables in universal metric space, Proceedings of 7th International symposium "Computer at the University", Cavtat, 518 1-6.

Gredelj M., K. Momirović and V. Dobrić (1986), Some relations between canonical covariance analysis and principal component analysis, Proceeding of 8th International symposium "Computer at the University", Cavtat, 5.03 1-5.

Lužar V. (1985), Inference in quasicanonical correlation analysis, Proceeding of 7th International symposium "Computer at the University", Cavtat, 518 1-6.

Lužar V. (1986), Inference in robust discriminant analysis, Proceeding of 8th International symposium "Computer at the University", Cavtat, 5.08 1-9.

Momirović K., V. Dobrić and Ž. Karaman (1983), Canonical covariance analysis, Proceedings of 5th International symposium "Computer at the University", Cavtat, 463-473.

Momirović K., V. Dobrić and M. Gredelj (1985), Quasicanonical relationships of partial image variables, Proceedings of 7th International symposium "Computer at the University", Cavtat, 522 1-6.

Momirović K. and V. Dobrić (1985), Diagnostic efficiency of a robust method for discriminant analysis, Abstracts of 1st European Biometric Conference, Hungarian Academy of sciences, Martonvasar, p. 101.

Momirović K. and V. Dobrić (1985), Some relations between canonical and quasicanonical correlation analysis, Proceedings of 4th International symposium "Data analysis and Informatics", Paris, Posters, 101-104.

Prot F., K. Bosnar and K. Momirović (1983), An algorithm for assymetric redundancy analysis without orthogonality constraints, Proceedings of 5th International symposium "Computer at the University", Cavtat, 485-490.

Štalec J. and K. Momirović (1983), Some properties of a very simple model for robust regression analysis, Proceedings of 5th International symposium "Computer at the University", Cavtat, 453-461.

Robust Recursive Estimation and Detection of Shifts in Regression

By E. Kuh and A. Samarov, Cambridge

Robust recursive estimation provides considerable computational advantage over iterative robust regression estimation, especially for large and ordered (e.g., with time) data sets. The robust recursive estimates are less sensitive than recursive least squares to the outliers and structural shifts, and produce residuals which are more effective in constructing tests for detecting a shift. In this paper we consider a problem of detecting a shift in regression when it is masked by outliers, and summarize results of a simulation study comparing several tests and estimates of the change point.

Keywords: robust recursive estimation, structural shifts in regression, robust testing.

1. Introduction. Robust and bounded-influence regression were developed to limit the effects that heavy-tailed noise distribution or gross outliers could have on the resulting coefficient estimates. Robust estimation procedures, including M-estimation, are nonlinear and require iteration to find the estimates. When the data are collected periodically, e.g., for time series, one may want to simply update the old estimates as the new data arrive instead of repeating the whole iteration again. The related problem of iterative M-estimation is that all past data, which are possibly very numerous, must be stored in order to obtain new robust estimates that also incorporate the new data.

These considerations point in the direction of robust recursive (RR) estimation. The term recursive estimation is used here to describe procedures where coefficient estimates for the current period are calculated only from current data and previous estimates. These methods of updating estimates have considerable computational and storage advantages over the standard iterative procedures.

Recursive methods have been used in estimation for a long time, especially in engineering for estimation of dynamic models (see, e.g., Jazwinski (1970)). More recently, Young (1974), Hannan (1977), Harvey and Phillips (1979), and Ljung and Soderstrom (1983), among others, emphasized again the usefulness of recursive estimation for linear regression and time series models. Robust recursive estimation for static and dynamic linear models was discussed by Martin and Masreliez (1975), Polyak and Tsypkin (1979), and West (1981).

Recursive residuals, which result from recursive estimation, have frequently been suggested for checking model assumptions and fit (e.g., in Phillips and Harvey (1974), Brown, Durbin, and Evans (1975), Galpin and Hawkins (1984)) because, unlike ordinary residuals, they do not spread the problem in one part of the data over all the residuals. They are particularly useful for detecting the structural change in the model because they show the effect of successive

Compstat 1986

updating of the fit with the new data, so that only the residuals after (and not before) the change will be affected.

In the next section we introduce the RR estimators and discuss their various modifications, properties, and applications. In Section 3 we consider the problem of detecting the structural shift in the linear regression model and estimation of the change point when the possible change is masked by heavy-tailed disturbances. We summarize the results of a simulation study which show that RR estimation improves the performance of the well-known tests (like CUSUM) based on the least-squares recursive residuals. We also consider regression stability tests, based on RR estimation, which are more resistant toward outliers.

The estimation and testing procedures discussed in this paper are implemented as part of the software package RECUR which is currently operational in fully interactive form in a large statistical modeling system TROLL (Kuh, Samarov, Shell (1986)). RECUR can be viewed as a robust analog of TIMVAR, a package for detecting instability in regression described in Brown, Durbin and Evans (1975). RECUR has a number of useful capabilities for estimation, testing, and forecasting, and provides a considerable amount of convenient graphical output to aid the interpretation of results.

2. Robust Recursive Estimation.

We consider the standard linear regression model

$$y_i = x_i'\beta + e_i, \quad i = 1,\dots,n, \tag{1}$$

where y_i is a response variable, x_i is a p vector of explanatory variables, β is a p vector of unknown coefficients, and distrubances $e_1,\dots,e_n$ are i.i.d..

The robust recursive estimates can be obtained using the first-order Taylor approximation of the difference $b_k - b_{k-1}$ between the M-estimates (e.g., Huber (1981)) based on the k and k-1 observations, and can be written in the following form:

$$b_k = b_{k-1} + B_k^{-1} x_k \psi\left(\frac{y_k - x_k' b_{k-1}}{\sigma_{k-1} c_k}\right) \sigma_{k-1} c_k, \quad k = s, s+1, \dots, n \tag{2}$$

where s is the number of observations used to compute the starting coefficients, and

$\psi(\cdot)$ - the robustifying psi-function,

c_k - the prediction interval inflation factor,

σ_k - the scale at "time" k, and

B_k - the p x p updating matrix.

The RR estimates vary in the choice of ψ and of c_k, σ_k, and B_k, which are also computed recursively. In particular, when $\psi(u) = u$ and $B_k = \Sigma\, x_i x_i'$, (2) gives the familiar recursive ordinary least squares (OLS) estimator.

A bounded psi-function (e.g., Huber's ψ, Huber (1981)) limits the influence of outliers by trimming large residuals in (2), while psi-functions with a finite rejection point (like Hampel's or biweight, Huber (1981)) ensure also that gross outliers will be completely rejected and will not affect the estimate (2) at all. The last feature of a psi-function will also result in the estimator being resistant against patches of outliers. When $c_k = 1$ the psi-function is applied to the scaled prediction residuals which trims to a greater extent outliers corresponding to leverage points. To make trimming more homogeneous the choice $c_k = (1 + x_k'B_{k-1}^{-1}x_k)^{\frac{1}{2}}$ could be used, in which case the psi function is applied to the recursive residual $rres_k = (y_k - x_k'b_{k-1})/c_k$. Yet another choice of c_k, involving a distance measure in the x-space, results in the recursive analog of the bounded-influence Krasker-Welsch estimator (Krasker and Welsch (1982)).

The scale σ_k in (2) can be estimated with various robust estimators of scale (see, e.g., Huber (1981)) which update σ_k in parallel with b_k in (2). In RECUR program we use the recursive versions of the interquartile distance and of the Huber proposal-2 scale estimator (see Huber (1981)).

The choice of matrix B_k in (2) is very important for the behavior of the RR estimates. After experimentation with different versions we have chosen to update B_k as follows: $B_k = B_{k-1} + WTS_k \cdot x_k x_k'$, where WTS_k is the weight associated with the kth observation which depends on the form of the psi-function and the choice of c_k.

The large sample properties (consistency, asymptotic normality) of the RR estimates can be obtained using methods of stochastic approximation (see, e.g., Polyak and Tsypkin (1979)). It turns out that, under standard regularity conditions, the asymptotic properties of the RR estimates coincide with those of the corresponding fully iterated "en bloc" M-estimates. In a number of moderate sample-size examples we have run, the RR estimates gave results very close (within 10% of the standard error) to their "en bloc" iterative prototypes while running 2 to 3 times faster. To get an idea of how stable the RR procedures are and how close they are to their iterative counterparts, it is useful to "recycle" RR estimators (i.e., to repeat recursions starting with the final estimates of the previous cycle). The effect of recycling is essentially equivalent to that of iteration.

As an empirical measure of an estimator's sensitivity at time k we consider

$$S_k = (b_k - b_{k-1})' \, Cov^{-1}(b_{k-1})(b_k - b_{k-1}), \tag{3}$$

where $Cov(b_{k-1})$ is the covariance matrix of b_{k-1}. The influence of an observation as measured with S_k depends on how early or late it appears in the sample, so that early outliers will be tolerated to a larger extent by this measure. Those early outliers can be detected, however, with backward estimation, which is another very useful capability in RECUR. Running the recursions backwards not only helps in detecting outliers that appear early in the sample, but also is most useful as graphical diagnostics for detecting parameter changes over time. The time plots of forward and backward coefficients, standardized coefficient changes, recursive residuals, as well as of other statistics, are indispensible for assessing the stability of regression relationships.

Another useful feature in RECUR is a capability not only to update but also to downdate a current estimation, i.e., to drop observations. These "moving window" estimators will better "track" the changing coefficients over time and will by very useful for detection of regression shifts and forecasting, especially when more than one abrupt change occurs over the sampling period.

3. Detection of Regression Shifts. The advantage of recursions for the analysis of regression stability was already mentioned: in the model with a structural change only recursive residuals after the change will be affected. For the least-squares regression with normal errors, Brown *et al.* (1975) proposed several procedures based on the recursive residuals to test structural change (see also Schweder (1976) and Dufour (1982)). However, since the OLS estimates are very sensitive to discrepant observations, the OLS recursive residuals may be greatly distorted a) immediately after a change in regression occurred, b) by outliers within stable segments, and even more so c) by the combination of a) and b).

The RR estimates are, by design, less sensitive to outliers than the recursive least-squares estimates, and we can expect that statistics based on recursive residuals from the RR fit will detect the shift more readily than their least-squares counterparts, especially in the presence of heavy-tailed noise or gross errors.

We have conducted a moderate size Monte Carlo study trying, in particular, to verify the above claim. Below we give a brief summary of our findings leaving the detailed account for another paper.

We considered a bivariate linear regression with 80 data points, and normal and contaminated normal errors. The tests based on cumulative sums of recursive residuals: CUSUM and CUSUMSQ (Brown *et al.* (1975)), and maximum CUSUM (Sen (1982)) showed a) dramatic decrease of power for contaminated normal errors with OLS estimation and b) much larger power of detecting a shift when used with the RR estimators than with the recursive OLS. The increase in power was especially significant for contaminated normal errors. The CUSUM average run lengths, which measured how early the shift in regression was detected, was 2 to 3 times shorter with the RR estimation. These improvements were paid for with some increase in the significance level which was not, however, very large.

To keep the significance level under control while maintaining the high power and short average run length we have to not only rely on RR estimation but also robustify test statistics. One such test is the cumulative sum test based on the "trimmed" recursive residuals $\psi(rres_k/\sigma_{k-1})\,\sigma_{k-1}$, which were used in the RR estimation (2), cf. recursive residual M-tests, Sen (1985). Simulation performance of this test, as well as of the recursive rank test proposed by Sen (1985), will be reported in the follow-up paper

The recursive estimation is useful not only for on-line, quasi-sequential testing discussed so far, but also for fixed sample, two-sided testing. The two-sided tests use information on both sides of the potential change and, therefore, would be expected to detect change better than one-sided tests when a linear model holds on both sides. The standard regression stability tests include Quandt (1958) likelihood ratio test and Chow (1960) F-test which in the case of unknown change point and robust fit would require enormous amount of computation. At the same time, only two, forward and backward, recursive estimation runs are needed to compute these test statistics for all potential change point periods, which gives a considerable computational advantage.

When the change point is known and the errrors ε_i are i.i.d. normal, Chow F-test has various attractive properties. However, a very few isolated outliers may easily mask the abrupt systematic change in the model since F-statistic inherits the outbounded sensitivity of the OLS estimates to outliers. Schrader and Hettmansperger (1980) and Ronchetti (1982) considered a class of robust tests of general linear hypothesis based on M-estimation of regression. Motivated in part by their tests and by the form of F-test we propose the following test statistic:

$$RF_k = \frac{1}{p}(b_{1,k} - b_{k+1,n})'(\mathrm{Cov}(b_{1,k}) + \mathrm{Cov}(b_{k+1,n}))^{-1}(b_{1,k} - b_{k+1,n}) \qquad (4)$$

where $b_{1,k}$ and $b_{k+1,n}$ are the forward and backward RR estimates (2) for the indicated range of time, and $\mathrm{Cov}(\cdot)$ denote an estimate of corresponding covariance matrix. (We calculate RF_k for $s \leq k \leq n\text{-}s$ with s not too small, at least 2 or 3 times p, so that it still makes sense to estimate p coefficients robustly. Note that for the standard F-statistic it is usually assumed that $p \leq k \leq n\text{-}p$.)

RF_k is a natural extension of F-statistic except that it estimates the scale of the errors in each regression segment separately. RF_k directly estimates the "relative change" of the coefficients in two segments and inherits the robustness properties of the RR estimates it is based upon. If a) the change point $k = k_n$ is such that $\lim k/n = \lambda$, $0 < \lambda < 1$, b) the regularity conditions guaranteeing the asymptotic normality of $b_{1,k}$ and $b_{k+1,n}$ are satisfied, and c) $k \cdot \mathrm{Cov}(b_{1,k})$ and $(n\text{-}k) \cdot \mathrm{Cov}(b_{k+1,n})$ consistently estimate the asymptotic covariance matrix of coefficients, then the statistic RF_k has asymptotic χ_p^2 distribution.

Naturally, we would like to use $RF_{k^*} = \max RF_k$ as the test statistic and k^* as the estimator of the change point. Unfortunately, even in the case of normal errors, the distributions of the corresponding statistics are not available (Hinkley (1970), Beckman and Cook (1979)), so that we again have to appeal to simulation.

For the estimation of k^* our simulation results show that a) the estimates of k^* based on the standard Quandt and Chow test statistics calculated with the RR estimates have approximately the same bias as those calculated with the OLS estimates when the errors are normal and have 3 to 4 times smaller bias when the errors are contaminated normal; and b) the estimates of k^* based on the statistics RF_k have dramatically smaller bias than all other (one- or two-sided) competitors for both normal and contaminated normal errors, e g . smaller by a factor of 10 compared with the estimates based on Chow statistics with the OLS fit when errors are contaminated normal.

References

Beckman, R.J. and R.D. Cook (1979), "Testing for two-phase regression," *Technometrics*, 21, 65-69.

Brown, R.I., J. Durbin, and J.M. Evans (1975), "Techniques for testing the constancy of regression relationships over time," *JRSS*, B37, 149-192.

Chow, G. (1960), "Tests of the equality between two sets of coefficients in two linear regressions," *Econometrica*, 28, 561-605.

Dufour, J-M. (1982), "Recursive stability analysis of linear regression relationships," *Journal of Econometrics*, 19, 31-76.

Hannan, E.J. (1977), "ARMAX and state space systems and recursive calculations, in M.D. Intrilligator, ed. *Frontiers of Quantitative Economics-111A*, Amsterdam: North-Holland, 321-336.

Harvey, A.C. and G.D.A. Phillips (1979), "Maximum likelihood estimation of regression models with autoregressive – moving average disturbances," *Biometrika*, 66, 1, 49-58.

Hinkley, D. (1970), "Inference about the Change-Point in a Sequence of Random variables," *Biometrika*, 57, 1-17.

Huber, P. (1981), *Robust Statistics*, J. Wiley and Sons, New York.

Jazwinski, A.H. (1970), *Stochastic Processes and Filtering Theory*, Academic Press, New York.

Krasker, W. and R.E. Welsch (1982), "Efficient bounded-influence regression estimation," *JASA*, 77, 595-604.

Kuh, E., A. Samarov and M. Shell (1986), RECUR: A program for robust recursive estimation, stability testing, and forecasting, Technical Report No. 62, CCREMS, MIT., Cambridge, MA.

Ljung, L. and T. Soderstrom (1983), *Theory and practice of recursive identification*, MIT Press, Cambridge, MA.

Martin, R.D., C.J. Masreliez (1975), "Robust estimation via stochastic approximation," *IEEE Trans. on Information Theory*, IT-21, 3, 263-271.

Phillips, G.D.A. and A.C. Harvey (1974), "A simple test for serial correlation in regression analysis," *JASA*, 69, 935-939.

Polyak, B. and Ja. Tsypkin (1979), "Adaptive estimation," *Automation and Remote Control 3*, 71-84.

Quandt, R.E. (1958), "The estimation of the parameters of a linear regression system obeying two separate regimes," *JASA*, 53, 873-880.

Ronchetti, E. (1982), "Robust testing in linear models: the infinitesimal approach," Ph.D dissertation, ETH, Zurich, 1982.

Schrader, R.M. and T.P. Hettmansperger (1980), "Robust analysis of variance based upon a likelihood ratio criterion, *Biometrica*, 67, 93-101.

Schweder, T. (1976), "Some 'optimal' methods to detect structural shift or outliers in regression," *JASA*, 71, 491-501.

Sen, P. (1982), "Invariance principles for recursive residuals," *Annals of Mathematical Statistics*, 10, 1, 307-312.

West, M. (1981), "Robust sequential approximate Bayesian estimation," *JRSS*, B, 43, 2, 157-166.

Young, P. (1974), "Recursive approaches to time series analysis," *Bull Inst. Math. App* **10**, 209-224.

How Robust is One Way Anova with Respect to within Group Correlation

By M. Schader and F. Schmid, Hamburg

Summary:
The paper shows that One Way ANOVA is highly nonrobust with respect to positive within group correlation because the true level of significance α^* of the procedure is much higher than the prescribed level α. Non robustness is also stressed by power considerations.

Key Words:
Robustness, One Way ANOVA, Central and Noncentral F-Distribution, Within Group Correlation.

1. Introduction

The robustness of ANOVA procedures has been investigated with respect to departures from normality and departures from the assumption of equal variances (see e.g. Scheffée (1959), p.331 ff). Robustness with respect to departures from the assumption of independent observations, however, is also of interest. The unsophisticated user often runs program packages containing ANOVA procedures in which dependencies which may have existed during data collection are neglected. It is therefore important to investigate the influence which dependencies have on the properties (i.e. level of significance and power) of the procedure.

In order to get precise numerical results the form of the dependence must be specified. In this paper we consider equi-correlation within groups: it is assumed that every two observations within one group have the same coefficient of correlation ρ. Dependencies of this sort may be caused by temporal or spatial effects.

The results of our investigation are given in paragraph 3. They indicate that one way ANOVA is highly nonrobust with respect to positive within group correlation and the results strongly underline Pearce´s observation (see Pearce (1982)) that independence is most important and that the analysis of variance in its usual form should be avoided if independence is not given.

Compstat 1986

2. Properties of Normally Distributed Random Variables in the Presence of Equal Within Group Correlation

Let $X_{ij} \sim N(\mu_i, \sigma^2)$ for $i=1,\dots,k$ and $j=1,\dots,n$ where k denotes the number of groups and n is the number of observations per group. Let $\underline{X}_{i.} = (X_{i1},\dots,X_{in})'$ the column vector of observations in the i-th group. The covariance matrix of $\underline{X}_{i.}$ is assumed to be $\text{Cov}(\underline{X}_{i.}) = \sigma^2 V(\rho)$ for $i=1,\dots,k$ where

$$V(\rho) = \begin{pmatrix} 1 & \rho & \cdot & \cdot & \cdot & \cdot & \rho \\ \rho & 1 & & & & & \cdot \\ \cdot & & \cdot & & & & \cdot \\ & & & & \cdot & & \\ \rho & & & & & 1 & \rho \\ \rho & \rho & \cdot & \cdot & \cdot & \rho & 1 \end{pmatrix} \quad , \; -1/(n-1) < \rho < 1.$$

Thus, for $i=1,\dots,k$ we have $\underline{X}_{i.} \sim N(\mu_i \underline{1}, \sigma^2 V(\rho))$, $\mu_i \in \mathbf{R}$ and $\sigma^2 > 0$ with $\underline{1} = (1,\dots,1)'$ denoting a column vector of n ones. We assume that $\underline{X}_{1.},\dots,\underline{X}_{k.}$ are independent (i.e. there is no correlation between groups).

Let $\bar{X}_i := \frac{1}{n}\sum_{j=1}^{n} X_{ij}$ for $i=1,\dots,k$ and $\bar{X} := \frac{1}{nk}\sum_{i=1}^{k}\sum_{j=1}^{n} X_{ij}$. Then it is possible to show: (see Rao (1973) for hints of a proof)

1. $\bar{X}_i \sim N(\mu_i, \frac{(1+(n-1)\rho)\sigma^2}{n})$

2. $$\frac{\sum_{i=1}^{k} n(\bar{X}_i - \bar{X})^2}{(1+(n-1)\rho)\sigma^2} \sim \chi'^2_{k-1}\left(\frac{\sum_{i=1}^{k} n(\mu_i - \bar{\mu})^2}{(1+(n-1)\rho)\sigma^2}\right)$$

   where $\chi'^2_m(\lambda)$ denotes a noncentral χ^2- distribution with m degrees of freedom and noncentrality parameter λ and $\bar{\mu}$ is the mean of $\mu_1,\dots,\mu_k$.

3. $$\frac{\sum_{i=1}^{k}\sum_{j=1}^{n} (X_{ij} - \bar{X}_i)^2}{(1-\rho)\sigma^2} \sim \chi^2_{k(n-1)}$$

   where χ^2_m denotes a central χ^2 - distribution with m degrees of freedom.

4. The quantities given in 2. and 3. are stochastically independent and we have

   $$\frac{\sum_{i=1}^{k} n(\bar{X}_i - \bar{X})^2 (1-\rho)\sigma^2 k(n-1)}{\sum_{i=1}^{k}\sum_{j=1}^{n} (X_{ij} - \bar{X}_i)^2 (1+(n-1)\rho)\sigma^2 (k-1)} \sim F'_{k-1,k(n-1)}\left(\frac{\sum_{i=1}^{k} n(\mu_i - \bar{\mu})^2}{(1+(n-1)\rho)\sigma^2}\right)$$

   where $F'_{m,h}(\lambda)$ denotes a noncentral F- distribution with m and h degrees of freedom and noncentrality parameter λ.

3. Level of Significance and Power of One Way ANOVA in the Presence of Within Group Correlation

We want to test

$$H_o: \quad \mu_1=\mu_2=\ldots=\mu_k \ .$$

The test statistic is

$$Z = \frac{\sum_{i=1}^{k} n(\bar{X}_i-\bar{X})^2 k(n-1)}{\sum_{i=1}^{k}\sum_{j=1}^{n}(X_{ij}-\bar{X}_i)^2(k-1)}$$

and H_o is rejected if $Z > c$, where c is the $(1-\alpha)$-quantile of the central $F_{k-1,k(n-1)}$-distribution and α is the desired level of significance. The true level α^*, in the presence of within group correlation coefficient ρ, can be computed (using the results of 2.) as follows:

$$\alpha^* = P(Z > c \mid \mu_1=\mu_2=\ldots=\mu_k,\sigma,\rho)$$

$$= P(Z\frac{1-\rho}{1+(n-1)\rho} > c\frac{1-\rho}{1+(n-1)\rho} \mid \mu_1=\mu_2=\ldots=\mu_k,\sigma,\rho)$$

$$= P(F_{k-1,n(k-1)} > c\frac{1-\rho}{1+(n-1)\rho}) \ .$$

($F_{m,h}$ denotes a central F-distribution with m and h degrees of freedom.)

Table 1 gives the true levels of significance α^* for $\alpha = 0.05$ and various combinations of k,n and ρ. The numbers given there clearly show that positive within group correlation strongly increases the level of significance of the test whereas negative within group correlation renders the test conservative. It follows that one way ANOVA is highly nonrobust with respect to positive within group correlation. This conclusion is stressed by power considerations:

Using again the results of 2. we have for the power

$$G(\mu_1,\ldots,\mu_k,\sigma,\rho) = P(Z > c \mid \mu_1,\ldots,\mu_k,\sigma,\rho)$$

$$= P(F'_{k-1,k(n-1)}(\lambda) > c\frac{1-\rho}{1+(n-1)\rho})$$

where $\lambda = \delta\frac{1}{1+(n-1)\rho}$ and $\delta = \sum_{i=1}^{k} n(\mu_i-\bar{\mu})^2/\sigma^2$ denotes the noncentrality parameter in the case $\rho = 0$.

In Figure 1 a comparison of power functions (as a function of δ) for $\rho \neq 0$ and $\rho = 0$ (dashed line) is given for $\alpha = 0.05$ and various combinations of k,n and ρ.

		ρ =								
		-0.01	-0.001	0.01	0.05	0.1	0.2	0.3	0.4	0.5
k=3	n=10	.037	.049	.064	.131	.223	.396	.538	.650	.740
	n=25	.020	.046	.090	.266	.442	.652	.767	.838	.887
	n=50	.003	.043	.135	.433	.628	.798	.873	.915	.942
	n=100	.000	.036	.224	.617	.779	.890	.933	.956	.970
k=4	n=10	.035	.048	.067	.151	.271	.492	.656	.772	.853
	n=25	.017	.046	.098	.327	.546	.773	.875	.928	.958
	n=50	.002	.042	.156	.535	.750	.899	.949	.972	.984
	n=100	.000	.034	.272	.739	.885	.959	.981	.990	.994
k=5	n=10	.034	.048	.069	.169	.315	.572	.745	.852	.918
	n=25	.014	.045	.106	.381	.630	.852	.933	.968	.984
	n=50	.001	.041	.175	.618	.832	.949	.980	.991	.996
	n=100	.000	.032	.315	.822	.940	.985	.994	.998	.999
k=7	n=10	.032	.048	.073	.202	.394	.696	.860	.938	.974
	n=25	.011	.045	.119	.475	.754	.938	.981	.994	.998
	n=50	.000	.039	.209	.742	.924	.987	.997	.999	1.0
	n=100	.000	.030	.391	.917	.984	.998	1.0	1.0	1.0

Table 1 : True levels of significance α^* for $\alpha = 0.05$

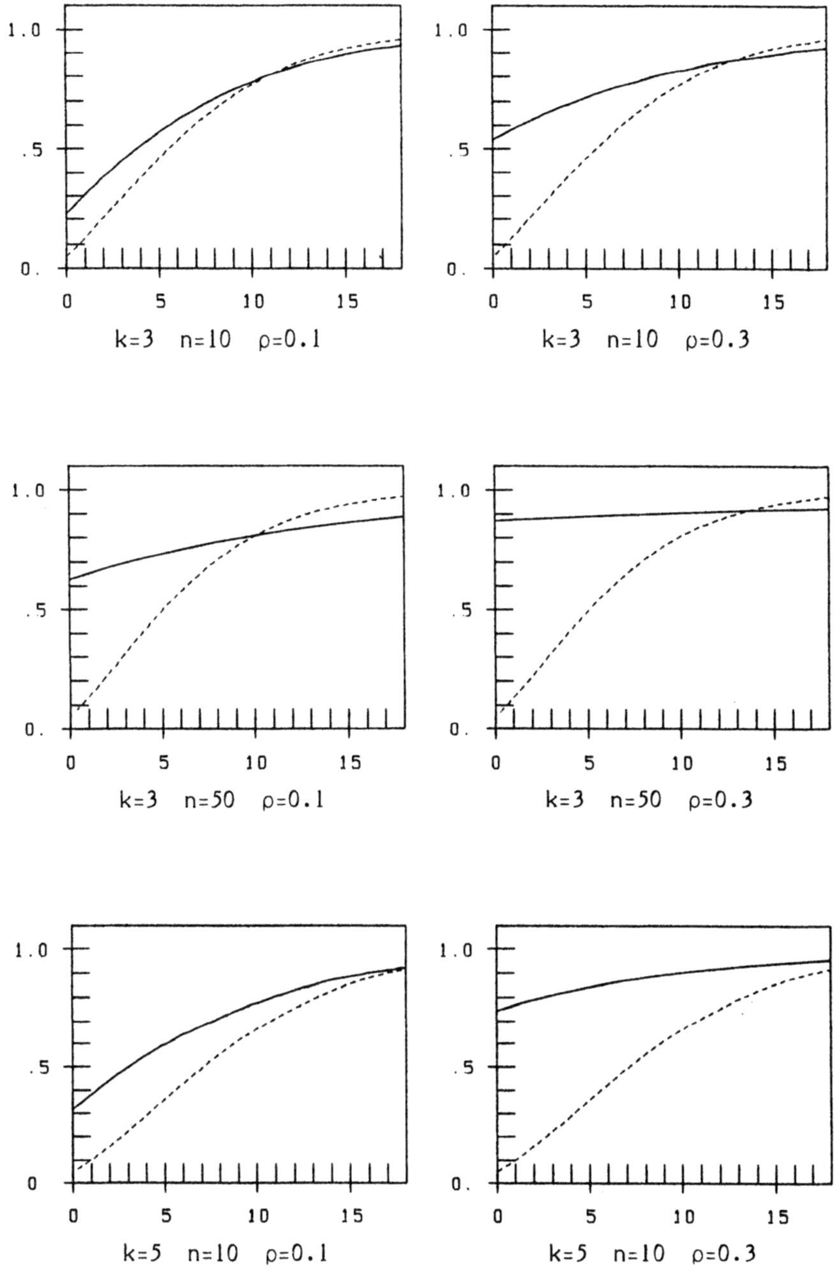

Figure 1: Power Functions for $\rho \pm 0$ and $\rho = 0$ (dashed line) with $\alpha = 0.05$

All computation for this paper was carried out on a UNIVAC 1100/82 computer (OS 1100, FTN 11R1 compiler). The values of the distribution function of a noncentral F-distibution were obtained by the subroutine DFNCF given in Schader & Schmid (1986).

4. References

S.C. Pearce: Analysis of Variance, in: S. Kotz & N.L. Johnson (eds.), Encyclopedia of Statistical Sciences, Vol. 1, Wiley, New York, 1982.

C.R. Rao: Linear Statistical Inference and its Applications (2. ed.), Wiley, New York, 1973.

M. Schader & F. Schmid: Signifikanzniveau und Gütefunktion von Parametertests für μ und σ^2 bei identisch normalverteilten, aber symmetrisch abhängigen Beobachtungen, Allg. Statist. Archiv 2, 1985.

M. Schader & F. Schmid: Distribution Function and Percentage Points for the Central and Noncentral F-Distribution, Statistische Hefte 26, 1986.

H. Scheffée: The Analysis of Variance, Wiley, New York, 1959.

Computer Graphics and Data Presentation

An Interactive Graphic System for Designing and Accessing Statistical Data Bases

By G. Di Battista and R. Tamassia, Roma

Research partially supported by Progetto Finalizzato Trasporti, CNR, Italy.

SUMMARY

An integrated graphic computer system is presented that supports the whole life cycle of a statistical data base. The system provides design and documentation tools for creating and mantaining a statistical data base, and a user friendly interface for the interactive retrieval of statistical data.
Two types of schemas are adopted: Entity-Relationship schemas for elementary data, and special schemas for aggregated data. For each schema, a diagrammatic representation is provided, so that the user can operate on schemas through the corresponding diagrams.

1. INTRODUCTION

The advantages of data base organization on conventional file systems for the management of statistical data are widely recognized, see for example [Sho]. The first step in the design of a data base is the creation of a conceptual data schema, which is an abstract description of the data of interest and their relationships. The schema plays a fundamental role in the rest of the life cycle: in the physical design phase, it represents a stable reference, independent from the storage technique adopted; in the maintenance phase, it allows a clear understanding of the information contents of the data base; finally, in the production phase, it can be used as a user friendly interface.
In this paper, we describe an integrated system for the design

Compstat 1986

and documentation of statistical data bases, named GIOTTO/S. Moreover, the system can be used for accessing and manipulating data.

One of the main features of the system is a graphic user-friendly interface. Schemas are represented by means of diagrams, which provide a pictorial description of data and their relationships, so that the user can operate on schemas through the corresponding diagrams. In the design phase, diagrams are created and updated on a high-resolution graphic terminal. During the production phase, queries are formulated by selecting paths on the diagram, corresponding to classes of data of interest. For documentation purposes, a hierarchically organized library of schemas (diagrams) is mantained by the system.

Two types of schemas are adopted: Entity-Relationship schemas [Chen] for elementary data, and GRASS schemas [RR] for aggregated data.

The next section describes the characteristics of the data models. Section 3 contains an outline of the capabilities of the system.

2. DATA MODELS

For elementary data, an extension of the Entity-Relationship (ER) model [Chen] is adopted. In the ER model two types of classes of objects are considered: entities and relationships. Entities represent sets of objects of the real world sharing common characteristics; relationships represent classes of aggregations between objects. In an ER diagram, entities are represented by means of rectangles, and relationships by means of diamonds. A connection between an entity E and a relationship R indicates that objects of the class E are involved in the relationship R.

For aggregated data, the GRASS (GRaphical Approach for Statistical Summaries) model is adopted, which describes the logical structure of a statistical data base by means of a graph. In the diagrammatic representation of a GRASS graph, circles and straight lines represent nodes and arcs, respectively. A GRASS graph is marked, labelled, directed, acyclic, and has exactly one root. For a complete description of the model, see [RR].

The system can support the methodology described in [DFB]: during conceptual design of statistical databases, the designer should be able to describe data at different levels of aggregation, find suitable strategies for building the different levels starting from user requirements, check the completness, coherence and minimality of the conceptual schemas at the different levels. The methodology fragments the overall process into the following activities:

1. Design of the schema of elementary data
2. Design of the schema of aggregations
3. Coherence check between schemas

For each step it provides indications that allow the designer to incrementally specify the schemas, integrate partial design products, restructure the schemas to improve their quality.

3. OUTLINE OF THE SYSTEM

3.1 The user interface

Graphic devices used for interacting with GIOTTO/S are: an high resolution display, a keyboard, a mouse, and an optional hardcopy device. The user is provided with a set of screen pages. Each page is divided into four work areas (viewports) as shown in fig. 1.

3.2 General features

The system manages a set of schemas, that are either ER or GRASS. In every viewport it is possible to operate on a schema. Conversely, a schema can be assigned to any set of viewports. A trasformation performed on a schema is visible in all the viewports that are assigned to that schema. At a time, only one viewport of a certain page is available to work. This viewport is called the active viewport of the current page.

3.3 Architecture of the system

The functions of the system are divided into eight modes, and at any time, each viewport can be in only one of them. A distinct set of commands can be issued in every mode, as follows:

1. SYSTEM: management of the schema library (similar to the Unix file system). In this mode it is possible to integrate two schemas into a third one.
2. EDIT: creation and update of a particular schema. An automatic layout function is also available, that generates diagrams according to a set of aesthetics for a better readability.
3. DISPLAY: this mode controls the graphic features of diagrams. It is possible to selectively display subsets of symbols in the diagram.
4. DOCUMENT: production of textual and graphic documentation, at several levels of detail.
5. QUERY: supplies an user friendly interface to access information stored in the data base.
6. ER-TO-GRASS: provides an interactive support to the methodology for conceptual design described in [DFB]. In this mode the user is able to design a GRASS schema starting from an ER one. It is possible to formulate aggregations on data using a query language.

The obtained query are a formal expression of aggregations, and are suitable to obtain semi-automatically a description of summary data in a GRASS schema.

The system performs also a coherence check between the informative content that is necessary to perform an aggregation, and the information content that is available in the ER schema. It is possible to enhance the second, adding new concepts and relationships.

The commands allows the user to: SELECT the part of the ER schema involved in the aggregation, SIMPLIFY it to enhance its readability, perform a QUERY to formally describe the aggregation, and TRANSFORM the query into a GRASS subschema.

Commands can be either entered through the keyboard, or selected with the mouse from the menu area.

These commands are common to every mode: the active viewport and the current page can be changed with the command:

CHANGE-VIEWPORT <page>,<viewport>

while the command

ADJUST-SCREEN

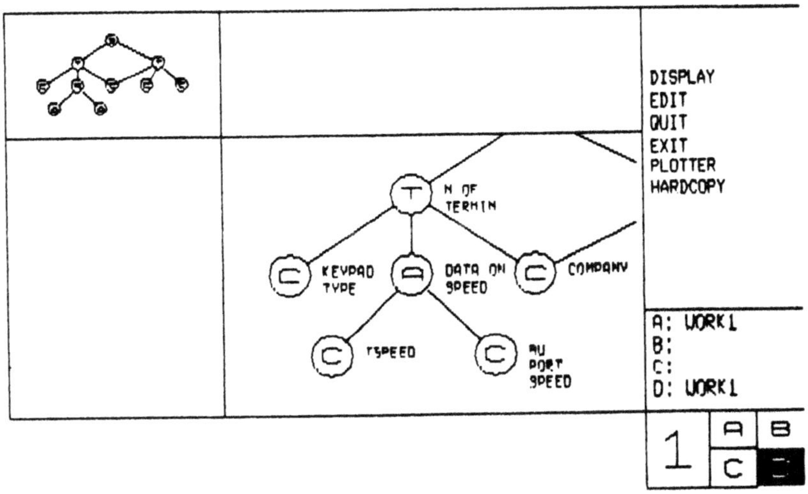

Fig.1 - The scree of GIOTTO/S

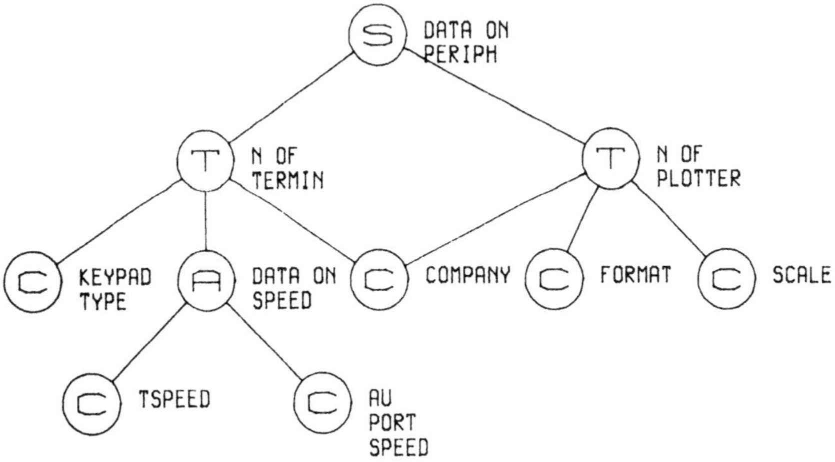

Fig.2 - A diagram produced in the DOCUMENT mode of GIOTTO/S

redefines the extension of the viewports.

4. CONCLUSIONS

We have shown the main features of an interactive graphic tool for designing and accessing statistical data bases, named GIOTTO/S. The system is under implementation at Dipartimento di informatica e sistemistica, University of Rome. All the figures that appears in this paper are made with the tool.
Future improvements will be focused on the QUERY mode. If both aggregate and elementary data are contained in the data bases, the answer will be obtained through a query optimizer, it will decide first if it is possible to answer using aggregate data, if it is not possible a query will be performed on elementary data.

References

[BL] Batini C., Lenzerini M., A methodology for data schema integration in the Entity Relationship Model. IEEE Transactions on Software Engineering, vol.se-10,no.6,november 1984.
[BNTT] Batini C., Nardelli E., Talamo M., Tamassia R., Gincod: a graphical tool for conceptual design of data base applications. in: Computer aided data base design: the Dataid project, North Holland pp. 33-51, 1985.
[DFB] Di Battista G., Ferranti G., Batini C., Design of statistical data bases: a methodology for the conceptual step. Accepted at COMPSTAT 86.
[RR] Rafanelli M., Ricci M., Proposal of a logical model for statistical data base. Proceedings of the second international workshop on statistical data base management, Los Altos 1983.
[Sho] Shoshani A., Statistical databases: characteristics, problems and some solutions. 8th International Conference on Very Large Data Bases, 1982.

Non-Standard Graphical Presentation

By J. Gordesch, Berlin

Summary: The use of general affine, projective, and curvilinear coordinates for graphical presentation is considered. In addition, geometric techniques other than point representation (general properties of algebraic manifolds, projection from three- or multidimensional spaces) are treated.

Keywords: Computer Graphics; Statistical Graphics; Computer Geometry

1. Introduction

Usual graphical display for analyzing data or communicating information is mainly based on Euclidean geometry, and even more restrictive, uses almost exclusively Cartesian coordinates. New ideas aim at aesthetic novelties, or endeavour to pay more attention to human factors (like Chernoff's face graphs). This paper, however, intends to give a survey on certain types of graphs based on techniques well-known in classical geometry or already used in some fields of non-statistical application, and to prompt the use of an enlarged set of geometric tools. First, affine, projective, and curvilinear coordinates are considered. In addition, some useful geometric properties other than point representation are proposed (e. g. qualitative properties of algebraic manifolds as dealt with in catastrophe theory). Finally, projections from three- and multidimensional spaces are treated.

2. Coordinate Systems

Cognition is impossible without a system of reference as music is a futile effort without scales and cords. The basis of mathematical description is an appropriately chosen coordinate system. Coordinates are defined as an ordered set of numbers or other elements which define geometric objects like points, lines, circles, etc. Parameters of a system of equations may represent completely different geometrical objects (in Cartesian coordinates e. g. a linear equation represents a line, in polar coordinates however an Archimedian spiral).

2.1 Affine Coordinates

For the following, we adopt tensor notation and summation convention, i. e. summation takes place whenever the same name is used for a superscript and a subscript of a tensor or tensor product.

Compstat 1986

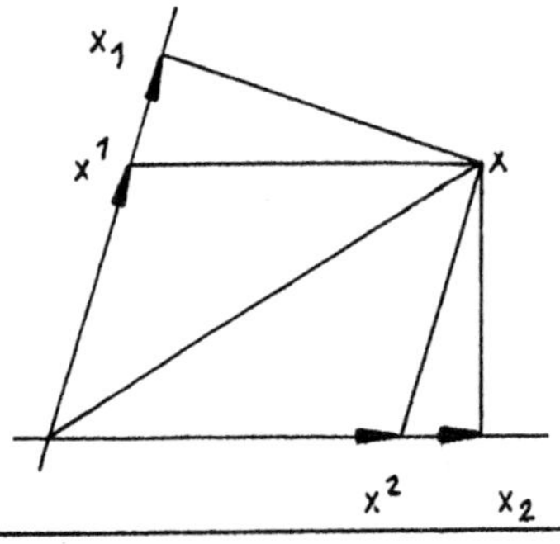

Fig. 1: Affine Coordinates

x_1, x_2 covariant coordinates
x^1, x^2 contravariant coordinates

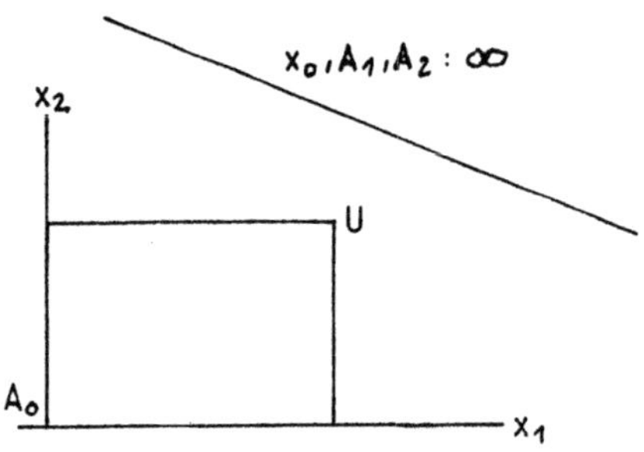

Cartesian Coordinates

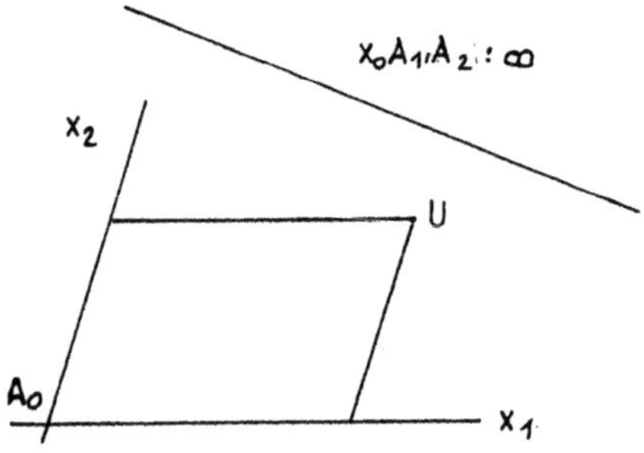

Oblique Coordinates

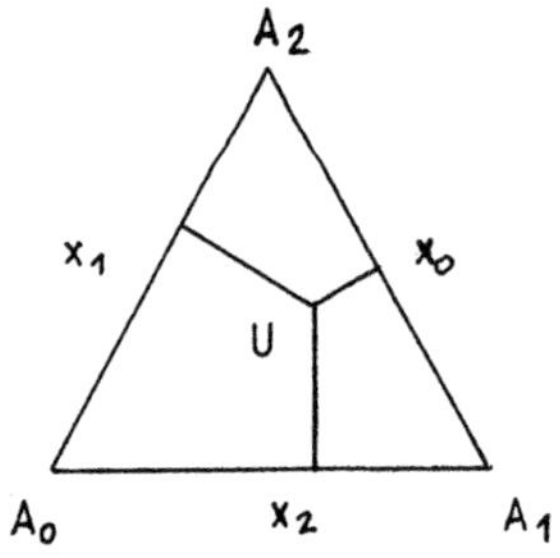

Triangular Coordinates

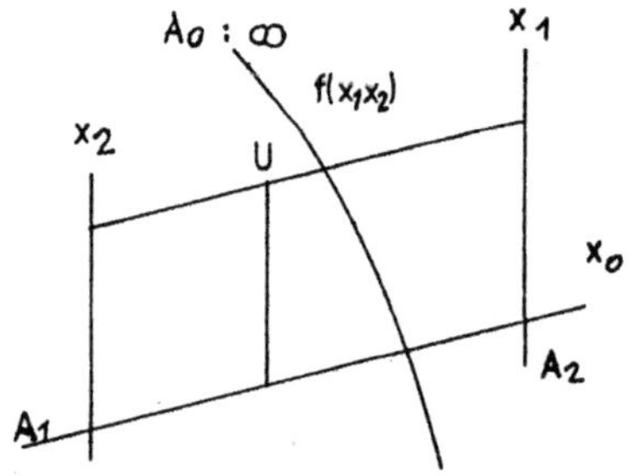

Alignment Chart

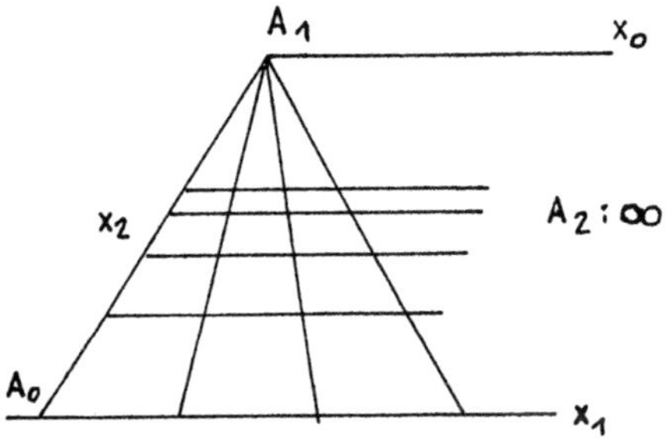

Projective Network

Fig. 2: Projective Coordinates

A_i vertices of coordinate triangle
x_i axes
U = (1,1,1) unit point

In a vector space where an inner product is defined (a Euclidean vector space) with basis $E = (e_{(1)}, \dots e_{(n)})$, the contravariant coordinates x^i of a vector x with respect to E are defined by

$$x = x_i e_{(i)} \ ,$$

and the covariant coordinates x_i of x with respect to E are defined by

$$x_i = x e_i$$

(cf. Fig. 1). The x_i's are also the contravariant coordinates in the dual space with basis E^*, and using the metric tensor $g^i{}_j$, we get

$$x_i = g^i{}_j x^j \ .$$

Firstly, this provides us with a new and possibly better presentation of a set of points, nonetheless it means nothing but the switching over to a new coordinate system. The interpretation however may be quite different as it is seen in the case of factor analysis (interpretation before and after rotating the factor structure). Secondly, a Euclidean space E_n may be identified with its dual space E_n^* by use of its inner product because of the one to one mapping of x_i and x^i. This leads to the generation of geometrical objects not by moving a point the trace of which is the geometrical object, but by moving a hyperplane and thus constructing an envelope (line or plane coordinates, etc.). This type of geometry is, so to speak, the geometry of the blind who perceive solids by the sense of touch and feel the shape using their palms like a tangent plane enveloping the surface under investigation.

2.2 Projective Coordinates

An n-dimensional projective space $P_n(E)$ related to a vector space E over a commutative field F is the quotient space of E - {o} where collinear vectors ($y = \alpha x$, α not 0 and $\alpha \in F$) belong to the same class of equivalence. To put it in a more geometric fashion, in projective geometry no geometric objects are distinguished as "being at infinity". Coordinates are determined only up to a common factor: Points in P_n can be represented by lines in an affine E_{n+1}. In particular, points in P_2 are represented by lines in an affine E_3. These may be cut by an optional plane and thus a particular coordinate triple will be singled out.
A coordinate system in the projective plane is determined by three points forming a not-degenerate triangle, and a fourth point (unit point) not lying on any of the sides of the triangle. The coordinates are normalized by

$$e = a_{(0)} + a_{(1)} + a_{(2)} \ .$$

Thus projective coordinates x^i can be defined by

$$x = x^0 a_{(0)} + x^1 a_{(1)} + x^2 a_{(2)} \ .$$

We pass from affine coordinates u_k to projective coordinates ("homogeneous coordinates") x_i by

$$x_i = A_i{}^k x_k \ ,$$
$$i,k = 0,1,\dots n, \ x_0 \equiv 1, \ \text{rank} \ (A_i{}^k) = n \ .$$

We may also divide e. g. equations 1 to n by equation 0, or multiply an equation by any constant, just to show that coordinates are determined only up to an arbitrary factor, or equivalently, only the ratios of the projective coordinates are determined uniquely.

The concept of projective coordinates is of enormous versatility; it comprises Cartesian as well as oblique coordinates, triangular coordinates, barycentric coordinates, and even alignment charts as well as projective nets (cf. Fig. 2). Moreover, the principle of duality doubles the number of possible coordinate systems (line coordinates, etc.). Projective geometry is more general than affine geometry, for it includes also the geometry at infinity. In practice, this means that "the very large" may be approximated by "infinite", which is often less cumbersome. Unfortunately, ordinary distance is not invariant under projective transformations, whence for instance concepts as variance or other measures of dispersion cannot be used any longer. A special metric, based on the projective invariant of cross ratio, has to be designed and used for measurement. For particular applications cf. Gordesch (1983).

2.3 Curvilinear Coordinates

With curvilinear coordinates, the straight axes are replaced by curves, whence results the connection with non-linear transformations of Euclidean coordinates. They are most versatile, and enable the reduction of complex structures to simpler ones. Almost only locally Euclidean spaces are of practical value: for each point exists a neighbourhood in which a Euclidean coordinate system can be established. As a result, statistical estimation using discrete data is possible. Distance in Riemannian geometry and additionally direction in Finsler spaces change from point to point. Well-known instances such as polar or elliptic coordinates are of little use in statistics, whereas general mappings as the fisheye transformation (Fig. 3) or the inversion with respect to a sphere may be of much more advantage. There are still many other examples of coordinate systems, as Graßmann's coordinates of projective subspaces, and each special kind of geometry leads to its own system of reference, possibly not with axes but with other geometrical objects. Since their use necessitates special training, they are more convenient for theoretical purposes.

3. Algebraic Geometry

Algebraic Geometry describes general characteristics of geometrical objects rather than it does construct these objects point by point. The algebraic theory of singularity is a very useful tool, which has been developed by Thom and others (cf. Lu (1976)) under the auspices of catastrophe theory, i. e. the theory of unstable behaviour. It enables classification of singularities up to diffeomorphisms. For example, the first order differential equation

$$y' = (x-a)y/(c_0+c_1x+c_2x^2)$$

leads to the well-known Pearsonian curves. The discriminant of the quadratic denominator yields a criterion for judging in advance what type of curve is appropriate to a proposed frequency distribution.

4. Descriptive Geometry

4.1 Description of Relationship

The purpose of descriptive geometry is to describe geometrical relationships among

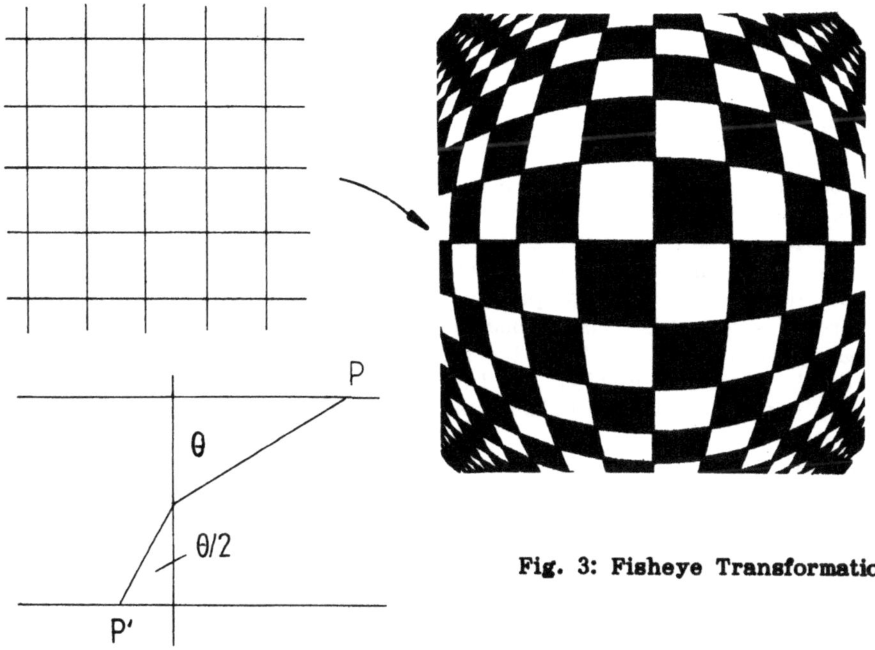

Fig. 3: Fisheye Transformation

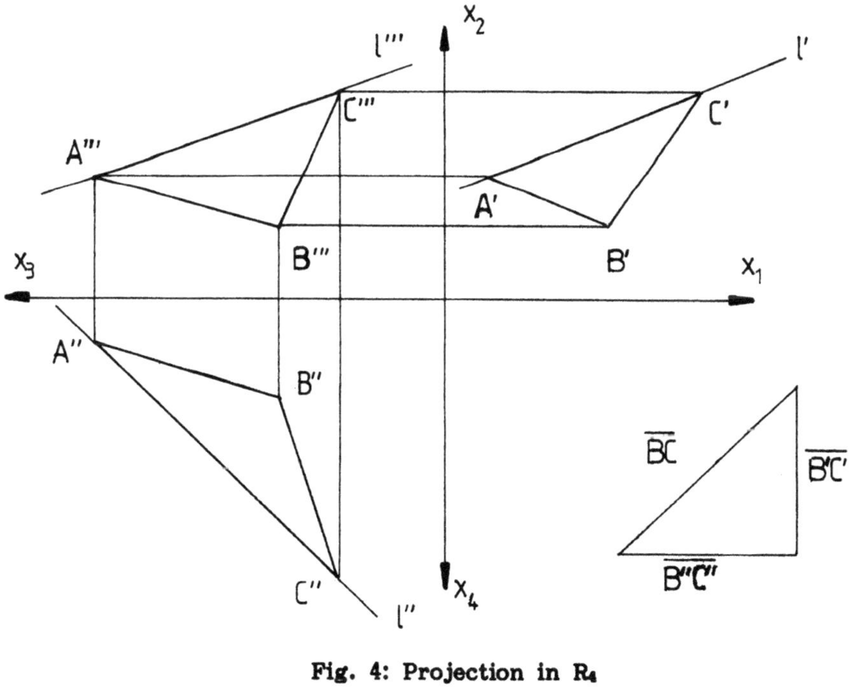

Fig. 4: Projection in R_4

three- or even multidimensional structures on a plane surface. For simple presentation, either cutting with a coordinate plane or topographic (coted) projection are usual means. Hence emphasis is laid here on relationship. As an example, a heuristic technique for multidimensional confidence regions (and interval estimation) is outlined in the sequel.

(1) The theoretical confidence region is set up according to a certain theory (conventional symmetric or unsymmetric intervals, fiducial intervals, Bayesian intervals, ...).
(2) The surface bounding the confidence region is approximated by Bézier surfaces, B-splines, etc.
(3) The surface, then, is represented using the Cartesian product of orthogonal projections onto a threedimensional hyperplane. Finally, these three-dimensional figures are depicted in one-view projection, preferably in topographic projection.
(4) The special shape of the region has to be determined by constraints originating from real world conditions. Thus the indeterminacy in selecting the confidence region can be removed. Finally, the two-dimensional projections are transformed accordingly (transformation into similar or congruent figures, figures of the same content, etc.).
(5) Eventual statistical conclusions are drawn.

As an example take the classical regression equation in n unknowns. The simultaneous confidence region for n regression coefficients is reduced to a set of simultaneous confidence regions of two regression coefficients (ellipses centered about the estimated regression coefficients and giving $(1-\alpha)$ confidence to include the true values).

4.2 Stereoscopic Views

Stereoscopic views are based on the composition of two centrally perspective pictures taken from slightly distant centres of projection. Anaglyphs are stereoscopic pictures printed in superimposed complementary colours. Three-dimensionality gives a much better impression of spatial relationship than do ordinary two-dimensional drawings. They require, however, some type of stereoscope (at least two-coloured spectacles), and its use therefore is a little bit awkward. The effort necessary for construction is considerable if done by hand, but lessens much if satisfactory computer graphics equipment is available.

REFERENCES

Bertin J. (1983), Semiology of Graphs, Madison, WI.
Fowler J. (1984), The IBM PC/XT Graphics Book, Prentice-Hall, Englewood Cliffs, NJ.
Gordesch J. (1983), Projective Scales, Mitteil. Schwerp. Methodenl. 6, Freie Universität, Berlin.
Gordesch J. (1983), Projective Metric in Cluster Analysis and Multidimensional Scaling, Mitteil. Schwerp. Methodenl. 6, Freie Universität, Berlin.
Grundzüge der Mathematik II, Geometrie, Part A (1967), Part B (1971), ed. by Behnke H. et al., Vandenhoek & Ruprecht, Göttingen (Foundations of Mathematics).
Hoschek J. (1969), Mathematische Grundlagen der Kartographie, Bibliogr. Institut, Mannheim (Mathematical Foundations of Cartography).
Klein F. (1928), Vorlesungen über nicht-euklididsche Geometrie, revised by W. Rosemann, Springer, Berlin, reprint 1968 (Lectures on Non-Euclidean Geometry).
Lu Yung-Chen (1976), Singularity Theory and an Introduction to Catastrophe Theory, Springer, New York.

Computer Graphics and Data Presentation, a First Step Toward a Cognitive and Ergonomic Analysis

By I. W. Molenaar, Groningen

Summary

After a brief review of some major papers on statistical graphs and tables, this paper sketches some general principles for choosing a display form and layout suitable for compact and yet faithful transformation of statistical information. An example is provided. It is argued that cognitive task analysis and controlled experimentation are required for understanding how people read (and sometimes misread) tables and graphs. Both improved software and a better training of students are necessary.

Keywords: display of statistical information, tables and graphics, task analysis, experiments in graphical perception, graphical software.

1. Introduction

Computerization has led to a rapidly rising stream of detailed information that transcends the processing capabilities of the human mind. Hence there is an increased demand for compact but faithful condensation of statistical information. Both the act of condensation and the consumption of the condensed information are no longer limited to experts: everybody reads graphs and tables, and almost everybody has access to some spreadsheet program or statistical package that produces them.

Judged by professional standards, or even simply by goal effectiveness, a majority of these tables and graphs are suboptimal, or even worse. Only a few researchers are familiar with the design recommendations given by e.g. Chambers et al. (1983), Cox (1978), Ehrenberg (1975), Schmid and Schmid (1979), Tufte (1983) and Tukey (1977). Moreover, such recommendations are

Compstat 1986

mostly based on common sense, not on controlled experiments nor on a psychological theory of the human understanding of tables and graphics (Kosslyn, 1985).

Bertin (1967) provides at least an orderly discussion of the concepts underlying graphics design. His system of display methods is based on the number and the measurement level of the variables to be simultaneously displayed. His list of "visual variables" to be used obviously starts with the two dimensions of the plane. At each position determined by the two coordinates one can choose between symbols (triangle, circle, star, etc.), each of which can have different size, color, darkness, and density of hatching. Bertin argues that coordinate positions are best for information at the interval level of measurement. Sizes and shades are good for showing a limited number of ordered classes. Color and form, for which no natural ordering is perceived, should be reserved for nominal categories. In a sequence of related graphs, the assignment of display formats to variables should be kept constant. Violation of such rules forces the conscientious viewer to repeated consultation of the legends, and triggers wrong inferences in the less conscientious viewers.

Tables can be improved by curing "decimalitis". Only two-digit numbers can be held in short term memory during comparison (Ehrenberg, 1975). Blanks for unimportant entries also facilitate the transmission of the main information (Boomsma, 1983; Molenaar, 1986). Another important tool is the reordering of rows and/or columns. Wainer "is not interested in Alabama first" and shows how alphabetic order may be less effective than ordering by region, population size, or some other important variable.

2. Some general principles

As discussed in Molenaar (1986), there are no uniformly superior display formats for summarizing data. The summarization should enable the viewer to easily obtain the answers to his/her main questions about the data. The optimal format depends on the nature of these questions, and on the amount of statistical knowledge and training of the viewer. Another disturbing factor is that the performance of each display method changes when a different variant is chosen: reversing the roles of the horizontal and vertical dimension, choice of scale units or transformations, or change of labeling often have a substantial influence on the effectiveness of the display.

Here are some general procedures that may help to find a good method for the transmission of statistical information in each particular case. First it should be established what types of question the persons looking at the display will probably seek the answer to. Next, the designer should assess the amount of statistical knowledge that can be assumed: a QQ-plot or an Andrews curve are useful only to those familiar with them. Then, some tentative choices of a display method and format can be made. Next they should be tested by establishing for each important question, how observation of the display followed by reasoning can lead the reader to the desired answer. If difficulties can be expected to arise at either the viewing or the reasoning stage, the designer should try to modify the display in order to remove or mitigate them. Finally it is recommended to validate the task analysis by asking a few persons representative of the future readers to look at the display and explain what they see.

Many authors cannot resist the temptation to cram a maximum of information into each table or graph. This is desirable only if it does not obstruct the essential observations and reasonings, and serves to answer important additional questions. These two conditions are rather restrictive, and very often a simple display should be preferred. "Details are available from the author on request" is a solution for the small minority of the readers interested in them. Great care should be taken, however, that all essential information is easily available from the table or graph with its legend.

Some modern computing facilities and software packages make it rather easy to try out several variants of display method, layout, labeling and legend. There should be a well chosen default option (adequately described in the program documentation) combined with easy ways to replace it by variants chosen by the user. Unfortunately a majority of the marketed software still has little flexibility, and fails to meet the standards of ergonomical design of graphs and tables. For major improvements, a joint research effort of statisticians, psychologists and computer scientists is required, as well as a better "graphicacy" training of students.

3. An example

Broersma and Molenaar (1985) report an experiment on the effectivity of box plots and stem-and-leaf displays (Tukey, 1977) for univariate frequency distributions. After a summary of the experiment, the procedures proposed in section 2 are illustrated by discussing how to present its results.

Eight samples A,B,...,H of size 173 were generated with two levels of standard deviation (sd), skewness (sk) and kurtosis (ku). For eight out of the 56 sample pairs, their two box plots were simultaneously shown for 50 seconds on an overhead projector. On an answer sheet, subjects had to fill in per pair whether the first sample had a larger, equal or smaller sd than the second, and similarly for sk and ku. The eight pairs were preceded by two training pairs and an explanation. Then this procedure was repeated for the same eight pairs using stem-and-leaf displays. The comparison tasks were purposively made difficult in order to get nontrivial error rates. Our 22 subjects were Dutch psychometricians familiar with statistics.

As there were no missing data, we have for 22 subjects (SU) for 8 pairs (PA) for 3 moments (MO=sd,sk,ku) for 2 display forms (FO) a right/wrong score. We seek a compact but faithful and informative data condensation.

According to section 2, it must first be established what questions the prospective reader wants to see answered. The main effect of FO on error rate (ER) plus the grand mean are of interest, and also the mean ER in each of the 3x2 cells of the MOxFO table. This was not only the idea of the authors; it was shared by 12 out of 15 students who knew the experiment but not the results. Some of them voted for a graph of y=ER versus x=MO with two lines for FO, or versus x=FO with three lines for MO. Most preferred the tabular form. Bertin would agree, as MO and FO are nominal and not interval. The report was to be published for readers probably familiar with all non-exotic display forms. It was decided to table ER as a percentage, not as an absolute frequency, sparing the reader the trouble of mentally dividing by total number per cell. Table 1 gives the MOxFO results, showing that the sd differences were far easier to detect than the sk or ku ones, and that the better performance of stem-and-leaf for sd is balanced by the superiority of the box plot for sk, with almost equal results for ku. Task analysis and experimental validation were skipped for this simple table.

Table 1. Mean percentage of correct answers per display form per moment across all 22 persons and 8 sample pairs (176 cases per cell, 1056 total).

	stand.dev.	skewness	kurtosis	total
stem-and-leaf	88%	39%	45%	58%
box plot	76%	49%	46%	58%
total	82%	44%	45%	58%

As the reader may wish to generalize the result both across persons and across sample pairs, it is desirable to add a display of average ER per person and per pair. If the latter includes the order of presentation one can also look for training or fatigue effects. First a table was designed with a line per person or per pair, and columns for stem-and-leaf, box plot and their difference, for each moment. A task analysis showed that it was unexpectedly difficult to answer the relevant questions from such a table, even after variation of its layout and ordering. Scatter diagrams of stem-and-leaf ER versus box plot ER, per person or per pair, turned out to give an idea of the univariate distributions and the correlation, but no clear picture of the format effect per pair/person. On the next page the finally chosen layout is illustrated. Note that Fig.2 identifies the pairs and their presentation order. The bipartite graph in Fig.3 does not identify persons, but emphasizes for each moment how many persons did better on each display format.

It will be clear that one experiment is insufficient to decide on the weak and strong points of box plots and stem-and-leaf diagrams. The conclusions may vary with the task, the layout, the distributions or the subjects. For more results, a task analysis and a discussion of the generalizability issue, see Broersma and Molenaar (1985) and Molenaar and Broersma (1985). The experimental validation of the final choice of the display forms is now in progress. The present summary intends to give an idea of both the impact and the limitations of this type of studies.

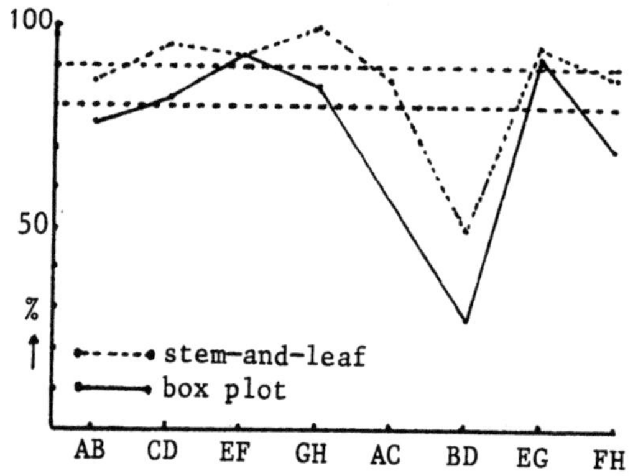

Fig. 2: Percentages of correct judgments about sd per pair in the order of presentation, with their median.

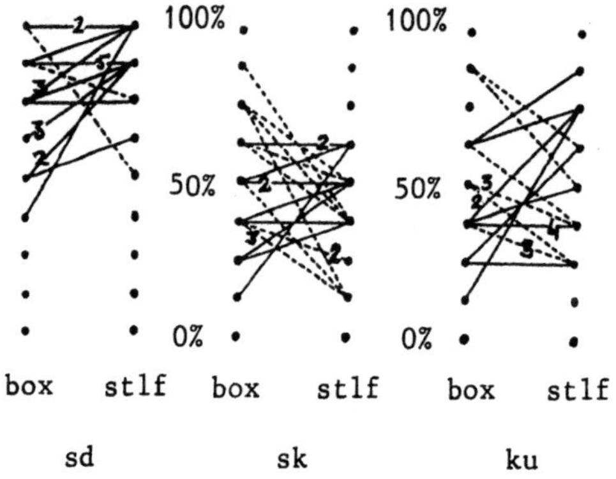

Fig. 3: Visual displays of the percentages of correct judgments per subject. Each edge connects a pair of vertices relating to the scores of one subject. Multiple edges have a multiplicity index.

REFERENCES

Bertin, J. (1967). Semiologie graphique. Gautier-Villars, Paris and Mouton, The Hague. English version Semiology of Graphs, translated by W.J.Berg, University of Wisconsin Press, Madison WI, 1983

Boomsma, A. (1983). On the robustness of LISREL (maximum likelihood estimation) against small sample size and non-normality (thesis University of Groningen). Sociometric Foundation, Amsterdam

Broersma, H.J. & Molenaar, I.W. (1985). Graphical perception of distributional aspects of data. Computational Statistics Quarterly 2, 53-72

Chambers, J.M., Cleveland, W.S., Kleiner, B. & Tukey, P.A. (1983). Graphical methods for data analysis. Wadsworth, Belmont (Cal)

Cox, D.R. (1978). Some remarks on the role in statistics of graphical methods. Applied Statistics 27, 4-9

Ehrenberg, A.S.C. (1975). Data reduction. Wiley, New York

Kosslyn S.M. (1985). Graphics and human information processing. Journal of the American Statistical Association 80, 499-512

Molenaar, I.W. (1986). Statistical information and human computer interaction. To appear in: G.C. van der Veer and G. Mulder (eds.), Human Computer Interaction. Berlin, Springer.

Molenaar, I.W. and Broersma, H.J. (1985). Exploratory Data Analysis : Psychology applied to the presentation of statistical information. In : Symposium Statistische Software 1985, Technisch Centrum Univ.of Amsterdam (ed.), 227-240.

Schmid, C.F. & Schmid, S.E. (1979). Handbook of graphic presentation, 2nd edition. Wiley, New York

Tufte, E.R. (1983). The visual display of quantitative information. Graphics Press, Cheshire (Conn)

Tukey, J.W. (1977). Exploratory Data Analysis. Addison-Wesley, Reading

Expert Systems

Expert Systems and Data Analysis Package Management

By J. Jida and J. Lemaire, Nice Cedex

SUMMARY : the possible connections between expert systems and data analysis softwares will be briefly developped, but with a little touch of pessimism. Then an example of such a connection shall be presented. It concerns the semantic analysis of a command file used in a statistical software on micro computers.

KEYWORDS : Expert Systems, Statistical software, Semantic analysis.

1. INTRODUCTION

Artificial Intelligence is now a fast developping field of research [Bar, Cohen, Feigenbaum (1982), Nilson, Nil (1980)]. The works performed in this field about knowledge modeling, automatic reasonning, learning processes..., have obtained succesful results as to their possible applications in very numerous areas of human activities. Expert systems [Sowa (1984), Waterman, Hayes-Roth (1978), Gondran (1984), Bonnet (1981)] are good examples of these results. Now, they are useful tools in various fields such as medecine, chemistry, financial help, formal calculus, oil prospection, teaching ...

Then it was pertinent indeed to try to make use of these systems in statistics [Gale (1985) for instance]. In the specific area of statistical packages, useful applications are obvious. "They are the only hope of connecting a slowly growing population of professional data analysts and statisticians with a rapidly growing population of data that requires analysis" says Tuckey (1985) in a comment about Hann (1985) 's paper.

The latter author has proposed three directions of applications :

- 1 "Computerized statistical answering and referral services"
- 2 "Expert guidance embedded in statistical programs"
- 3 "Automated statistical consultation and data analysis."

Before the presentation of a little work which lies within the scope of direction number two, we should like to report some remarks about the general use of expert systems in data analysis [Escoufier (1985)].

Compstat 1986

2. EXPERT SYSTEMS AND DATA ANALYSIS

Declarative programmation constitutes the main charm of expert systems in human knowledge modeling. In a field where analysis schemes are numerous - a little experimentation conducted by Y. ESCOUFIER at Montpellier (France) in 1985 was a good illustration of this variety of schemes ; he just submitted a little collection of different data bases to French "experts" in data analysis - it seems easy to represent expert knowledge by rules such as :

IF H and H' and H" and ... THEN C and C' and C" ...

and to do this for several experts. The declarative feature makes the representation of different schemes possible in the same rule basis. Unfortunately it can be observed that the useful realizations [for instance Malvache (1986), Loprieno (1986), Dambroise, Massotte (1986), Demonchaux, Quinqueton (1984) in france] are very rare in spite of numerous declarations of intentions. It can be interesting to find the reasons for this difficulty.

- In most cases (not only in statistics !) experts have not a clear vision of their knowledge.

- Perhaps, the present model (rules, frames ...) is not well adapted to represent complex knowledges. It may be the case for the expert as for the user of a statistical package who "frequently has only imprecisely stated question or end objective" [Hahn (1985)].

- Declarative programmation rapidly leads to the difficult problem of non contradiction of the rule basis. There are solutions in the case of zero order logic [for instance Hery, Laleuf (1985)]. But can we really speak of useful results in the case of the order one logic where the rule order is relevant ?

- Learning schemes which would make it possible to have an automatic adjustment of the rules are not yet efficient enough.

- The classical tools used by expert system conceptors are often too rough. Various research works have been achieved to improve the performances of inference motors, which constitute the core of an expert system. But the same cannot be said of the the interface part between user and inference motor. On the other hand, the actions in the consequent part of rules often requires complex calculus that need procedural languages such as PASCAL, FORTRAN ... and complex data structures. But are the interface tools really satisfactory ? I think they are not at the moment.

- Before undertaking an important action in computer science, it is necessary to take care of the future portability of the realized product. Unfortunately the normalization is not yet established in this field. PROLOG language and its extensions certainely constitute a first step, but not a sufficient one, when considering the problem of interfaces with other languages for instance.

- Computer performances are often too poor to offer the possibility of using important knowledge data bases. It is true for micro computers such as IBM-PC on which we have worked. Here the PROLOG interpretors [Giannesini, Kanoui, Pasero, Van Caneghem (1981)] are quite time-consumming languages. Perhaps TURBO-PROLOG (not yet distributed in France) will contradict this assertion. I hope so !

Because of all the facts mentionned above, we have prefered to adopt a non committal attitude in the third domain of Hahn (1985) 's paper. "Rather, we need, and can just find the strengh to construct, effective systems for individual areas" as it was proposed by Tuckey (1985). In this direction, we have chosen to restrict ourselves, for the first construction of an expert system in data analysis, to the second domain defined by Hahn (1985). Now we shall describe this investment.

3. EXPERT SYSTEMS AND DATA ANALYSIS PACKAGES

3.1. The software

In order to be concrete, we have worked on a French data analysis software named CHADOC VS (CHADOC means in French : CHaîne d'Analyse de Données à Organisation Cohérente). This software written in TURBO-PASCAL is being developped on IBM-PC. It constitutes a professional version of a previous software, named CHADOC too, which was written in BASIC, a few years ago with pedagogical objectives. It offers the main useful tools in multivariate data analysis. In fact it is not the software which is important but only its organisation which is rather common. So this work could be extended to other softwares with small modifications.

3.2 The general organisation of the software

It's described in the following scheme :

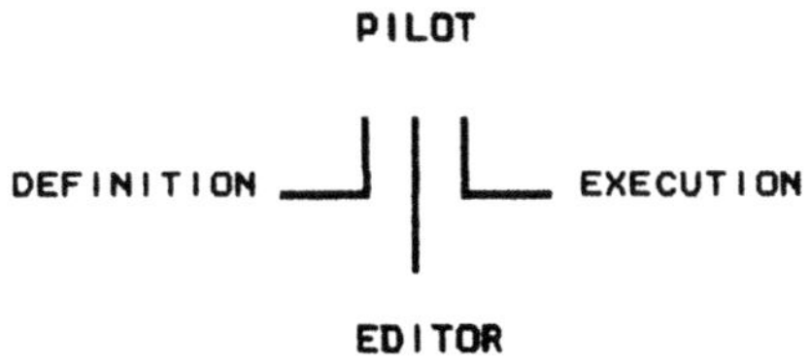

It contains three modules. The EDITOR is dedicated to the data bases input and their modifications. An example of such a data basis is given below :

Example of very, very small data basis

AGE : What is your age ?

SEX : What is your sex ?
MAN : Man
WOM : Woman

LIKE : Do you like data analysis ?
YES : yes
NO : no

	AGE	SEX	LIKE
0001	35	MAN	YES
0002	12	MAN	NO
0003	20		YES
0004		WOM	NO
0005	45	WOM	YES

The DEFINITION module helps the user in the definition of the analyses to be done and their corresponding parameters. It creates a parameter file which contains the following informations : analyses names and their corresponding parameters.

The EXECUTION module is used to execute the analyses requested. It is handled by the parameter file previously defined. This separation between the DEFINITION step and the EXECUTION step in the realization of a data analysis was proposed to avoid the user 's presence during the execution step which can be tedious when it is long (unfortunately rather frequent on these micro computers).

3.3 Management by an expert system

One can imagine using expert systems at different levels in this software:

- At the begining, when making choice of analyses and their options. Here we are in the field of automatic statistical consultation but with a restriction : in the conclusions of the rules, the actions to be executed are well defined ; they are the CHADOC VS treatments.

- At the end, with the interpretation of analysis results. Here, the hypotheses are more clearly defined. They are the outputs of CHADOC VS. But the conclusions of the rules may be more complex. For instance, in factorial analysis, the interpretation of axes depends on the context. On the basis of axe oppositions, the statistician has to produce an interpretation in connection with the specialist of the studied field.

- In the middle, between the DEFINITION step and the EXECUTION step. Here is the section of our work. The context is perfectly precised : the parameter file and the data bases in input. At this level, an expert system can be used to produce a semantic analysis of the commands contained in the parameter file before executing these commands. The advantages of such an analysis are discussed below.

3.4. Semantic analysis of a parameter file

Firstly, it can be noticed that this analysis is necessary in the EXECUTION step in order to avoid incorrect treatments and to produce error messages if necessary. So, in CHADOC VS, this semantic analysis has been already performed in this step, in PASCAL, and for each treatment module.

What should the improvement of an expert system be, then ? I think that it can offer many advantages. Its main purpose is to produce a rapid diagnostic on the correctness of the file parameters without doing effectively the required analysis executions. Then the user can immediately see if his parameter file is correct or not. This feature is important if he has to command a long serie of data analyses in which the slightest error can bring the computer to stop.

From the conceptor 's point of view, there is also a great interest in grouping all these verifications in order to explain, in a technical book, what constraints are involved by the use of his software. Moreover, in the case of a semantic analysis, declarative programmation is very useful. It is well known that parsing algorithms can be rapidly implemented, in a very straightforward manner, with languages such as PROLOG. Modifications don't generate problems ! The IF ... THEN ... rules are very clear and they can be completely presented in a book which explains how to use the software. Error messages can also be grouped. Then the user can learn not only what he has to do but also, and clearly, what are the common errors he has to avoid.

Before presenting the main features of this semantic analysis, I should like to make some remarks about its limits. In certain cases, it is not possible to present a complete diagnostic, from a semantic point of view, without executing the commanded analyses. For instance, if the user wants to cut a hierarchical tree, it is not possible, without this tree, to know how many clusters would be constituted. This number may be important in a following analysis such as a variance analysis, a cross validation analysis ... Then, it is not surprising if, at time, we have been modest in our ambition.

3.5 The details of the semantic analysis module

For each possible analysis (also called treatment in CHADOC VS), the specific part of the parameter file contains generally the following records :

- 1 name of the analysis
- 2 name of the treated data bases
- 3 parameters of the analysis
- 4 name of the created data bases

The part 3 generally begins with the series of active variables and active individuals (= observations) in the treated data bases. The first problem in this work was the selection of a substructure in the data bases so that most of verifications can be done quickly. We have chosen the following information named the <u>structure</u> of a data basis. It contains three lists which are :

- 1 the list of the individuals codes
- 2 the list of the variables codes
- 3 the list of their type (qualitative = ql, quantitative = qt)

For instance, for the little data basis of paragraph 3.2., the structure is composed of the following lists :

```
(0001.0002.0003.0004.0005)
(AGE .SEXE.LIKE)
(qt.ql.ql)
```

The main problem in the verification is that the commands of the parameter file can concern data bases that do not exist at the begining of the execution step. For instance, it is possible to command firstly a factorial analysis about an existing data basis and secondly (but in the same parameter file), a clustering analysis on the previously created data basis of the principal components. So it is necessary to simulate the structures of the created data bases named in the file parameters. For instance, a two-axe correspondance analysis where all the variables and the individuals are active, applied to a data basis whose structure is :

```
(0001.0002.0003.0004.0005)
(VAR1.VAR2.VAR3.VAR4)
(qt.qt.qt.qt)
```

will generate the following structure :

```
(0001.0002.0003.0004.0005.VAR1.VAR2.VAR3.VAR4)
(AXE1.AXE2)
(qt.qt)
```

It obviously is associated with the data basis of the principal components (for individuals and variables).

We have restricted our choice to these three lists for they are sufficient for most of semantic verification cases. For instance, we give a possible PROLOG rule for the stepwise regression module :

```
REGM                        {stepwise regression}

     {list of parameters}

     (*data_name.           {data basis name}
      *pred_var_sel.        {selection expression for predictive var.}
      *nb_max_pred_var.     {maximum number of predictive variables}
      *var_to_pred.         {variable to be predicted}
      *sel_ind.             {selection expression for individuals}
      *creat_var.           {created variable with regression coef.}
      *creat_data_name)     {created data basis name}

     {verifications to be done}

      -READ_DATA_STRUC(*data_name,*data)
      -VAR_SEL(*data,*pred_var_sel,*pred_var,qt)-NO_EMPTY(*pred_var)
      -POSITIVE_INTEGER(*nb_max_pred_var)
      -EXIST_VAR(*data,*var_to_pred,qt)
      -NOT_IN(*var_to_pred,*pred_var)
      -IND_SEL(*data,*sel_ind)-NOT_EMPTY(*active_ind)
      -NOT_EXIST_VAR(*data,*creat_var,?)
      -ADD_VAR(*data,*creat_data,*creat_var,qt)
      -SAVE_DATA_STRUC(*creat_data_name,*creat_data).
```

The verifications are specified in the IF part of the rule. They use general predicates such as READ_DATA_STRUC, VAR_SEL ... written in PROLOG or in PASCAL. These verifications are rather obvious :

- the data basis is searched for and its structure is put in the PROLOG variable *data;
- then the selected possible predictive variables are selected (they must be quantitative variables); a list of their codes is put in *pred_var;
- after we observe if the variable to be predicted exists and if it is not in the list *pred_var;
- then the active individuals are selected; the list *active_ind is constructed in doing this;

- finally, we create the structure of the new data basis by adding the prevision variable; before, we have to verify that its code does not already exist in the treated data basis.

Rules of this type have been written for each CHADOC VS treatments and we have constituted a library (embedded in a PROLOG module) with the general predicates. These rules are selected by a pilot rule which constructs, at the same time, the needed parameter lists. To end this paragraph, we give a little example of a correct parameter list for the stepwise regression treatment :

```
(test.(AAAA:CCCC.EEEE:).3.DDDD.****.XXXX.test1)
```

where test is the name of a data basis with the following structure :

```
L1 = (0001.0002.0003.0004.0005.0006.0007.0008.0009.0010)
L2 = (AAAA.BBBB.CCCC.DDDD.EEEE.FFFF.GGGG)
L3 = (qt.qt.qt.qt.qt.qt.ql)
```

In PROLOG, the instanciations will be (**** means everybody) :

```
*data_name = test
*data = (L1.L2.L3)
*pred_var = (AAAA.BBBB.CCCC.EEEE.FFFF)
*var_to_pred = DDDD
*active_ind = (0001.0002.0003.0004.0005.0006.0007.0008.0009.00010)
*creat_var = XXXX
*creat_data = (L11.L21.L31)
*creat_data_name = test1
```

with L11 = L1, L21 = (AAAA.BBBB.CCCC.DDDD.EEEE.FFFF.XXXX) and L31 = (qt.qt.qt.qt.qt.qt.ql.qt).

3.6 Some remarks to conclude

I hope that this type of rules appears very simple to you. This is one of our objectives in using expert systems in data analysis. When it comes to the choice of the programming language (here, PROLOG-P, achieved by the CNET in France, on IBM-PC like micro computers), we can observe now that it is well adapted to the handling of lists. But this choice is not critical; for instance, languages such as LISP, LOGLISP ... would do as well. I hope that TURBO-PROLOG will provide a better speed of verifications and a more simple interface with CHADOC VS (written in TURBO-PASCAL).

To be completely honnest, I have to give a last example where the chosen structure is not sufficient to make a complete verification. For instance, if a multiple correspondance analysis has to be performed on a data basis with qualitative variables, it is not possible to construct the structure of the created data basis. This poses a problem because the new data basis contains the principal components for the active individuals and the <u>modalities</u> of the active qualitative variables. Unfortunately their codes are not in the structure lists. One can think of systematically adding this information to the structure. This is not a good option because it generates more complicated calculus in the verifications step (here speed is necessary for this tool to be useful). It also creates new problems : for instance we cannot estimate the number of such modalities when cutting a hierarchical tree (not effectively). So, for the time being, we prefer to restrict ourselves to this simple choice of data basis structure and to admit that sometimes the verifications cannot be performed without doing the effective execution of the required treatments. This may be a future subject of research !

4. REFERENCES

Barr, Cohen, Feigenbaum (1982), The handbook of artificial intelligence, Pitman.

Bonnet A. (1981), Application de l'intelligence artificielle : les systèmes experts, RAIRO Informatique, 15, 4.

Dambroise E., Massotte P. (1986), Mise en oeuvre d'un système expert en statistiques : MUSE, Journées de statistique, Lille.

Demonchaux E., Quinqueton J. (1984), OURCIN : Manuel d'utilisation, Rapport technique no 36, INRIA, France.

Escoufier Y. (1985), Synthèse des résultats des groupes de travail mis en place par le GRECO "Analyse des données et informatique", Rapport CNRS.

Gale A. (1985), Knowledge representation in data analysis. Proceedings of the fourth international symposium "Data analysis and informatics", Versailles, France.

Gondran M (1983), Introduction aux systemes experts, Eyrolles.

Giannesini F., Kanoui H., Pasero R., Van Caneghem M. (1985), Prolog, InterEditions.

Hahn G.J. (1985), More intelligent statistical software and statistical expert systems : future directions, The Amer. Stat., February, 39,1.

Hery J.F., Laleuf J.C. (1985), Cohérence d'une base de connaissance : la convergence commutative en langage LRC, Note EDF-DER, HT 14/22/85, HI/5080-02.

Loprieno C. (1986), ADELLE : automatisation du dépouillement d'enquêtes par logiciel expert. Sixièmes journées internationales "Les systèmes experts et leurs applications", Avignon, France.

Malvache P. (1986), ADONIA : un système d'aide à la conduite d'une analyse de données. Sixièmes journées internationales "Les systèmes experts et leurs applications", Avignon, France.

Nilsson, Nil (1980), Principles of artificial intelligence, Tioga Publishing Company.

Sowa (1984), Conceptual structures information processing in mind or machine, Addison Wesley.

Tuckey J.W. (1985), Comment, The Amer. Stat., February, 39,1.

Waterman D.A., Hayes-Roth F. (1978), Pattern directed inference systems, Academic Press.

5. ACKNOWLEDGEMENTS

I would like to thank the Scientific Programme Committee and the Organising Commitee of the 7th Symposium on Computational Statistics (COMPSTAT) for having offered me an invited conference in ROME in september 1986.

NICE le 1 juin 1986.

Developing Intelligent Software for Non-Linear Model Fitting as an Expert System

By C. Berzuini, Pavia, G. Ross, Harpenden, and C. Larizza, Pavia

A prototype rational front-end for assisting users of the MLP program to input or generate data and formulate a model has been developed as an expert system using the shell EXPERT. Relationships between (a) MLP structure (b) selection of the task for the expert system (c) front-end design and (d) choice of programming tool are described.

Keywords: artificial intelligence in statistics, expert systems, Maximum Likelihood Program, non-linear model fitting

1. INTRODUCTION

Many projects in the area of Artificial Intelligence Research in Statistics focus on a particular task done by statisticians and investigate on whether and how this can be actively assisted by expert system techniques.

The program MLP (Ross, 1978), allows a flexible set of facilities for fitting models to data. This appeared to be a challenging background for the development by means of knowledge engineering techniques of a ***rational front-end:*** an interface providing MLP users with help and advice concerning the specification of input parameters, interpretation of results and in the adoption of a suitable strategy of analysis. The term "rational" indicates that the front-end contains a detailed representation, expressed as computer rules, of the decision and planning process used by a human expert in specifying MLP instructions to conduct a data analysis.

As a first stage in a joint project between the Dipartimento di Informatica e Sistemistica, University of Pavia, Italy, and Rothamsted Experimental Station, England, a prototype front-end capable of guiding the user in data input and model formulation by MLP has been developed *as an expert system*. The expert-system shell EXPERT, developed at Rutgers University, U.S.A. (Weiss & Kulikowski, 1979) has been used to implement the expert system and to translate it into a FORTRAN subprogram to be integrated into MLP. Rather than describing in detail the front-end performance, the present paper focuses on relationships between (a) MLP characteristics, (b) selection of the task for the expert system, (c) front-end design and (d) choice of the programming tool.

2. MLP

MLP stands for "Maximum Likelihood Program": a computer program that fits models to data. The models are either ***standard models*** selected by name and fitted by the most suitable method available for the particular combination of model and data, or ***general models***, defined by the user in a high-level language and solved by general algorithms. Standard models include various models for curve or distribution fitting, assay analysis, regression analysis and genetic models. The general models section supports non-linear regression, solution of differential equations and various methods of estimation, such as robust estimation. MLP is intended for a wide range of users: from the non-specialist with a particular model who requires a solution with the minimum of trouble, to the statistical research worker who wishes to study the relationships between models and data. The latter user usually requires detailed knowledge of the program control language to exploit the flexible investigative tools offered by MLP. A common wish across the whole spectrum of users is to be able to run the program without reference to the manual for relevant options and input parameters and without the need to be familiar with the program control language and job structure constraints. Non-specialist users may also benefit from suggestions about the strategy of use of MLP facilities, and at a different level, those concerning statistical strategy. A prototype MENU facility has been added to the language, and will be

Compstat 1986

available to users of MLP Release 3.09. This has provided valuable insights for developing the project described below, in which an approach by means of knowledge engineering tools is pursued to develop a system supporting robust and flexible guidance and advisory capabilities.

3. WHAT IS AN EXPERT SYSTEM BUILDING TOOL

This paper describes the development of a front-end to MLP using an expert system building tool (ESBT). ESBTs are software packages which do more than solve single problems: they help in *constructing new programs*, specifically, new expert systems (ES). The main difference with respect to traditional programming is that in an ES *facts* (specifically: facts about the data to be statistically analyzed, about the user aims, about the statistical package) are separated from *procedures* using them for problem-solving. The front-end designer can concentrate on stating facts and relationships among them (eg. relating a user's choice to a statistical package facility) in a direct way, by means of symbolic specifications, in a modular style resembling input of facts in a data base. The set of facts+relationships is called *knowledge base*. As in a data base data retrieval procedures are embodied in the DBMS and separated from recorded data, in an ES the messy algorithmic details of ordinary programming are hidden in a domain-independent *inference engine*, capable of using facts + relationships of the knowledge base to perform reasoning. Like data in a data base, a knowledge base can be easily modified, incrementally expanded by adding/removing modular symbolic specifications.

Prototype consultation models capable of running within exceptionally short time periods can be developed using an *expert system shell*, a programming environment with a pre-programmed inference engine, providing the user with an easy language for representing knowledge and tools for inspecting the program's reasoning while it is solving a problem, for dynamic tracing, explanation (eg. WHY facility in EMYCIN) and debugging.

4. AIMS OF THE PROJECT.

The following are important aspects of MLP front-end development:

(a) *front-end interaction with a novice user of the statistical package;* the front-end elicits necessary information from the user in dialogue form and then takes over the standard input instructions. Therefore the user needs no reference to a manual, irrespective of his knowledge of the input language;

(b) *computer representation of relevant elements of statistical knowledge;* eg. if the user does not know what kind of statistical model to select the front-end may provide him with help in choosing the right tools from the facilities provided by MLP;

(c) *developing techniques for tutoring ;* the front-end evaluates the user's need for statistical tutoring at each step of the process, alters its view of the user dynamically and uses this view to determine its own behaviour; for example, instead of simply presenting MLP options, it may in addition discuss with the user statistical reasons for selecting the options;

(d) *interfacing the front-end to MLP;* nowadays the majority of important statistical programs possess a well-designed and sophisticated language interpreter, and AI language compilers are becoming available. This allows front-end - program communication via text strings. A different approach has been followed here, in which the front-end bypasses the language interpreter by directly passing control parameters to MLP modules.

Robust support of points (a)-(c) may well depend on the following design aspects:

(i) *generate suggestions that take into account user knowledge*

(ii) *rhetoric capability:* choosing the appropriate level of detail of explanation for the current user

(iii) *unobtrusiveness:* the MLP user is in complete control of the statistical package, the front-end being only in a background advisory mode.

5. OVERALL DESIGN OF THE MLP FRONT-END.

The front-end is meant to be integrated into MLP as a FORTRAN subroutine. The MLP user should be able to communicate with MLP in two modes: via standard control language through the MLP interpreter, or by invoking the front-end module, He will not be able to switch mode during an analysis. When invoking the front-end, all the user has to do is answering questions to enable the front-end module to take over standard MLP input instructions. It was decided that the project should not involve substantial rewriting of the existing MLP code. The first step in the design was to organize the process of interaction with the data and of model formulation via MLP into the following set of subtasks/operations, called **nodes** (of a network across which the user moves to carry on the analysis) , in which the user:

NAMES provides, if needed, non standard names for operands appearing in calculations and/or model formula

DATA enters data to be analyzed, or edits previously entered data to correct errors, delete units, or reorganize the data array (include/delete variates, change mode of addressing it)

PROBLEM specifies the statistical problem to be solved (if a standard model is requested, he is asked further details)

MODEL enters equations for a non-standard model

FIT fits a model and specifies output options

TRANSFORM derives new variates computed from original ones, or generates systematic variates

DISPLAY asks for data plots or printout

The general front-end **control strategy** consists in prompting the user to select a node, then assisting him in the subtask characterizing that node. This cycle is repeated many times. The assistance at a node takes form of a question and answer dialogue, with the front-end asking for necessary information, setting MLP control parameters and calling appropriate MLP routines into action; moreover, it acts like a "background" consultant, unobtrusive, but capable of conveying to the user knowledge and suggestions at a very precise level if necessary. The dialogue can assume a highly variable form depending on previous interaction , previous nodes , on user's knowledge of statistics and of MLP, on properties of data being analysed. Basically, at any moment the user can go from any node to any other node, with some constraints (of which examples are given later in this section). In particular, the user can at any moment return to any previously visited node for correcting previous answers or to elicit tutorial information.

Access to a node is allowed conditional on the "state" of other nodes or on information obtained at them. Sometimes the user is not inhibited in pursuing his will, but receives warnings. Two examples: if node DATA has not been "switched on", that is if data input has not been successfully carried out, access to nodes TRANSFORM, FIT or DISPLAY is obviously not allowed: any attempt causes the front-end to stop the dialogue and suggest the user to go to node DATA for data input. If the user attempts PROBLEM (distribution fitting) -> NAMES -> DATA, as soon as he requests node NAMES he is advised that distribution fitting does not expect non-standard operand names to be specified, except in non-standard situations such as when the distribution is to be artificially generated prior to fitting.

The front-end keeps a **history** of the interaction: a record of sub-tasks carried out and an abstraction/summary of *(a)* information obtained from the user at previous nodes, and *(b)* results of previous model fittings. Updating the history involves *adding* facts (record that "node DATA has been switched on"), or *modifying* them ("the current model is no more regression, but probit") or *deleting* them (erase flag indicating that "the user needs data reorganization for fitting the new selected model"). The different propensity of such facts to "survive" reminds of the idea of a short-term and a long-term history memory. Complex long-term history management may enhance front-end capability of effective suggestions. As an example, differences between successive model fits may be recorded in history so that they may eventually be plotted to show graphically the effect of changing the model. The front-end uses the history to judge at any moment the appropriateness of the next step attempted by the user (eg. the next node selection) and to issue state-of-the-process driven questions and suggestions. For example, assume the user visits the following nodes in sequence: PROBLEM (specifying a type of curve fitting) -> DATA -> MODEL; the front-end at node DATA is able to guide him in providing the correct data layout expected by the MLP procedure for fitting curves. In fact standard models in MLP expect to find the data they require in a certain order, and to compute the dimensions of the arrays from the stored total length. For example, when using the MLP control language the instruction DATA 1 3 4 6 8 9 21 31 42 51 62 66; reads a list of 12 items in the data array; the directive FIT CURVE interprets this list as two arrays of coordinates, first the x values, then the y values, and calculates that there are six units. Other standard models require more complicated data organization. Change the order of nodes into DATA -> PROBLEM(curve fitting) -> MODEL and the front-end has the option to help the user at node PROBLEM to recognize the appropriate curve from the shape of the data (eg. from the ranges of x and y, classifying each data pair into classes and then assigning classes to possible curve models). A further example: consider a user moving along DATA->PROBLEM(standard model A)->FIT->PROBLEM(standard model B); at the last node he is warned whenever the data layout needs to be reorganized before fitting model B, and he is invited to move to node DATA where he may ask for help in carrying out the reorganization.

6. FRONT-END CONTROL STRATEGY.

The front-end uses two primary components: the **temporary memory** (TM) and the **working memory** (WM). The former records user's answers at the current node, the latter records the history. Each time a new task (node) has been carried out, the WM is updated and the TM is cleared and becomes ready to accept user's answers to questions that will be asked at the subsequent node.

The front-end adopts the following control strategy:

(a) *ask user for the node he wishes to enter (including the possibility to quit)*

(b) *determine from history whether conditions of admissibility for the selected node are satisfied*

(c) *if the node is not admissible go to (f)*

(d) *if the node is admissible, ask user for relevant information to be recorded in the TM*

(e) *interpret the data-base of facts (TM+WM): reason from facts to determine and carry out appropriate actions (eg. options settings, MLP routine calling, WM updating, recommendations to the user), then return to (a)*

(f) *explain why the requested node is not admissible, then return to (a);*

For the purpose of implementing such strategy using an expert system approach, we needed to define in a declarative fashion relationships among facts and between facts and conclusions. An idea is to identify **functional kinds of inference rule** linking TM + WM to conclusions. Here are some examples:

NA **node admissibility** rules, checking node admissibility conditions:

IF *TRANSFORM is the selected node*

THEN *if at least one of the following history facts is true: 'node DATA is switched ON', 'random variables have been generated', then switch ON admissibility flag for node TRANSFORM*

ELSE *explain why node TRANSFORM is not currently allowed*

Q **questionnaire** rules. They determine "askable" questions keeping current node, history and user's answers into account: *IF node DATA is NOT the current node OR has not been previously switched ON THEN do not ask the user whether he wants to correct errors in the data or to reorganize the data array*, or, as another example: *IF node DATA is switched off OR node PROBLEM is not the current node OR distribution fitting is not the current problem OR discrete distribution is not the user's intention THEN do not calculate poisson distribution index from the data.*

UM **user model** rules; they evaluate the user's degree of naivety on the basis of his/her answers. The user is assigned to categories of expertise, and this assignment is recorded in the TM.

H **help** rules. They check whether the user has not been able to indicate a choice at the current question. In such case they trigger explanations of the meaning of the current question. This can be done via examples.

A **action** rules. They heuristically match the current state-of-analysis as defined by previous history and by user's answers given after the last node selection to a set of possible actions to be carried out by MLP, possible options to be set, possible recommendations to the user:

IF *the admissibility flag (IPROB) for node PROBLEM is ON*

THEN *if at least one of the following facts is true: (a) the negative binomial (NEBI) is one of the selected distributions , (b) the random variable represents the number of intervals before a random event (NINT), then do the following action (OP254): inspect the option vector from element 2 to 5 and set the first found free equal to 4;*

if distribution fitting is a current selected problem (TM fact DISTR), and nodes DATA and PROBLEM were previously switched ON (history facts WDATA, WPROB) and distribution fitting (recorded in history as WDISTR) was never previously selected , then activate hypothesis RD warning that the data array may need to be reorganized;

WM **working memory updating** rules. They update the working memory

C **consistency** rules. They check for inconsistencies in the informations recorded in the working memory and resolve them.

7. BUILDING THE "EXPERT" MODEL OF THE FRONT-END

EXPERT is a rule language developed for building expert systems based on classification problems (Weiss and Kulikowski 1979). In EXPERT terminology, a knowledge-base is called a *model*. Such model distinguishes between "findings", or **facts** to be interpreted, and a taxonomy of "hypotheses", or **conclusions** that the model may infer. In the application here described facts are stored in the WM and TM, and conclusions represent possible front-end *messages*, *action recommendations*, *warnings*, *help explanations*, *actions*. **Intermediate conclusions** may be included, representing higher level abstractions of facts: eg. user-categories of expertise inferred from user's answers, node admissibility flags, indicators of currently requested node, and, more in general, intermediate logical steps of the reasoning. In an EXPERT model, conclusions are simply listed with associated identifiers, eg:

NVAR	*Print "give names of variates"*
GDATA	*Print "Go to Data Input node"*
MS15	*Print "Control mortality and immunity may be supplied in CONSTANTS"*
CGENE	*Call MLP routine GENERATE*
CHECK	*Check that length is a multiple of the number of variates*

WM facts, first, and TM facts, after, are simply listed as in the following example:

**Checklist*		**Multiple Choice*	
Switched on nodes (WM)		*Type of operation on data (TM)*	
WNAME	*node NAMES is switched on*	*CORR*	*Correct errors*
WDATA	*node DATA is switched on*	*REOR*	*reorganize the data array*
WFIT	*node FIT is switched on*		

The example above shows *logical* facts grouped as part of a *multiple-choice* (none or only one may be true, they may be all false or unknown) or of a *checklist* (any number of them may be true) entry. For what concerns *numerical* facts (eg. "number of variates in the analysis") in the current EXPERT implementation no real data structure except the scalar is available, which might prove to be a serious shortcoming when reasoning from complex objects such like MLP output data structures.

In the application here described the order of questions to be posed to the user and then stored in the TM can be stated in advance during front-end design. With EXPERT, by declaring questions to be organized as a *questionnaire*, the model designer eliminates the need of a complicated control mechanism. Control of the questioning about TM facts is handled by simple EXPERT implementation of Q rules The Q rule example in sec. 6 may have form *F(WDATA,F) & F(DATA,T) -> F(CORR:REOR,U)* where *F*, *T* and *U* within brackets stand for "false","true" and "forced to be unknown". *F* otherwise stands for "fact".

Intermediate conclusions may be used to "trigger rules", so that an *opportunistic* search (combining fact- and hypothesis-directed inference in a flexible way) may be carried out such as, for example, adapting the front-end behaviour both to the state of the analysis as described by facts and to the user's degree of knowledge inferred as an intermediate conclusion. As another example, the appropriate subset of NA, H and C rules to be triggered at one node may be selected by conditioning on an intermediate conclusion representing the node admissibility flag, as in the A rule example in sec. 6 (the "/" indicates newline):

**IF / H(IPROB,0.9:1.) / *THEN / [1:F(NEBI,T),F(NINT,T)]->H(OP254,1) / F(DISTR,T)&F(WDATA,T)&F(WPROB,T)&F(WDISTR,F)->H(RD,1) / *END*

where *H* stands for "hypothesis" (or conclusion). Note that each hypothesis is qualified by a priority coefficient ranging from -1. to 1. In a (THEN-END) block as many *antecedent* -> *consequent* rules as desired may be contained. The IF - THEN structure causes them to be evaluated on the data base of facts only when the IF part is positively checked. A careful use of such facility may allow a decisive enhancement of the front-end efficiency.

Instead of using a backward- or forward-chaining mechanism for rule evaluation, EXPERT evaluates rules in an ordered fashion that has been prespecified by the model designer. This seems not a limitation in designing a front-end model using the approach here proposed: the node defines the active rule set, within which Q rules are evaluated before NA rules, which are evaluated before H and A rules, and so on. Within each rule subset, if a rule *R1* sets a conclusion appearing in an antecedent of rule *R2* , care must be taken in writing *R1* before *R2*.

8. FRONT-END DEVELOPMENT PHASES

Three EXPERT modules were used in the development stages of the final front-end version: program XP for compiling a consultation model, program EXPERT for executing the compiled representation and program BASE for automatic translation of the consultation model into FORTRAN.

At the beginning a standard text editor is used to create a file that contains statements in the special purpose EXPERT symbolic programming language (see previous section) describing what we call the **EXPERT model of the MLP front-end.** This language is very simple, so that the front-end model is very readable , easy to correct, expand and refine.

Program XP allows checking the model for syntactic errors and translating it into an efficient internal representation. The model may then be executed (program EXPERT). At this stage the working memory is transparent to the user. The model is debugged by simulating the front-end behaviour in a variety of data analysis situations by directly altering the WM, and of user's behaviour patterns by entering different kind of answers into the TM.

When a satisfactory EXPERT model is obtained, it is translated into a very efficient FORTRAN routine, capable of autonomously running as a front-end, in which a clear separation between rule-base and inference engine is not recognizable, and the data base of facts has the simple structure of a FORTRAN vector. We call this routine **intermediate front-end.** The intermediate front-end is elaborated into the **final front-end** by working on the FORTRAN code. The final front-end is ready to be incorporated into MLP as a subroutine. Appropriate structures of MLP are made accessible to it via FORTRAN COMMON blocks. The working memory is made opaque to the user, previously simulated actions such as controlling MLP options and parameters and calling MLP modules are actually carried out, any useful FORTRAN routine can be inserted, a SUMMARY command is added allowing the user to request printout of the history at any moment.

ACKNOWLEDGEMENTS

The authors are grateful to Marta Landini for helpful discussions and programming assistance.

REFERENCES

Ross G.J.S. (1978), Curve fitting using the Rothamsted Maximum Likelihood Program, in : Jacobs D.A.H. (ed.), *Numerical Software - Needs and Availability*, Academic Press, 293-323.

Weiss S.M., Kulikowski C.A. (1979), EXPERT: a system for developing consultation models, *IJCAI*, 6, 942-947.

Express - An Expert System Utilizing Standard Statistical Packages

By F. Carlsen and I. Heuch, Bergen

Summary: EXPRESS is an expert system based on the idea that complex chains of statistical analyses can be simplified using a standard package for each set of intermediate computations. This principle is illustrated by two-sample testing for equal location parameters, relying on various BMDP programs. Distributional assumptions are decided on after preliminary runs of BMDP. The system each time generates control language for the package, and then extracts relevant information from the print files produced. It is argued that such a stepwise procedure, considering basic results before proceeding to more complex analyses, is close to the actual behaviour of most statisticians. However, this approach raises fundamental questions concerning overall statistical characteristics of the strategy built into the system. It is suggested that these characteristics may be assessed through simulation.

Key words: Statistical expert systems, statistical packages, two-sample problems

1. Introduction

Despite the recent general interest in artificial intelligence, there are at present relatively few successful intelligent systems functioning in the area of statistical software. In most cases these systems fall into either of two categories, (i) systems with built-in procedures for all computations or relying on other specialized software not always readily available, or (ii) systems working more or less as catalogues or preprocessors, ultimately referring the user to separate outside packages for the actual data processing.

On an experimental basis we have constructed an expert system EXPRESS which attempts to combine these approaches. EXPRESS (EXPert system Relying on Existing Statistical Software) takes advantage of the capabilities for convenient computation offered by well-known and proven statistical packages. At the same time it draws on information in a knowledge base, emulating a practising statistician who evaluates results provided by a package in a stepwise manner. In order to illustrate this approach, EXPRESS has been specificially designed for the problem of comparing location parameters in two samples, utilizing methods available in BMDP. However, the general issues

Compstat 1986

raised have much wider implications. EXPRESS is written in Fortran, and it is implemented on a Sperry 1100 computer.

2. General ideas motivating the design of EXPRESS

Early development of expert systems in statistics concentrated on programs giving advice about selection of statistical methods (e.g., Jones, 1980). The dialogue between user and system led to recommendations influenced by characteristics of the data only if the user conveyed such information to the system. However, more recent work (e.g., Pregibon and Gale, 1984) has also taken into account information derived immediately from the data. In fact, this approach seems to be more related to the way statisticians behave in practice. Thus, in a complex series of statistical analyses it is customary to combine at each stage information extracted from the observations with prior outside knowledge before proceeding to the next stage. The basic idea behind EXPRESS was to build this technique into an expert system, assuming that the statistician at each step in his analysis would be using facilities offered by a standard package. Prior knowledge could be incorporated by allowing the user to set aside any decision made by the system.

As a very simple example retaining the important feature of stepwise reasoning, we chose to study testing for identical location parameters in two independent samples. In such a situation a statistician with limited prior knowledge may begin by considering simple statistics as skewness or kurtosis to determine whether normal distributions can provide a reasonable fit. Alternatively, he may prefer just to have a look at various simple plots. Rejecting the normality assumptions, he may wish to explore effects of different transformations. He might then settle for a t-test, applied to the original or transformed data, with or without an assumption of equal population variances. The decision on variances could be based on a separate F-test. By contrast, if all attempts to obtain two normal distributions failed, the data could be subjected to a non-parametric procedure such as the Mann-Whitney test.

An experienced statistician, with any large package at his disposal, can easily set up instructions for all intermediate and final results required by such a procedure. However, decisions made during this process will depend on results produced by the package, so it is impossible to write down once and for all the complete set of instructions. An expert system adopting a similar approach, must be able to initiate execution of the package, interpret the output, store essential results, and then decide on the next step. This is the general strategy followed by EXPRESS.

3. Structure of EXPRESS

EXPRESS consists of four main logical components, (i) a common data base with space allocated to data sets, to results from computations, and to conclusions drawn by the system, (ii) a tree structure describing steps in the chain of statistical analyses, and indicating what information is needed at each step, (iii) an interface with BMDP, and (iv) a user interface. The system is menu-driven. The user may initially address the problem of testing location parameters, or he may start by investigating distributional shape. Every time a particular question is asked, the system checks if the information requested has already been written into the appropriate record in the data base. If not, the system attempts in accordance with the tree structure to generate the relevant information. This is done by combining other information stored previously, by executing a BMDP program, or by requesting special information from the user. Basic specifications, such as variable names or labels of external files containing data, must always be supplied by the user.

If the system must run a BMDP program, it first checks if other information is available in the data base ensuring that all assumptions for using this program are satisfied. Otherwise the system goes on recursively to generate the information needed. The interface with BMDP must necessarily depend on the computing environment. In our implementation, adapted to Sperry 1100, the execution of another completely independent program is carried out by producing in our Fortran program a separate instruction file. These instructions must redirect output to a temporary print file, initiate execution of BMDP, provide BMDP control language, and finally restart EXPRESS at the correct point. The print file is scanned by EXPRESS, and the results are inserted into the data base, along with other information from BMDP corresponding to blank records in the data base. The results transferred may vary from single numbers to quite comprehensive histograms or normal plots.

4. Discussion

On a trial basis we have studied how results from a standard statistical package can be incorporated into an expert system. The main novelty lies in the repeated use of the package, with decisions being based on results from earlier runs. The choice of the two-sample problem as the statistical setting was motivated by the relatively straightforward procedures involved. This situation was also considered by Jones (1980) in connection with expert systems providing guidance for selecting methods.

Chambers (1981) suggested that expert systems could take into account repeated cycles of inference based on conventional statistical methods. An advanced system as REX (Pregibon and Gale, 1984) exploits many techniques for dealing with actual data, but it relies on the language S for computations. Although S is a powerful tool, it is not widely used among statisticians. One of the characteristics of expert systems is their ability to explain decisions made. We feel that it will be easier to introduce intelligent statistical software if explanations can be related to such standard tools as packages like BMDP. Of course the design of EXPRESS would have been simpler if a more subroutine-like system could have been used. We utilized BMDP because of our greater familiarity with this package, but we envisage no problems using systems like SPSS or SAS in a similar way, or for that matter any combination of packages. These systems have the important feature in common that they are written as separate units with no possibility at all for the user (or the designer of an expert system) to modify the software. (User-supplied internal subprograms may be partly allowed, but do not help very much in this connection.) Similar problems adhere to a large amount of the existing statistical software in specialized areas. Separate complex Fortran programs are often available, but it is not feasible for anyone except the author to modify the source text. Our experiences with EXPRESS indicate that it may still prove worth while to develop expert systems based on established programs in such areas. As pointed out by Hahn (1985), such narrow fields of application may initially offer the best opportunities for creating successful expert systems.

It has previously been recognized that preprocessors to statistical packages can be useful in selecting program options and setting up control language (Kuik, 1984). The idea that the system should also initiate execution of the package and analyse results seems a natural extension. Prat et al. (1985) describe a system for time series analysis which can run various packages and extract information from the output. They indicate that a future version may be expanded to storage of information in a data base, permitting automatic diagnostics of results.

Because of external restrictions it was necessary to construct EXPRESS using Fortran. However, EXPRESS incorporates features of more general expert systems (as backward chaining for generating information and forward chaining for proceeding to more complex problems). The tree structure is incorporated in the Fortran program itself. We expect that many of the practical problems could have been handled more easily by a language specifically adapted to artificial intelligence. It may even be justified to design a general "shell" for setting up statistical expert systems in any area utilizing a number of established statistical programs.

Repeated use of the same data set with interactive statistical software may seriously invalidate such conventional measures as significance levels (Ader et al., 1985). Expert systems with stepwise data processing may create similar problems. In our opinion this should not prevent anyone from using such systems, but it is important to investigate the extent of these problems. Even with only slight complexity in the chains of statistical methods applied to the same data, simulation seems to offer the only practical method of evaluating overall performance of an expert system. To illustrate this approach, we have carried out simulations with EXPRESS, submitting each time to the system 4000 or 5000 replicates of two samples, each consisting of 20 observations, generated according to prescribed distributions. The procedure used in the simulation is somewhat simplified compared with actual use of EXPRESS, since user interaction is not included. The test for identical location parameters can now be specified as follows: (i) Consider skewness and kurtosis with corresponding standard errors as found by BMDP2D. Assign conventional P-values for deviations from normality. If $P<0.15$ for either skewness or kurtosis, normality is rejected. (ii) With rejection for both samples, and positive skewness values, similar tests are carried out for lognormal distrbutions. (iii) If normality is accepted for both samples, or lognormality, a Levene test for equal variances is performed (as in BMDP3D), with a nominal level of 0.10. A t-test (for equal or unequal population variances) is then carried out. (iv) If any other conclusions are reached about the distributions involved, a Mann-Whitney test is done (by BMDP3S).

Table 1 shows the simulation results, according to three distinct null models and one alternative model. For the overall strategy built into EXPRESS the most important figures are the attained significance levels given in the rightmost column. With nominal levels of 0.05 (in t-tests and Mann-Whitney tests) the system evidently produces close to correct values. In fact, all complete null distributions of P-values agree fairly well with the uniform distribution. Nonetheless, the relative frequencies of the procedures selected internally in EXPRESS clearly reflect different underlying models.

Table 1. Relative frequencies of test procedures selected by EXPRESS in simulation[1], with overall frequencies of rejection of null hypothesis[2]

Distribution of samples[3]	original data: t-test pooled variance	original data: t-test separate variance	logtransformed: t-test pooled variance	logtransformed: t-test separate variance	orig.d. Mann-Whitney test	Relative freq. of rejection
Both N(μ=5,σ=1)	0.746	0.098	0.001	0.000	0.155	0.053
Sample 1: N(μ=5,σ=0.5), sample 2: N(μ=5,σ=1.5)	0.006	0.847	0.000	0.000	0.147	0.060
Both Cauchy, median 5, int.quart.range 0.674	0.048	0.014	0.020	0.005	0.913	0.048
Sample 1: N(μ=4.75,σ=1), sample 2: N(μ=5.25,σ=1)	0.753	0.096	0.002	0.000	0.148	0.336

[1]5000 replicates, for Cauchy distr. 4000. Every replicate includes two samples, each with 20 obs. [2]Nominal level is 0.05. [3]All distributions truncated at 0.

It may be objected in this situation that the complete procedure is based on methods known to be well-behaved. On the other hand, even with the simple rules given above, the effect of interdependence between successive steps is very hard to predict. Thus, after rejection of normality or lognormality, the use of the Mann-Whitney test is not really correct for samples with unequal dispersion. Indeed, the simulation with different variances produces too high observed levels for this test, but this does not affect overall results very much. There may be serious obstacles to carrying out similar simulation experiments in more complex expert systems, in particular if unpredictable user interaction is important. Still, this approach may give a useful performance test, especially for studying effects of changes in internal parameters of the system, provided that realistic classes of models can be specified. An alternative is to measure practical performance of system and user combined, but such experiments will be very time-consuming and perhaps not too reliable.

References

Ader, HJ, Kuik, DJ, Opperdoes, E, Schriever, BF(1985): The use of conversational packages in statistical computing. Statist. Softw. Newsl. 11, 106-116.

Chambers, JM(1981): Some thoughts on expert software. In: Proc. 13th Symp. Interf. Comp. Sc. Statist., New York: Springer, 36-40.

Hahn, GJ(1985): More intelligent statistical software and statistical expert systems: Future directions. Amer. Statist. 39, 1-8.

Jones, B(1980): The computer as a statistical consultant. BIAS 7, 168-195.

Kuik, DJ(1984): QUESTOR, a conversational preprocessor to BMDP. In: COMPSTAT 1984, Vienna: Physica-Verlag, 329-324.

Prat, A, Marti, M, Catot, JM(1985): Incorporating expertise in time series modelling: The STATXPS System. Statist. Softw. Newsl. 11, 55-62.

Pregibon, D, Gale, WA(1984): REX: an expert system for regression analysis. In: COMPSTAT 1984, Vienna: Physica-Verlag, 242-248.

Muse: An Expert System in Statistics

By E. Dambroise and P. Massotte, Montpellier

ABSTRACT : This paper is aimed at describing an approach and some methods to be used in defining the field of applications and the structure of an expert system in statistics. These concepts were applied to develop the expert system M U S E (MUltivariate Statistical Expertise). This expert system is intented to be implemented in an industrial environment. The main features and spécifications of M U S E are also described in this document.

KEY-WORDS : Expert System, Data Analysis, Multivariate Statistical Analysis, PROLOG, Industrial Environment.

I - INTRODUCTION

Decision Support Systems are particularly useful in the high technology industries where decisions, strategic options and choices have an impact on the long term investments, and involve a large amount of expenses. In the same way, the complex manufacturing lines are controlled by more and more advanced information systems. They have a higher computational and storage capabilities, involving larger volumes of various collected data, available to many users.
The increasing complexity of these information systems implies use of sophisticated tools for decision supporting. They have to enable analyses of large volumes of data and they need a wider capability of synthesis. Building a Decision support System in an industrial field useful for naive users, seems to be an ambitious project : the industrial constraints require a userfriendly, reliable tool available to a great number of people. This conducts to introduce the Expert System (E.S) concept in such an environment. Due to the new aspects of this approach, a specific methodology has been set-up in the design as well as in the development of the E.S .

II - PRELIMINARY PHASE

An investigation has been performed at management level in order to determine the needs belonging to the decision support field. They can be broken down into the following :

Compstat 1986

1 - Diagnosis and prediction.
2 - Definition, study, analysis of a set of indicators / parameters.
3 - Modelling, simulation.

Points 1 and 2 could be satisfied in involving multidimensional data analysis. At the issue of our preliminary study, it was decided to emphasize on these statistical techniques. It must be noted that people very often have a right idea about the data to be collected in a data analysis. However, they generally do not know what are the adequate statistical methods to be carried out and how to process and interpret the results. For these reasons, developping an E.S was necessary either for diagnosis or for an educational purpose.

III - M U S E SPECIFICATIONS

Each E.S development requires thinking about the process used by the experts of the application domain. In the statistical field, five stages have been pointed-out :

1 - Identification of the statistical expertise field.
2 - Data collection, data checking, and data cleaning / updating.
3 - Determination of the user's problems and needs.
4 - Choice of the statistical methods.
5 - Carrying out of the methods, interpretation of the results.

This approach of the statistical expertise, has been the basis of the M U S E architecture design.

III.1 - M U S E ARCHITECTURE

MUSE consists of five modules (figure 1), each module is split into three parts. The first one represents the name of the module. The second one describes its functions. And the last one determines the different links towards other modules.

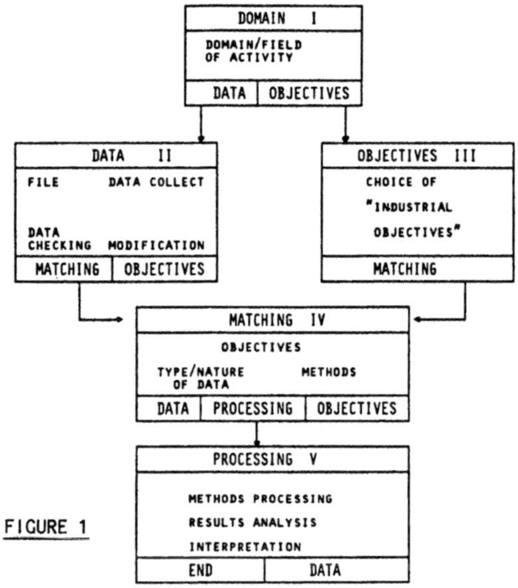

FIGURE 1

1 - "DOMAIN" MODULE

It is devoted to the initialization of the expertise : providing user's identification and determination of the application domain. This module is also intended to be used in determining the data to be collected. The user may have either his own data file or a direct access to the data bases. The knowledge representation is in the form of production rules.
example : "If the domain is the Personnel Department THEN the name of the data base is PERSO".
The information related to the data bases is encoded in a "frame" form."Frames are data structures used to represent a stereotype of objects or concepts. This data structure is made up of "slots" which contain information about the various aspects of these objects or concepts". Each of these frames contains information about the description of the data base (address, extraction process).
This information about the domain of application is transferred towards both the "DATA" and "OBJECTIVES" modules.

2 - "DATA" MODULE

After extracting the data base, the "DATA" module builds up the data file.
It checks when possible, the missing data and the outliers, and updates the data base. At the end of this stage, the data types are characterized. The knowledge necessary for data type determination is encoded in the form of production rules. This consists in two steps :

1 - The first one is aimed at determining if a character is either a variable or only an indentifier. These diagnosis rules are based upon the percentage of unique items of a character set.

2 - The second one acheives, from the results of the previous step, the determination of the data type. When a character is viewed as an identifier, it will be dropped out. Four data types are possible : quantitative, nominal, ordinal, dichotomic. The rules consist in computing the percentage of real values, character or binary items, missing or outlier values.
Examples of rule :
"If the percentage of real values is greater than 10 % THEN the variable type is quantitative".
"If the percentage of missing values is greater than 5 % THEN the variable must be dropped out".

All the results of these investigations are shown to the user for validation

and/or modification.
The data type may take a great part in the determination of both the industrial objectives and the statistical methods or strategies. Thus, connections towards the "OBJECTIVES" and "MATCHING" modules are set-up.

3 - "OBJECTIVES" MODULE

This step of the expertise is one of the most important and one of the most difficult of an E.S in statistics. The aim of this module is to do an easy determination of the industrial objectives. The domain of application being bounded to "IBM Montpellier", this target must be reached. During the exploratory analysis, many needs and industrial objectives were pointed out. Several elementary objectives were also defined. These elementary objectives were common to several industrial objectives and correspond to statistical methods.
Consequently, the industrial objectives may be viewed as a sequence of these elementary objectives. The concept of "SCRIPTS" is used in order to supply a better representation of this kind of concept.
"Scripts enable to describe and to represent a sequence of stereotyped situations or objects".
Example of script : 'opinion survey script' : (Data pre-processing, bidimensional analysis, situation analysis).
Thus, several scripts were built up and coded in this module.
The user can have either a list of scripts or a list of elementary objectives.
For a faster and a more user-friendly determination of the industrial objectives, two techniques related to the Artificial Intelligence field were used :

1 - Plausibility factors.
2 - Meta-rules (rules which specifically deal with how knowledge is to be manipulated).

The plausibility factors are associated to each script, and are based upon their frequency of use. They are updated after all the run sessions of the E.S. The scripts are proposed to the user according to their decreasing plausibility rank.
Two kinds of meta-rules have been coded in this module. These meta-rules manage the selection rules of the scripts. The first one enables only selecting of the scripts having specific elementary objectives. The second one forbids selection of certain scripts.
Examples : "The 'opinion survey' script can only be selected IF the applica-

tion domain is the Personnel Department and IF the data are p ordinal variables and q nominal variables".

"IF the domain is the Test Engineering Department THEN select scripts which contain the elementary objective : 'search of analogies'".

The scripts are selected in function of their plausibility, weighted by the meta-rules which can remove their pre-selection rank.
If the user refuses the predefined scripts, he can build up his own scripts through an other kind of meta-rules. These meta-rules act on the selection of the elementary objectives : a specific user or a specific domain requires specific elementary objectives.

Thus, this process of choosing a statistical method can be well suited after every environment change.

4 - "MATCHING" MODULE

Once the data type and the industrial objectives have been defined, the third component of a statistical expertise : the choice of the statistical methods or strategies, is activated. This choice is a function of two factors :

1 - The types of the data.
2 - The user's objectives.

This stage is performed by matching the objectives, the data types and the statistical methods. A good knowledge representation for the statistical methods is the frame one. It is well suited as the methods can be viewed as stereotyped objects with some attributes and properties. Each frame consists of 3 kinds of "slots" :

1 - Slots describing the conditions of application related to the method.
2 - Slots defining the aims of the method.
3 - Slots enabling the carrying out of the method.

MUSE selects all the possible methods whose slots match with the initial conditions of the expertise and displays them to the user : he can choose any one of them, and by default the first one.

5 - "PROCESSING" MODULE

After completion of calculations made with the methods selected in the previous module, the results are stored in order to be used in further studies. The interpretation of the results depends upon the statistical methods, so that specific sub-knowledge bases are built-up. At present time, the interpretations of both the Principal Components Analysis and of the Correspondence Analysis are under a validation phase. Their interpretations are broken down into 3 parts :

1 - Choice of the number of axes.

The rules are based on either the percentage of inertia or on the cross validation method generalized by JUNCA-HOLMES (1985).

2 - Stability.

3 - Description of the graphics.

The co-ordinates and the contributions of each points are translated into PROLOG lists, and MUSE manages these lists in order to determine the relationships between the points and enable finding typical patterns in the population of items :

- significant items.
- clusters of items.

III.2 - IMPLEMENTATION

MUSE is used in VM/CMS environment in order to have benefits of all the VM facilities and then, to make its building-up easier. The inference engine and the knowledge base are written in VM/PROLOG. Moreover this PROLOG enables usage of different tools like GDDM (Graphical Data Display Manager) for the graphics. The statistical programs and the mathematical applications are written in APL. Then, a link has been established to enable exchanges of information and computer capabilities.

REFERENCES

GALE A. (1985) : Knowledge representation in data analysis. Proceedings of the Fourth International Symposium, Data Analysis and Informatics, Versailles France.

HAND D.J. (1985) : Statistical expert systems : lecessary attributes. Journal of Applied Statistics, vol. 12, n° 1.

HAHN G.J. (1985) : More intelligent statistical software and statistical expert system : Future directions. The American statistician, February, vol 39, n° 1.

JUNCA-HOLMES S. (1985) : Outils informatiques pour l'évaluation de la pertinence d'un résultat en Analyse de Données. Thèse 3ème cycle, USTL, Place E. Bataillon, Montpellier, France.

Building Expert Systems with the Help of Existing Statistical Software: An Example

By P. L. Darius, Heverlee

Summary

This paper discusses the implementation of an expert system shell written completely in SAS (*). The features of the shell and its interface with the statistical procedures are described . The system provides a flexible tool for executing sequences of statistical analyses guided by a knowledge base in the form of rules.

(*) SAS is a registered trademark of SAS Institute Inc, Cary, NC, USA.

Keywords

expert system, intelligent statistical software, SAS

Introduction

Computational statistics can take advantage of artificial intelligence technology in many ways (see Hahn (1985) and the accompanying discussion). One of the possibilities is to combine existing statistical software with expert systems tools (Hayes-Roth et al (1983)).

Statistical software is currently available in many forms, ranging from libraries of independent subroutines to complete integrated environments. On the other hand, expert systems software requires knowledge encoded in a specific way (rules or frames), or expressed in a typical AI language like Lisp or Prolog. Few statisticians are familiar with these languages.

Compstat 1986

To combine expert systems technology with existing statistical software, two approaches can be taken :

1. write the expert system in an AI language and interface with more or less independent statistical routines
2. build the expert system within the environment provided by the statistical package (if this environment is rich and flexible enough).

Examples of the first approach are REX (Pregibon and Gale (1984), with Lisp and the S-system) and ADONIA (Malvache and Reissian (1986), with Prolog and data analysis programs in Fortran). In this paper we discuss a realization based on the second approach: an expert system shell written completely in SAS.

Production rules and statistical knowledge.

In many expert systems, knowledge is represented in the form of (production) rules. In our system, rules have the following format :

```
IF (attribute) IS (value)
 AND (attribute) IS (value)
 ...
 AND (attribute) IS (value)
THEN (attribute) IS (value)
```

An example :

```
IF type of test IS parametric t-test
 AND equality of population variances IS accepted
THEN test IS pooled t-test
```

In a set of rules, the clauses usually form a tree structure. Fig 1. shows such a structure: it is a small part of a rule set for doing t-tests (Darius (1984)). The system processes the tree in a way known as "backward chaining" (see Thompson and Thompson (1985) for a detailed description). It starts with the attribute designated as the goal of the search ("test" in fig 1) and tries to infer its value from other facts (the values of "type of test" and "equality of population variances"). To determine the latter, it continues its search upward, until it arrives at a clause which cannot be inferred from any rule ("purpose" or "equality of population variances"). The value of the attribute is then determined by asking the user or by computing.

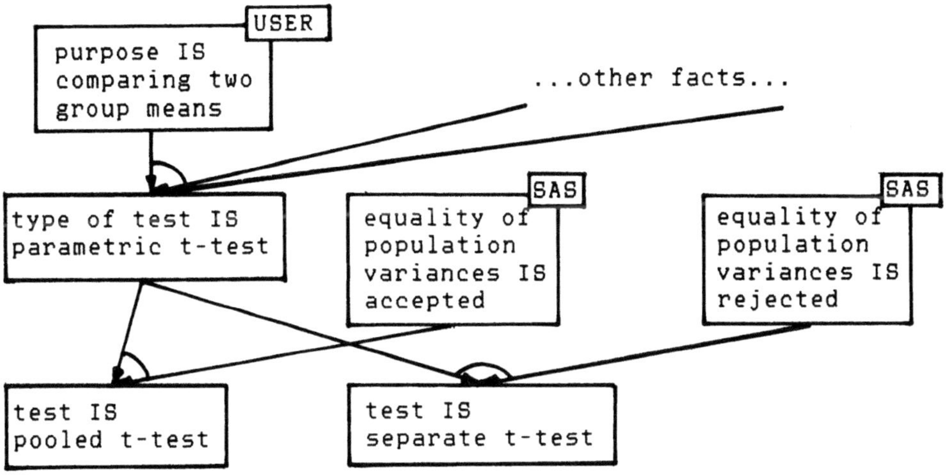

Fig 1 A graphical representation of some rules concerning t-tests

In such a system there are three classes of attributes:

1. Those for which the value should be inferred from the values of other attributes (e.g. "type of test" in fig 1).
2. Those for which the user must provide the answer (e.g. "purpose").
3. Those for which the statistical software must provide the answer (e.g. "equality of population variances").

Hence the software needed to build this kind of system will have three parts:
- an inference machine, guiding the search through the tree
- an interface to the user, asking questions, preferably in a user-friendly format
- an interface to the statistical software, able to put the software to work on a specific program and to interpret the results.

The next section describes how these parts have been implemented in SAS.

An inference machine written in SAS.

The inference machine is a piece of SAS code that operates on three user-provided SAS-datasets (respectively containing rules, the goal and prompts). It also maintains a number of utility SAS-datasets.

Currently, two versions are operational. The first one runs under release 82 of SAS and makes use of PROC MATRIX for its inference strategy and PROC FSEDIT for its interface with the user. There are also a number of DATA steps for initialization and writing of a final report. The whole is driven by a macro structure which activates each part at the appropriate moment. The code for the whole system has about 250 SAS statements.

In the versions of SAS following release 82, PROC MATRIX is no longer supported (SAS now markets an expanded version of PROC MATRIX as a separate product). A second version of the inference machine was written in the form of a DATA step with about 200 statements. It is currently operational under SAS Version 6 on a microcomputer. The new version makes extensive use of the windowing capacities in SAS Version 6. The user can see 4 windows simultaneously on the screen:

- the question window (where he can input the answer to a question asked by the system)
- the log window (which reports e.g. the facts that have been inferred, the user's answers to the questions and the reasons why they were asked... ; the accumulated content of the log window forms a report that can be consulted after the session)
- the output window (with output from the SAS procedures described in the next section)
- (optionally) a trace window (with the current state of the search through the tree).

The knowledge base needed to feed the inference machine consists of 3 SAS-datasets. Although SAS-datasets have a rather rigid structure (each observation must have the same variables and takes the same amount of space), they can be used to store the information needed here. E.g. in the rules dataset, each clause is coded as one observation, with one numeric variable (the number of the rule) and 3 character variables of fixed length: TYPE (with values IF, AND and THEN), ATTRIB and VALUE (with the names of the attribute and the value). The inference machine also maintains a SAS-dataset named CONTEXT, where it stores the values it has already found.

Connecting the inference machine with the statistical procedures.

The user interface within the inference machine's DATA step can cope with the attributes designated (see fig 1) to be asked to the user. Questions asked to SAS require the execution of other SAS steps, and hence must be handled by a separate piece of SAS code: the interface code.

The prompts dataset specifies, for each attribute, the questions to be asked. For questions asked to the user, it contains the question that must appear on the screen. For questions asked to SAS, it contains the piece of SAS code that should be executed, plus additional instructions.

The SAS code in the prompts dataset can be very simple (e.g. to do a linear regression) or more complex (e.g. the code to answer the question "is this design balanced"), and may contain an arbitrary number of steps. Before execution it is preprocessed, allowing substitutions based on the facts available in the context dataset. As one of the ways to examine the output, simple instructions are provided for extracting a string with a location relative to a known string (e.g. the value of the attribute "probvalue" is the string immediately to the right of the string 'PROB<W' in the output). All input and output is available for inspection afterwards.

Applications

The inference machine on itself can be used for any knowledge base that only asks questions to the user. It has been used for developing prototype knowledge bases in fields outside statistics (Maes et al (1986)).

Currently two prototypes of statistical applications are under development. One provides guidance with some aspects of regression analysis, using a strategy somewhat similar to the one used in REX (Pregibon and Gale (1984)). The other application mainly aims to protect users of ANOVA techniques against serious errors. Among other things, it checks if the design is balanced and whether there are empty cells, it helps in avoiding non appropriate comparisons of means, and it automatically generates plots for 1 and 2 way analyses.

Conclusions

The expert system shell consisting of the inference machine and the interface code forms a highly flexible and transparent tool for executing sequences of statistical analyses. It operates in an environment familiar to the user of the statistical package, and it is portable to all the computer environments that run the package.

References

Darius P.L. (1984), Expert Systems and Statistics, Proceedings SEAS Spring Meeting, Knokke Belgium April 9-13 1984.

Hahn G.J. (1985), More Intelligent Statistical Software and Statistical Expert Systems: Future Directions, The American Statistician, 39, 1-16.

Hayes-Roth F., Waterman D.A. and Lenat D.B. (1983), Building Expert Systems, Addison Wesley, Reading, MA.

Maes J. and Darius P., Vereecken H. (1986), Knowledge Processing in Land Evaluation, Workshop on quantified land evaluation, Washington D.C. April 1986.

Malvache P. and Reissian S. (1986), ADONIA: un systeme expert d'aide a la conduite d'une analyse de donnees, The 6th International Workshop on Expert Systems and their Applications, Avignon 28-30 April 1986.

Pregibon D. and Gale W.A. (1984), REX: An Expert System for Regression Analysis, Proceedings COMPSTAT 84.

Thompson B.A. and Thompson W.A. (1985), Inside an Expert System, BYTE April 1985, 315-330.

Statistical Data Validation and Expert Systems

By J. M. Dickson and M. Talbot, Edinburgh

SUMMARY

Automated procedures for validating experimental data as they are recorded are described. We outline the development of these procedures as an expert system for data validation.

KEYWORDS: data capture; data quality; data validation; expert systems.

1. Introduction

Data in many areas of research are now recorded automatically: observations are transferred directly from the measuring device in a computer-readable form. The use of programmable computers as data capture devices makes possible automated data scrutiny. Data can be validated as recorded, when errors are most easily rectified and unusual observations can be explored further.

The removal of the human element in data recording eliminates some errors, but other sources of error become important (eg. failure to stabilise a balance when weighing). These errors are more difficult to detect without the checks for reasonableness that a conscientious human recorder might apply. Such checks need to be replaced by automated procedures which can identify unusual observations.

Ideally, automated data validation should bring together the experience of recorders and of data analysts. The validation procedures should be dynamic, developing and learning from information recorded and from the operator's responses. In effect, the automated procedures should serve as an expert system.

Compstat 1986

2. Automated data validation

2.1 Purpose

A data validation system can be expected to monitor data for one or more variables as they are recorded, drawing attention to unusual values and, where possible, helping to identify the reason for discordancy.

Automated systems are likely to be required in circumstances where :

* long sequences of data are being recorded directly on a computer;
* the volume of data or its complex structure is such that it is not possible for an operator to easily identify discordant values by eye;
* an operator is present to investigate any unusual values which might arise;
* the discordant values have to be identified immediately if corrective action is to be taken;
* the consequences of error are non-trivial.

Such circumstances might arise in a laboratory of a research station where samples from many experiments are analysed. The samples from an experiment may be grouped by treatment factors or may represent readings taken over time on the same experimental units. Validation must take account of such structuring to make checks worthwhile.

2.2 Predicting new observations

To validate data as they are recorded, one needs a prediction function to give an idea of the value to be expected, and a credibility function to assess how unusual the observation is. Both these functions will be based on experience with past data.

The prediction function will estimate the value to be expected of a new observation. The appropriate function will depend on the structure of the data that are recorded. For data from a designed experiment with treatment factors, a predictor of a new observation might be the mean of all previous observations with the same treatment. Where no such observations are available, the predictor might be the mean of all previous observations. A prediction function for observations recorded over time might be derived from weighted linear regression of past observations, with weights decreasing exponentially into the past (Healy,1983).

2.3 Measuring the surprisingness of an observation

A credibility function will estimate the extent to which the recorded value may depart from the expected value without being considered unusual. The distribution of departures of observed from predicted will often take the normal form. However it should be possible to incorporate other functions. It is stressed that the credibility function is not being used as a test for outliers, but simply to measure the surprisingness of an observation.

The credibility function must allow for improvement in knowledge about variability as recording progresses. Account must be taken of the short-term experience of the current dataset, and of the longer-term experience of many datasets. This may be done in several ways. One is to use estimates of variance which are a weighted function of both long and short-term variation, eg.

$$s^2_i = s^2 + (i/c)\,(s^2_{i-1} - s^2)$$

where :

s^2 is the long-term variance of observed minus predicted,

s^2_{i-1} is the variance for the current dataset after i-1 observations, c is a large constant not less than i, but equal to i when there are sufficient observations in the current dataset on which to base estimates of variability.

When several variates are recorded on each unit, eg. a person's height and weight, it is desirable to look at all variates jointly to assess their credibility. One method is to use a discriminant function based on the long-term variances and covariances.

3. Expert systems and data validation

An expert system for data validation will comprise some of the following elements :

* rules for identifying discordant values and their possible causes, ie. the knowledge base;
* methods of reasoning from the rules to deduce likely causes of discordancy;
* a way of explaining its reasoning to the user;
* a means of gaining more information about the causes of discordancy and the frequency of occurrence.

The knowledge base is the prediction and credibility functions, details of the way in which errors may occur, and estimates of the a priori likelihood of their occurrence.

Methods of reasoning from rules to identify errors are likely to be relatively straightforward : low values of the credibility function will indicate an unusual observation. Occasionally it may be possible to and use the credibility function to point to possible sources of error. For example, when several variates are recorded and the joint credibility of the observations is acceptable but the credibility of an individual variate is low, then the individual variate can be pointed out as suspect. If the joint credibility is low then there is more reason to suggest that the unit of observation has been wrongly identified.

In some circumstances it may be possible to apply tests to identify specific sources of error. For example, the dry matter content of herbage can be assessed by weighing a sample before and after drying. If several samples are taken, the dried weights can be expected to vary linearly with the fresh weights. If two samples from the same treatment give very different dry matter percentages, then the sample which shows the largest deviation from the dry weight / fresh weight relationship is likely to be the most suspect.

Explanation of the reasoning can be given by linking credibility values to descriptive phrases, eg. "the reason for this is / almost certainly / likely to be / may be / unlikely to be / ...".

4. Implementation of an expert data validation system

The data capture facilities described by Dickson (1984) are being extended to include some of the features of the expert data validation system described above. Programming is in BASIC to permit implementation on small-scale computers. The validation system consists of several program modules, which can be linked with programs already written for data capture. Some modules are concerned with the prediction of observations from field crop experiments and with data recorded over time. Others deal with the assigning of credibility.

For each recording operation, stored information about the measures will include the expected mean, range, variation within and between treatments, and covariation between measures. For each of the possible causes of error the program holds an estimate of the probability that the event will occur, pointers to the checks that can confirm or deny the event, and a textual explanation of the cause.

At the start of a recording session the program retrieves details of the data, including experimental plan or previous observations. As recording proceeds the program, having identified the unit to be recorded next, estimates its expected value and compares this with the value recorded to obtain a measure of credibility. A coded form of the credibility measure is printed alongside the recorded values. At any stage the operator can ask the program for more details on the reported credibility measure.

5. Conclusion

A tacit assumption in data analysis is that data are what they purport to represent. Validation aims to provide assurance. However, until recently, the statistician could *"do no more than act as a detective, identifying the flaws and assembling the evidence that something is wrong, but not correcting "* (Finney,1974). One of the benefits of automated data collection is that statistical expertise can now be used positively in the prevention of errors. If these potential benefits are to be realised, statisticians must contribute to future developments in small systems computing and expert systems.

References

Dickson J.M. (1984), Data capture and validation using portable terminals, Proceedings of COMPSTAT Symposium, 1984.
Finney D.J. (1974), Problems, data and inference, J.R.Statist.Soc., A, 137, 1-23.
Healy M.J.R. (1983), A simple method for monitoring routine statistics, The Statistician, 32, 347-349.

Knowledge Base Supported Analysis of Longitudinal Data

By K. A. Froeschl and W. Grossmann, Wien

It is the aim of the paper to describe an expert system supporting the - at least moderately experienced - user in the analysis of longitudinal data. The system's knowledge about typical statistical processing strategies for such kind of data is used interactively to derive an action plan suitable for a given situation. The basic design features of the system are discussed along with some considerations on the actual decision processes. A brief example and a few implementation issues are presented, too.

Keywords: Analysis of longitudinal data; statistical expert system; computer aided data analysis.

1. Introduction

This paper presents the concepts of an automated statistical consulting system designed for the analysis of longitudinal data (LDA) emphasizing nonparametric techniques. LDA is a well-suited example to fit into the context of expert system technology for several reasons :

- (R1) statistical methodology is sufficiently complex to justify qualified guidance
- (R2) this complexity encourages the exploration of dedicated techniques for the design of a statistical expert system
- (R3) formalization of analysis strategies is a vivid statistical research topic.

The creation of a computer program doing statistical data analysis satisfactorily requires a careful investigation of the spectrum of occurring statistical problems as well as the ways and means statisticians actually solve these problems. It basically turns out (also cf. Hahn (1985)) that very often the statistician is faced with a client's rather vague problem statement which has to be expressed in precise statistical terms. A basic statistical structure and an appropriate computation procedure must also be determined. In order to answer the original problem the statistician then performs the analysis and

Compstat 1986

interpretes its results.

The benefits a statistical expert system can offer to the applied statistician are therefore :

(B1) drawing attention to (tacitly) introduced model assumptions which may influence the significance of results

(B2) giving advice to improve quality of analysis (e.g. by reporting bad data quality, extreme skewness, ...)

(B3) drawing attention to peculiar results.

The following paragraphs describe an attempt to develop a system for LDA meeting these design criteria.

2. Statistical Design Issues

The formalized statistical expert knowledge contained in the proposed program comprehends a finite set of "analytical models" (AMs). Each AM represents a typical scheme of statistical analysis applicable to both a particular type of observation data, and particular statistical assumptions. From this predefined set of alternatives one AM has to be chosen in order to answer a given statistical problem. The selection of an appropriate AM is effected by comparing the problem's characteristic features to the corresponding attributes of AMs. These attributes are grouped into five partitions, or submodels :

(M1) investigation model : defines the appropriate type of statistical problem formulation, i.e. summary (estimate), comparison (testing), or relationship between variables

(M2) data model : defines the structure of supplied data in rather qualitative terms

(M3) statistical model : defines a conceptual model, its mathematical setting, and error structure

(M4) computational model : defines a sequence of computing procedures to carry out the calculations

(M5) interpretation model : defines the mode of displaying and commenting statistical results.

The submodels (M1) through (M3) generate a "combinatorial lattice" of possible "basic models" (BMs). With each BM is associated a small set of algorithmic procedures which are complete sequences of analyses including interpretation of results.

In case of application a set of attribute values corresponding to the various attributes within each of the five submodels has to be determined and compared to the typical patterns of attribute values of the candidate AMs. Thus by stipulating a perfect specification of an AM this method resembles the procedure of taxonomic object identification: the "total" space of analyses is divided into several classes by means of attributes; within each class of analyses there exists a recommended strategy of analysis which is represented by an AM. The system´s goal then is to assign any particular analysis to one of the predefined classes, i.e. to select a recommended AM.

The critical part of system design is the decision structure for finding an adequate BM. In the development of the system´s current version the adopted attitude is :

(A1) substantial knowledge for determination of a BM is supplied by the client directly instead of deriving it by logical or statistical inferences, or data inspection, respectively, as it is done e.g. in the REX system (Pregibon, Gale (1984))

(A2) it is reasonable to suppose that in any practical situation at most one AM can be chosen unambiguously.

These severe restrictions are imposed for pragmatic reasons. Later on they will be replaced by more sophisticated methods for selecting a proper AM, e.g. by employing a fuzzy logic calculus, Baysian rule, or some kind of scoring mechanism to evaluate the appropriateness of AMs, as decision procedure as well as by extending the nesting depth of the inference network.

3. Decision Process and Plan Selection

Within the described system a complete data analysis consists of two major stages : determination of a BM, and performing of computations associated with the chosen BM.

After program initialization the system calls for a dataset as primary input. This dataset may be contained in a prepared datafile, or entered interactively. In a next step the dataset is submitted to a validation process, i.e. mainly a syntactic analysis of data structure, similar to data checking in conventional statistical software. After data validation a trimmed nonredundant dataset is available.

Now the system tries to settle a BM assigned to a valid combination of an investigation model, a data model, and a statistical model. The BM is determined partly by asking the client, partly by direct inspection of the prepared dataset to gather several attribute values (cf. section 2).

The relation between the BMs and the algorithmic procedures is mapped into a kind of IF ... THEN ... - rules. The conditional part of each rule is a conjunction of characteristic attribute values for one BM as defined by the combinatorial lattice mentioned above. The THEN-parts of the rules refer to the computational/interpretation models comprising the sequences of executable statistical programs. As an example consider the following verbalized rule :

```
IF   principal_task_of_analysis is estimation_of_curve
  AND data contain replications
  AND aggregation_of_replications is requested
  AND number_of_aggregated_observation_data exceeds 50
  AND computing_model is nonparametric
  AND functional_model is standard
  AND error_model is standard
THEN use kernel_estimate_type_2
```

In order to infer a proper computational model the system searches for a feasible rule, i.e. one with satisfiable conditions. In general there are few, but more than one, computational models tied to a rule. Again the choice among these competing models involves asking additional questions or, occasionally, performing some simple tests on input data. Thereupon the tuning parameters ("options") of the statistical algorithms have to be initialized interactively, thereby yielding a precise action plan to carry out. Finally a compatible interpretation model is selected analoguously to the computational model. After setting its tuning parameters the obtained interpretation plan which is actually a self-contained program is applied to the numeric results already computed.

4. An Example

In order to illustrate the way the system communicates with a client in an example session, a rough sketch of posed questions is given. With exception of posed questions and requested explanations any other system output is suppressed. Submodel references are inserted for clarification.

```
(investigation model)
  1. type of statistical problem ?   "estimation of curve"
(data model)
  2. should replications be aggregated ?      "yes"
(statistical model)
  3. type of computing model ?       "nonparametric"
  4. type of functional relationship ?  "normal regression"
  5. kind of error structure ?   "independent homoscedastic"
(validation of selected BM)
  6. accept suggested kernel estimate ?        "why?"
     -- if you wish to estimate a fitting curve
        using nonparametric regression under
        standard statistical assumptions and
        the number of observations exceeds 50 then
        the recommended procedure is kernel estimate; ... ? "yes"
(computational model)
  7. preferred kernel function ?               "none"
(validation of selected computational model)
  8. accept suggested quadratic kernel function
     incl. automatic calculation of band-width
     by means of cross-validation ?               "why?"
     -- if the fitting curve (without any
        derivatives) is to be estimated alone
        and the number of observations is greater
        than 50 and less than 200 then a
        quadratic kernel is recommended; ... ?   "yes"
(set parameters of selected computational model)
  9. type of result display ?                   "graphical"
(interpretation model)
 10. summaries to display ?   "mse-estimation, bias-estimation,
                               plot of confidence band"
 11. explain any statistic ?  "confidence band"
     -- the confidence was calculated as a local
        .95 band; the variance was calculated
        according to (Rice, 1984); increase of
        confidence band-width at the boundaries
        of observation interval is typical and
        should not injure your interpretation;

 ........
```

5. Implementation

Up to now the primary research efforts have been dedicated to the development of an adequate logical system structure, whereas the numerical algorithms are not yet fully integrated. The logical part of the program system, i.e. the inference and decision making procedures, will be realized in a carefully tailored expert system framework currently developed in PROLOG. The various numerical tasks the system takes charge of are solved by means of a collection of FORTRAN77 subroutines borrowed mainly from the SPASP-package (Grossmann, Pflug (1982)). The knowledge base of the system is therefore composed of

(K1) a set of logical sentences expressing the intrinsic statistical knowledge for deciding what to do and

(K2) a library of numeric routines virtually realizing the calculations of statistics.

The integration of logical and numerical components demands precise interface definitions and fine tuning of algorithmic procedures to enable satisfactory cooperation of these componentsand to support the system's explanation facilities. That will be the focal point of further work.

References :

Grossmann, W.; Pflug, G.Ch. (1982)
"SPASP - A Statistical Program for the Analysis of Stochastic Processes", in: H. Caussinus et al. (eds.), "Proceedings of COMPSTAT 1982", Part I, Physica Wien, pp. 254-259

Hahn, G.J. (1985)
"More Intelligent Statistical Software and Statistical Expert Systems: Future Directions", in: The American Statistician, February 1985, Vol. 39/1, pp. 1-8

Pregibon, D.; Gale, W.A. (1984)
"REX: An Expert System for Regression Analysis", in: T. Havranek et al. (eds.), "Proceedings of COMPSTAT 1984", Physica Wien, pp. 242-248

How to Assist an Inexperienced User in the Preliminary Analysis of Time Series: First Version of the Estes Expert System

By P. Hietala, Tampere

Summary: In this paper we describe the first version of a statistical expert system, called ESTES (Expert System for TimE Series), intended to assist in the preliminary analysis of time series. The ESTES system provides guidance for an inexperienced time series analyst in detecting the following essential properties of time series: trend, seasonality, level shifts and outliers. The system can utilize user's own knowledge of the time series being considered and can also give instructions to the user if he doesn't know or is not sure what is good way of dealing with that particular time series.

Key words: Expert system; Time series; Trend; Seasonality; Level shifts

1. INTRODUCTION

Statistical software packages are nowadays widely available, but a statistically unsophisticated user has a danger to use these packages incorrectly because of the minimum amount of guidance offered by the packages. The necessity of more user-friendly, and above all, more intelligent statistical software is widely recognized. For example, Chambers (1981) expressed this as follows: "Here is the moral imperative: Statistical software in its present form, made widely available by cheap computing, will participate much uninformed, unguiding and simply incorrect data analysis. We are obliged to do something to help. ... I have argued here that expert software for data analysis would respond to an important need."

Several expert systems for data analysis have been developed after Chamber's ideas. The best known include REX (e.g. Pregibon and Gale (1984)) and STATPATH (Portier and Lai (1983)). The REX system deals with a specific statistical problem (a linear regression analysis), while the STATPATH system is developed to solve a more general problem: the selection of a statistical method. However, in spite of some promising results in these expert systems for statistical data analysis, this has not lead, at least to our knowledge, to commercially available statistical expert systems.

In particular, for *time series analysis* there are very few expert systems. The STATXPS system (Prat, Marti and Catot, 1985) is "an intelligent interface between the final user and some statistical packages, which guides, instructs, and helps the final user in analysing data and in the interpretation of the most relevant statistical results, ...". STATXPS system includes a host computer, a microcomputer and several general statistical software packages.

However, an inexperienced user will most probably prefer to use a stand-alone personal microcomputer, instead of a complicated computer configuration. Unfortunately, most of the

Compstat 1986

currently available statistical software for personal microcomputers is but a slight modification of packages for large mainframes, not actually making use e.g. of the up-to-date interface capabilities provided by the personal microcomputers, or the advanced methods in statistics, e.g. statistical expert systems.

On the other hand, it is evident that there are benefits in providing an uniform interface for general statistical software (as the STATXPS does) if the user is statistically sophisticated. Then the user has a lot of methods available through an uniform interface but he does not have to learn to use each and every statistical software package included. However, for a *statistically unsophisticated user* the advantage of such system is not so evident because he does not need so much a heavy repertoire of methods but more often only instructions on a very basic level.

The use of general time series packages (see e.g. TSPACK Chaghaghi (1985)) often places severe restrictions on the form of the analyzed time series and presupposes also detailed statistical knowledge from the user. For example, if the user intends to build an ARIMA-model (McLeod (1983)), the time series must be stationary (i.e. there should be no trend, no seasonality, etc.). However, in "real life" it is common that time series are not so well-behaved. Furthermore, the user of a general time series package must be aware of relevant terminology (e.g. the concept of trend, seasonality, etc.). These systems also require that the user is able to discover the essential features of the time series by himself. After that he must know how these features influence the model building process and what to do with them (e.g. how to remove trend or seasonality). Note that for an experienced time series analyst the preliminary analysis is not very difficult: usually a graphical representation of the time series can reveal most of essential points needed and will often indicate the best way to proceed with the analysis.

In this paper we describe the first version of a statistical expert system ESTES intended to attack the questions described above. First of all, the system assists in the *preliminary analysis of time series*. Moreover, our system is operating independently on a commercially available personal microcomputer (Apple Macintosh™) without other equipments or other software. In this paper we focus on the up-to-date user interface and explanation capabilities of the ESTES system, which in our opinion are essential features especially for an inexperienced user. This first version of the ESTES system is currently being extended to comprise a more full-fledged expert system.

This paper is organized as follows. In section 2 we give an overview of the system, after which we discuss in more detail its two-dimensional user interface. In section 4 we focus on the explanation capabilities of the system, especially on the concept of visual lexicon. We conclude with a few remarks on the implementation status of the system.

2. OVERVIEW OF THE ESTES SYSTEM

Our basic assumption is that the user of the ESTES system is statistically quite inexperienced: he doesn't know much about methods used in time series analysis, but he might have non-statistical knowledge concerning of environment of the model building problem. First of all we suppose that the user doesn't want to have a "blackbox system" with input and incomprehensive output. We hope that he likes to learn something about model building process and want to use his own knowledge. These assumptions about the user lead us to implementing the system without any automatical features: system analyzes the data and gives instructions, but does not make any real decisions. We agree with Velleman (1985) in that "the goal of an expert research assistant program would be a system that would serve as an aide or a colleaque to the data analyst without taking control of the analysis".

The main idea behind the ESTES system is to provide assistance in dealing with trend, seasonality, level shifts, and outliers. The basic procedure of handling each of these four properties is as follows (will be illustrated with more detail in sections 3 and 4 of this paper):

(1) The user selects the property he is interested in. Next the system tries to find out what the user knows about this property corresponding to the time series being considered.
(2) The system analyzes the time series (the thoroughness of this analysis depends on the user's level of knowledge of the property). In all cases the system informs the user of the actual (or probable) situation, e.g. confirming user's views on the property, or suggesting more likely alternatives to him. The system also gives the user appropriate instructions on how to proceed. However, it is the user who eventually makes the decision how to continue the session.
(3) After completing the treatment of one property (this might include making the necessary transformations on the time series), the user continues the session from (1) with the next property or exits the session. In each of the above phases the user can request for an explanation of a term unfamiliar to him. For this a "visual lexicon" is provided: the ESTES system explains an unknown term with a short definition and displays (if possible) typical representations of the term in graphical form.

3. USER INTERFACE

When designing the ESTES user interface we have drawn heavily on the modern style of interaction with high-resolution bit-mapped displays exemplified by the XEROX Star™ and Apple Macintosh™. This "what-you-see-is-what-you-get" way of interaction (also called as the direct manipulation paradigm (Shneiderman (1983)) seems to suit very well an inexperienced user and is currently becoming a standard on modern personal microcomputers.

Figures 1 and 2 illustrate how the user interacts with the ESTES system. Assume that the user has already provided the system his data and possibly edited it to a suitable form (all inside the ESTES system). In Figure 1 he is analyzing this data (a time series) the values of which can be seen in a *window* on the screen named "data". A graphical representation of the time series is displayed in another window named "time series". At the upper edge of the screen there is a menu-bar which contains the basic *menus* of the ESTES system. The user has just pulled down the property menu and selected a menu item called "seasonality".

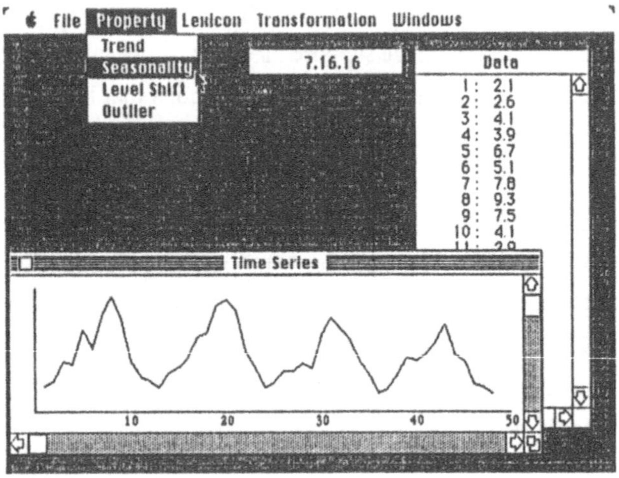

Figure 1. Windows containing two representations of the data.The user can resize or reposition windows at will.

The selection of an alternative out of the property menu creates a new window onto the screen. This "dialog" window is seen in Figure 2. Using the dialog window the user can convey to the ESTES system his current knowledge about the seasonality of the time series being considered. In our example the user knows or supposes that the time series contains a single seasonality with period 12 (he has clicked the alternative "single seasonality" and filled in "12" onto his dialog window). Note that also several periods can be given in the case the user selects either the option "multiple seasonality" or the option "other cyclic changes".

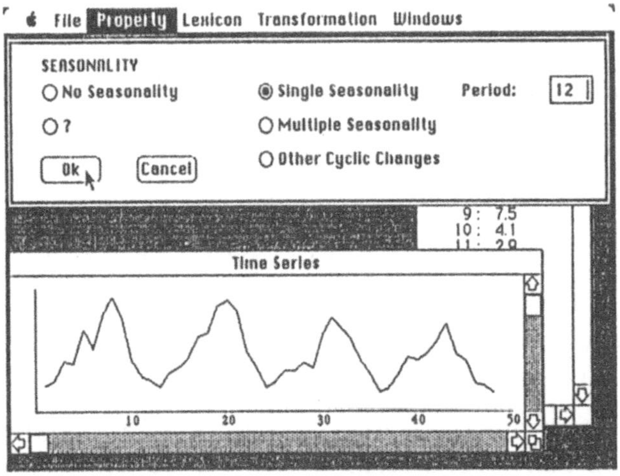

Figure 2. A dialog between the user and the ESTES system for explicating the knowledge of the user about his time series.

After the user has signalled the end of the dialog by clicking the soft key "OK" the system starts to analyze the time series taking into consideration the level of the user's knowledge about the property. For example, this means that in the session outlined above the analysis process will not be as thorough as if the user would have indicated that he knows nothing about seasonality of the time series being considered (the alternative "?" in the dialog window).

As an outcome of the analyzing phase the ESTES system displays to the user the results it has computed and how these relate to the knowledge the user possesses. However, the user has always the final choice whether to proceed further on using his own knowledge or to accept the changes suggested by the system. So the system just informs the user about how well the tests it has carried out on the hypotheses given by the user support these. We see this kind of procedure of making explicit as well as validating the acquisition of user knowledge very important but much neglected area in statistical software packages.

4. EXPLANATION CAPABILITIES

As our first approximation of explanation capabilities typical to expert systems we have included into the ESTES system a *visual lexicon* (Pregibon and Gale (1984)). In our system this means that the user can ask for an illustration of unknown terms. An unknown term may be an elementary concept such as "time series" or "seasonality", or it may refer to the results produced by the ESTES system. Explanation of a term includes a defition and a graphical representation of the phenomenon in question. Our main objective has been to design the lexicon so that it contains as little as possible precomputed information about the graphical representation. Only some generation rules are usually needed. This kind of data generation makes it possible to produce several different representations of a phenomenon.

Suppose the user has a window similar to that of Figure 1. If he does not know what the term "seasonality" means, he can choose from the "lexicon menu" an item called "basic item". Then the system displays a list of all the basic items out of which the user can select the term "seasonality" to be explained. Figure 3 illustrates the results of this query. There the ESTES system has generated an example seasonal time series with the period of 4. If the user desires also other examples of seasonal time series can be displayed. Moreover, it is possible for the user to save a time series generated this way and use it later on as if it were his own data. We think that these features are very important in supporting an inexperienced user in his work.

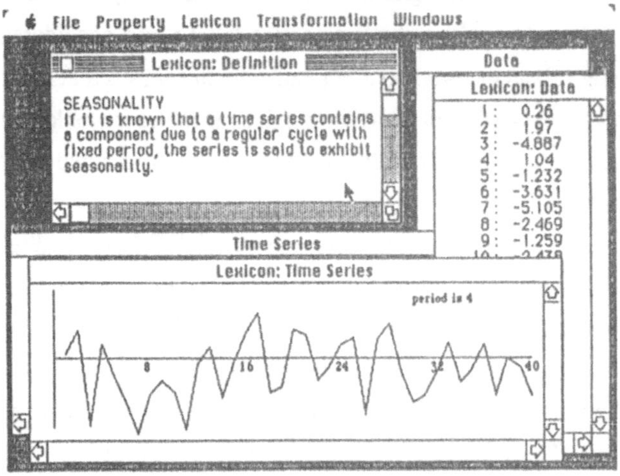

Figure 3. The visual lexicon approach to explain unknown terms to the inexperienced user.

We intend to extend the above described explanation capabilities in the future. For example, it would be useful if the system would be able to explain why a certain operation should be preferred in a particular situation.

5. IMPLEMENTATION STATUS

The current version of the ESTES system must be considered only as our first approximation of an expert system for time series. It will be further extended in the near future to include more complete domain knowledge (mechanisms to deal with time series) as well as enhanced inference and explanation capabilites. The initial part of the system described in this paper has been implemented using the language Pascal.

REFERENCES

Chaghaghi, F.S. (1985), Time Series Package (TSPACK). Lecture Notes in Computer Science 187, Springer-Verlag, Berlin.

Chambers, J.M. (1981), Some Thoughts on Expert Software. Proceedings of 13th Symposium on the Interface of Computer Science and Statistics, Springer Verlag, New York, 36-40.

McLeod, G. (1983), Box-Jenkins in Practise, Vol 1. G.McLeod and Gwillym Jenkins & Partners Ltd., Lancaster.

Prat, A., Marti, M. and Catot, J.M. (1985), Incorporating expertise in Time Series Modelling: The STATXPS SYSTEM. Statistical Software Newsletter, Vol. 11. No. 2, 55-62.

Pregibon D. and Gale, A. (1984), REX: an Expert System for Regression Analysis. Proceedings of Compstat 84, Physica-Verlag, Wien, 242-248.

Shneiderman, B. (1983), Direct manipulation: a step beyond programming languages. IEEE Computer, Vol. 16, No. 8, 57-69.

Velleman, P.F. (1985), Comment on G.J. Hahn's "More Intelligent Statistical Software and Statistical Expert Systems: Future Directions", The American Statistician, Vol. 39, No. 1, 10-11.

Object-Oriented Data Representations for Statistical Data Analysis

By R. W. Oldford, Waterloo and S. C. Peters, Berkeley

Summary

We discuss the design and implementation of object-oriented datatypes for a sophisticated statistical analysis environment. The discussion draws on our experience with an experimental statistical analysis system, called DINDE. DINDE resides in the integrated programming environment of a Xerox Interlisp-D machine running LOOPS. The discussion begins with our implementation of arrays, matrices, and vectors as objects in this environment. We then discuss an additional set of objects that are based on statistical abstractions rather than mathematical ones and describe their implementation in the DINDE environment.

1.0 Introduction

We have built an experimental statistical analysis system, called DINDE, that provides the professional statistician with new software tools for statistical analysis and its management. The analysis tools differ from the usual ones in that each contains information on how and when it is used in accepted practice. This information is then used by DINDE to provide unobtrusive guidance to the user and to allow the analysis to be meaningfully managed as it progresses (see Oldford and Peters, 1985, 1986ab, for details).

Such tools are most effective when they can also access the data information necessary for their successful interpretation. For example, if each point in a plot represents an individual in a dataset, then access to all information recorded on that individual should be available to the user, preferably by simple interaction with the displayed point. The additional information could be very important in interpreting interesting features of the plot. Conversely, if the plotted point is unusual in some respect, the user may wish to record such information directly on the individual, perhaps by marking that individual for highlighting in all plots, or by actually adding new information to that already recorded for the individual.

To provide easy and immediate access to such information requires a careful and flexible representation of the basic statistical data. In DINDE, we have implemented one such representation using object-oriented programming (e.g. see Stefik and Bobrow, 1985). The object-oriented programming language used here is LOOPS embedded in Interlisp-D (Bobrow and Stefik, 1983, and Teitelman and Masinter, 1981). However, the abstractions described here can be implemented in any object-oriented language that allows multiple inheritance (i.e. allows objects to inherit properties from more than one other object).

In what follows, we discuss the data abstractions and their current implementation in DINDE. The next section describes our implementation of arrays as objects, and how we think they should be used in a statistical analysis system. Sections 3, 4, and 5 present data representations based entirely on statistical, as opposed to mathematical, abstractions. The order of presentation is from the bottom up: from basic data abstractions in section 3 (**Individuals**, **Variates**, and **Datums**), to compositions of these in section 4 (**Batches**, **Cases**, **Factors**, and **DataSets**). Some implementation details are then given in the fifth section. In the last section, we make some brief general comments on the new data representations and possible extensions.

2.0 Arrays and TreeStructures

Arrays are typically used to store data in statistical systems. With each element corresponding to a data value, they provide an easy and unique way to reference the data. In DINDE, an array is implemented as an object, called **Array**. As such, each **Array** can respond to a number of messages that are appropriate to an array. For example, an **Array** may be sent the **Reexpression** message with an arbitrary scalar function as an argument, the **Reexpression** method applies this function to each element in the array, stores the result as an element of a new **Array**, and returns the new **Array**.

While matrices, vectors, and scalars are simply special kinds of arrays with 2, 1, and 0 dimensions, respectively, they are used enough to merit distinct representations as **Matrix**, **Vector**, and **Scalar** objects . We arrange for each of these objects to have the same behaviour as more

Compstat 1986

general **Arrays**, by explicitly declaring each to be a specialized **Array**. Figure 1(a) shows the resulting inheritance pattern (parent to child = left to right).

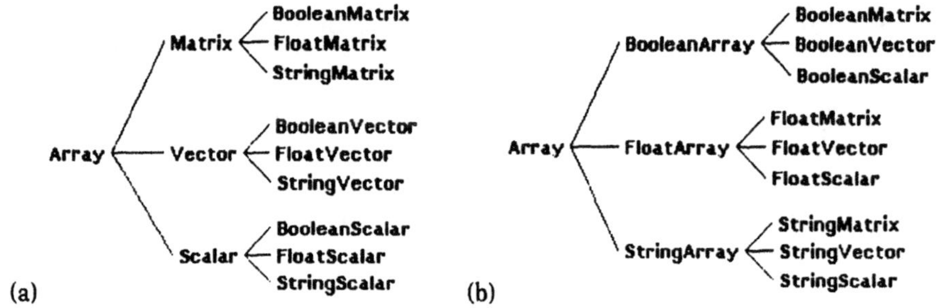

Figure 1. Array inheritance structure

Every message understood by **Array** is also understood by **Matrix**, **Vector**, and **Scalar**. However, any behaviours specifically attached to **Matrix**, say, will not be reproduced in **Array** - the inheritance flows in one direction only. Thus, **Matrix** is an **Array** having exactly two dimensions and the additional ability to understand new messages like Transpose (a somewhat ambiguous message for the more general **Array**).

Similarly, we distinguish arrays whose elements are booleans, numbers, or strings, by introducing three new specializations of **Array**: **BooleanArray**, **FloatArray**, and **StringArray**. Figure 1(b) shows the corresponding inheritance structure. Each has special behaviours that do not belong to generic array structures. **FloatArray**, for example, responds to a SQRT message by taking the square-root of each of its elements and returning a new **FloatArray**.

Together with **Matrix**, **Vector**, and **Scalar**, these new objects are used to construct the matrices, vectors, and scalars usually found in statistical systems, shown here as the rightmost nodes in either tree of figure 1. For example, **FloatMatrix** is both a kind of **Matrix** and a kind of **FloatArray**, and thus inherits behaviours from both.

FloatMatrix, **FloatScalar**, and **FloatVector** are especially important since they correspond to mathematical abstractions. As such, their behaviours are identical to their mathematical counterparts. For example, **FloatMatrix** can be multiplied by another to return a third. Via the messages they understand, these three kinds of **FloatArrays** give the user the basic functionality of a matrix language.

For some messages received by a **FloatArray**, it might be natural to return several objects. The SVD message of a **FloatMatrix**, for example, calculates three **FloatMatrix** objects as the singular value decomposition of the original. However, it returns a single object called a **TreeStructure** which has the three **FloatMatrix** objects as components. **TreeStructures** are entirely analogous to the general data structures of S (Becker and Chambers, 1984), and, as in S, can be used to build data structures of arbitrary complexity.

It is straightforward to incorporate statistical procedural information into the data representations. Statistical methods, like numerical methods, could be attached to the appropriate data objects and activated by sending messages. For example, a **FloatVector** could construct a histogram of its elements in response to a Histogram message.

After implementing this approach, we soon discovered shortcomings. For example, we thought it important to be able to label observations in histograms. But this information was not part of the numerical vector, rather it was contained in a separate **StringVector**. This led to a proliferation of logically related vectors that were unconnected in either our data abstractions or software.

Our first solution was to make these connections explicit by using **TreeStructures**, with **Arrays** as components, to represent the statistical data. To associate string labels with individual data values, one need only create a **TreeStructure** with two components: one called Data to point to a **FloatVector** and another called Labels to point to the matching **StringVector**. Alternatively, we could have specialized **Arrays** further, say to **StatisticalArrays**. Statistical methods could then be attached to the new objects as messages. Either way, the approach soon leads to complex structures that become difficult to manage.

It seemed to us that the difficulties were caused by using arrays as the basic building blocks. The vector abstraction failed to adequately capture important data characteristics like range, or distinctions between numbers that represent categories and those that represented measured values, and so on.

In point of fact, except for the mathematics, there is no statistical reason to work with arrays. Indeed, statistical information distinguishes a datum from a number. Hence, it would seem that such information naturally belongs attached to a statistical abstraction much smaller than an array.

Our present solution, which we now describe, begins with very small grain statistical abstractions and goes on to build larger grain ones by composition.

3.0 Basic elements of statistical data

The basic statistical data are the observations, where each observation consists of a value of a variate on some individual. Multivariate observations are just many of these recorded on the same individual. This suggests three basic elements: the individual on which the measurement was taken, the variate measured, and the measurement or data value itself. Each of these is a statistical abstraction that can be represented as a distinct kind of object - an **Individual**, **Variate**, or **Datum**.

An **Individual** object represents the individual on which measurements *can be* taken. In a clinical trial this might represent an actual person; in an economic model it might represent the year 1974, or a nation. Each **Individual** contains descriptive information peculiar to that individual (Name and DescriptiveText) and can have arbitrary information added to it at any time (AddedInfo). It does not include any measured values.

Similarly, a **Variate** object represents a variate which *can be* measured, not the actual recorded values. Like **Individuals**, **Variates** contain basic descriptive information (Name and DescriptiveText) and can have information added at any time (AddedInfo). Both inherit these abilities from a more generic object called **Identifier**. Figure 2 shows the complete inheritance structure.

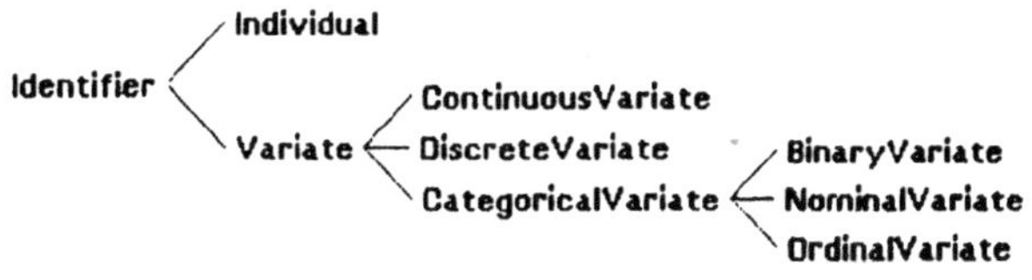

Figure 2. **Identifier** inheritance structure

Variates contain additional information specific to a variate, like the natural **Range** of the variate, and are specialized to represent the kinds of variates found in data analysis, namely **ContinuousVariate**, **DiscreteVariate**, and **CategoricalVariate**. Measured data values on either of the first two must be numbers, reals for **ContinuousVariates** and integers for **DiscreteVariates**; on the third they are categories. **CategoricalVariates** differ from the others in that their **Range** is a list of possible values (e.g. ("Favour", "Indifferent", "Oppose") or (1, 2, 3, 4)). **CategoricalVariate** is further specialized into **OrdinalVariate** and **NominalVariate**, depending on whether the order of the **Range** list is meaningful. **BinaryVariate** is included because of its frequent occurrence in practice.

A **Datum** object represents the value actually recorded. It is meant to contain all the information related to that recorded value but none of the information on the variate or individual on which it was measured. It represents only the measurement itself. As such, a **Datum** contains three distinct pieces of information: the DataValue (a number or a string), the Censoring (Uncensored, Left, or Right), and the number of SignificantDigits, if applicable. It might be argued that the units of measurement should also be attached to each **Datum**. However, for most datasets, the units do not change from one data value to another on the same variate (cf. significant digits). Therefore, like Range, the units of measure are attached to **Variate** rather than **Datum**.

As an example, suppose a person, X, is known to be taller than 1.71 metres. This information is recorded using three distinct objects. An **Individual** is created and given the name "X". On that object, we record whatever information we have about X - X's date of birth, address, and so on. A **ContinuousVariate** named "Height" is then created, where we record that this variate represents the height of humans, that the measurement is to be recorded in metres, and that the Range of measurements can be from 0.70 to 2.10. Finally, a **Datum** is created with DataValue = 1.71, Censoring = Right, and SignificantDigits = 3.

Note the economy of the representation. Should more variates be measured on X, no new **Individual** will need to be created. Similarly, if the height of a different person is measured, the same **ContinuousVariate** will do.

To be useful, a **Datum** must be associated with its **Identifiers**. This is accomplished by using an object called a **DatumAssociation**, which is a specialization of a more general object called an **Association**. **Associations** pair two arbitrary objects, identifying one of them as the Key, and the other as the Value, of the **Association**. A **DatumAssociation** requires that its Key be an **Identifier** and its Value a **Datum**.

Note that a **DatumAssociation** pairs a **Datum** with either a **Variate** or an **Individual**, rather than associating it with both its **Identifiers**. It turns out that this simple pairing provides a parsimonious way to build representations of more complex statistical abstractions.

4.0 DataRecords and DataSets

In practice, data are typically examined in the following conditional fashion: one kind of **Identifier** is held fixed, and then the **Datum** associated with each **Identifier** of the other type is examined in turn. In particular, given an **Individual**, we consider the **Datum** associated with each **Variate** measured on that **Individual**. This arrangement corresponds to a multivariate observation. The dual arrangement fixes a particular **Variate** and considers the **Datum** associated with each **Individual** on which the measurement was taken. This arrangement is just a sample of univariate observations.

The common aspects of these two data arrangements are collected together in a **DataRecord** object. As can be seen in figure 3, a **DataRecord** is a specialization of two other objects: a **Collection** and an **Association**. It is a **Collection** because each **DataRecord** contains a collection of **DatumAssociations** (either **Variate-Datum** or **Individual-Datum** pairs, but never both). From **Collection**, **DataRecord** inherits many useful behaviours, such as the ability to apply a function to each of its elements. It is special in that no two **DatumAssociations** are permitted to have the same Keys, so neither **Individuals** nor **Variates** may be repeated in a **DataRecord**. **DataRecords** are also special kinds of **Associations** - they associate a fixed **Identifier** (the Key) with a collection of **DatumAssociations** (the Value).

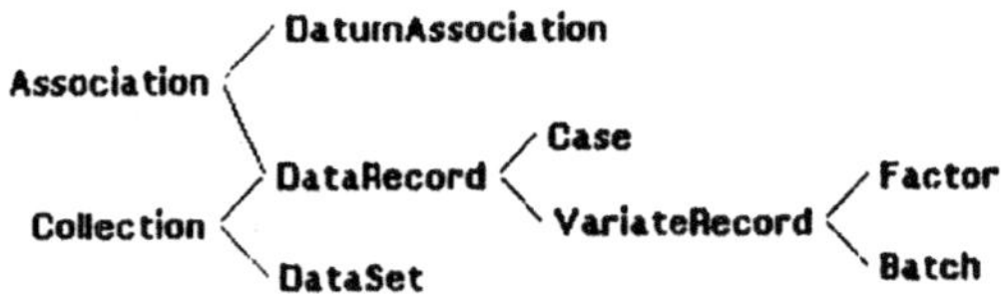

Figure 3. Inheritance structure for **DataRecords** and **DataSets**

DataRecords have many behaviours that are not inherited. First, they can be probed for information on the contents of every **DatumAssociation**. The message DataValueAt, with the appropriate kind of **Identifier** as an argument, will return the DataValue of the **Datum** associated with the specified **Identifier**. The message IdentifierAtDataValue causes the converse to happen - the first **Identifier** paired with a **Datum** having the specified DataValue is returned. There are many messages that operate in this fashion. Others, like HowManyCensored? and IncludesIdentifier?, perform simple tests. Second, an arbitrary function can be applied to every **DatumAssociation**, or **Identifier**, or **Datum**, or DataValue in the **DataRecord**. This allows one to easily craeate new behaciours for a **DataRecord**. Finally, **DataRecords** can convert themselves into some other kinds of objects. For example, the AsVector message will cause the appropriate kind of **Vector** object to be created - one whose elements are the DataValues of the **Datums** in the **DataRecord**. This means all the mathematical functionality of a **FloatVector** need not be redeveloped for a **DataRecord**.

A **Case** is a special kind of **DataRecord** that is meant to represent a multivariate observation. Its Key is an **Individual** and it contains only **DatumAssociations** that pair a **Variate** with a **Datum**. To this kind of object, behaviours that are appropriate for a single multivariate observation are attached. For example, a message which would cause a graphic like a glyph, or a cartoon face, to be constructed would be of statistical interest.

A **VariateRecord** represents a set of observations on a single variate. Its Key is a **Variate** and it contains only **DatumAssociations** that pair an **Individual** with a **Datum**. In addition to its inherited behaviours, a **VariateRecord** can access and set its Range, and might determine which **Individuals** have DataValues outside that Range. Before more statistically interesting behaviours can be added, the type of **Variate** needs to be specified.

We distinguish two different kinds of **VariateRecords** : **Batches** and **Factors**. The basic distinction is that the **Variate** of a **Batch** must be either a **ContinuousVariate** or a **DiscreteVariate**, while the **Variate** of a **Factor** is a **CategoricalVariate**.

Given this distinction, some statistical methods are now appropriate. For example, **Factors** might return all **Individuals** whose associated **Datum** had a specified DataValue, or the number of **Individuals** associated with each DataValue in the **Factor**. For **Batches**, there are many more possibilities, from the standard numerical summaries to displaying their contents graphically in histograms and boxplots.

Together, a number of different **DataRecords** constitute a dataset, which we represent by an object called **DataSet**. As figure 3 indicates, a **DataSet** is a special kind of **Collection**. A **DataSet** can be viewed as either a collection of **Cases** or a collection of **VariateRecords** - the difference should not be distinguishable to the user. Wherever possible, the same message should be used to interact with the **DataSet** as a collection of **Cases** or as a collection of **VariateRecords** - the arguments given with the message will themselves indicate which view the user intends. For example, the message AtPut, which is meant to return (or Put) information At a location in the **DataSet**, will return: a **VariateRecord** if the argument supplied is a **Variate**, a **Case** if the argument is an **Individual**, and a **Datum** if both a **Variate** and an **Individual** are given. Note also that, unlike arrays, there is no need to record missing values as NAs - they are simply indicated by their absence.

Like a **DataRecord**, a **DataSet** has behaviours which allow interaction with its contents (including building matrices from its DataValues) and statistical behaviours that describe its contents. For example, the statistical behaviours will include the following: (i) histograms, or boxplots, of all the **Batches**, or conversely, (ii) a display of all the **Cases** as cartoon faces, or glyphs, and to demonstrate joint relationships (iii) a dynamic scatterplot matrix showing all two-way scatterplots of **Batches** (the **DataSet's Factors** would be available for grouping points). Interactions with such graphical displays should allow interaction with the representative object - to access and record pertinent information. For example, touching a displayed point in a scatterplot should allow access to the **Case**, or **Individual**. represented by the point. Information that is learned from the plot could then be recorded directly on the **Case**, or **Individual**. For boxplots, the corresponding objects are **Batches** and **Variates**.

5.0 Implementation details

We now, briefly point out some details of our present implementation. First, the **Collection** we have implemented is based on that developed in SmallTalk - the behaviours and messages are essentially the same (see Goldberg and Robson, 1983, for details). Following SmallTalk, we also implemented some of their specializations of a **Collection**. In particular, we implemented **Sets**, **Bags**, and **Dictionaries**. The last kind of **Collection** are relevant here. A **Dictionary** is simply a **Collection** whose members are **Associations**. One looks up a Key in the **Dictionary** to find the Value of the **Association** which has that Key. Hence, Keys in a **Dictionary** are unique.

To implement DataRecords and DataSets, two special kinds of a **Dictionary** were implemented: **DataDictionary** and **CaseDictionary**.

A **DataDictionary** contains only **DatumAssociations**. We implement **DataRecords** as **Collections** whose contents are contained in a **DataDictionary**. The lookup **Keys** in a **DataDictionary** are **Identifiers**, either **Variates** or **Individuals**.

Similarly, a **CaseDictionary** contains only **Cases** (which are **Associations**) and **DataSets** are actually implemented as **Collections** whose contents are **Cases** collected in a **CaseDictionary**.

6.0 Concluding Remarks

Object-oriented programming mixes procedures and data in each object. These objects are most easily understood, and managed, when they represent clear and distinct abstractions. Distinguishing the mathematical abstractions of **Matrix**, **Vector**, and **Scalar**, from the statistical abstractions of **DataSet**, **DataRecord**, and **Datum**, is an attempt to provide such meaningful abstractions.

By building on the simple statistical abstractions of **Individual**, **Variate**, and **Datum**, the composite data representations provide promising and flexible alternatives to array-based representations. For example, if, in the course of an analysis, something noteworthy is learned about a particular **Individual**, then that information can be recorded directly on that **Individual** and accessed later as required. Further, these representations do not preclude mathematical manipulation of the data since conversion messages are available (e.g. AsVector, AsMatrix). Instead, a conceptual distinction between statistical and mathematical views of a collection of numbers is reinforced.

To be useful, our data abstractions must also be clearly distinguished from yet higher level statistical abstractions. For each composite data abstraction (**DataRecords** and **DataSets**), the only statistical behaviours that we allow are those which are descriptive, simply giving different views of the data. Thus producing a scatterplot matrix is appropriate, while producing all possible regressions of a selected variable on the remaining ones is not. The latter incorporates information - a mathematical model and a fitting routine - that is extraneous to the data. These statistical behaviours are better attached to richer objects where the purpose of analysis is more clearly stated. For example, a **MultipleLinearRegression** object, that takes a **DataSet** and the identity of the dependent variable as input, better separates the data from the proposed analysis. Such statistically richer objects are also implemented in DINDE.

Finally, the above data objects represent simple and generic data abstractions that are relevant in statistical practice. Other statistically interesting data abstractions (e.g. **SequencedDataRecords** for time series analysis) can be represented as specializations of the objects presented here. We recommend that the behaviour of such new objects be similarly restricted to preserve their distinction as data abstractions.

7.0 Acknowledgements

This research was supported in part by U.S. NSF grant IST-8420614, by the MIT Center for Computational Research in Economics and Management Science, by the MIT Statistics Center, and by an equipment grant from the Xerox Corporation. We also thank D.A. Belsley, E. Kuh, and J.A. McDonald for helpful discussions.

8.0 References

Becker, R.A. and J.M. Chambers [1984]. *S: An Interactive Environment for Data Analysis and Graphics*, Wadsworth, Inc., Belmont CA.

Bobrow, D.G. and M. Stefik [1983]. *The LOOPS Manual*, Xerox PARC: Palo Alto, CA.

Goldberg, A. and D. Robson [1983]. *Smalltalk-80: The Language and its Implementation*, Addison-Wesley, Reading, MA.

Oldford, R.W. and S.C. Peters [1985]. "DINDE: Towards more statistically sophisticated software", *CCREMS Technical Report # 55*, MIT, Cambridge, MA.

Oldford, R.W. and S.C. Peters [1986a]. "Statistically Sophisticated Software and DINDE", *Proceedings of Computer Science & Statistics: 18th Symposium on the Interface.*

Oldford, R.W. and S.C. Peters [1986b]. "Data Analysis Networks in DINDE", *Proceedings of the American Statistical Association: Statistical Computing Section.*

Stefik, M. and D.G. Bobrow [1985]. "Object-Oriented Programming: Themes and Variations", *The AI Magazine, 5*, pp. 40-62.

Teitelman, W. and L.M. Masinter[1981]. "The Interlisp programming environment", *IEEE Computer, 14*, pp. 25-34.

Statistical Software

Algorithms for the Beta Distribution Function

By H. O. Posten, Storrs

1. SUMMARY

The present paper provides a set of useful algorithms for evaluating the beta distribution function in easily programmed form. These algorithms are the most important of the currently available procedures. Each is expressed in recursive form and a discussion of the quality of each is included. Applications of these algorithms to the evaluation of t, F and noncentral F distribution functions is also included. KEYWORDS: beta distribution, beta algorithms.

2. TWO CONTINUED FRACTION ALGORITHMS

In an earlier paper Posten (1982), the author provided a survey of a number of effective algorithms for many of the classical distribution functions. The present paper provides a more complete development (with extensions) of the beta distribution function algorithms presented in the previous paper. The objective of the present paper is to provide for teachers and practicing statisticians a set of useful algorithms for evaluating beta probabilities in easily programmed form. In the forms presented, each of these algorithms can be programmed (with beginning programming skills) on a personal computer or on a larger computing system. Accuracy statements will refer to use double precision on IBM 370 systems.

Because the beta distribution function can also be used to evaluate t and F probabilities it is, next to the normal, probably the most important continuous distribution function requiring evaluation. For self programmed algorithms, of course, it is desirable to have more than one algorithm to provide independent checks on the quality of each algorithm.

One of the most effective ways to evaluate the beta distribution function

$$I_x(a,\ b) = \frac{\Gamma(a+b)}{\Gamma(a)\Gamma(b)} \int_0^x t^{a-1}\ (1-t)^{b-1} dt \tag{1}$$

is through a continued fraction expansion. Two such expansions are available The first is due to Muller (1931) and is given by Abramowitz and Stegun (1964), Bargmann and Ghosh (1963), Boardman and Kopitzke (1975) and Kennedy and Gentle (1980). It has the form

Compstat 1986

$$I_x(a,b) = \frac{\Gamma(a+b)}{\Gamma(a+1)\,\Gamma(b)}\, x^a\,(1-x)^{b-1}\, CF, \tag{2}$$

where,

$$CF = \frac{1}{1-}\;\frac{c_1}{1+}\;\frac{d_1}{1-}\;\frac{c_2}{1+}\;\frac{d_2}{1-}\;\cdots\,, \tag{3}$$

is a continued fraction with coefficients

$$c_k = \frac{(b-k)(a+k-1)\,x}{(a+2k-2)(a+2k-1)\,(1-x)}, \tag{4}$$

$$d_k = \frac{k\,(a+b+k-1)\,x}{(a+2k)(a+2k-1)(1-x)}.$$

The notation of (3) is shorthand notation for successive division and in practice (3) is evaluated in truncated form with the divisions terminated at a specific number of steps. This evaluation is efficiently done by a recursive form as given in Table 1.

TABLE 1. First Continued Fraction Algorithm for the Beta Distribution Function.

STEP 1: Input values of a, b, x and n = number of recursive steps to be used (effective number of divisions used in (3) is therefore 2n).
STEP 2: Evaluate $S_n = c_n/(1+d_n)$ using (4) to evaluate c_k and d_k.
STEP 3: For k = n-1, n-2,... ,1, evaluate $S_k = c_k/(1 + d_k/(1 - S_{k+1}))$ using (4) to evaluate c_k and d_k.
STEP 4: Evaluate $CF = 1/(1 - S_1)$ and $LA = \ln\Gamma(a+b) - \ln\Gamma(a+1) - \ln\Gamma(b)$
STEP 5: Evaluate $LF = LA + a\ln x + (b-1)\ln(1-x) + \ln CF$ and $I_x(a,b) = exp\ (LF)$

Bargmann and Ghosh indicate that this method is effective over a wide range of values of a and b, with accuracy of at least 10^{-8} using 80 divisions in the continued fraction (n = 40 in Table 1). However, Kennedy and Gentle recommend the use of this method for x < (a-1)/(a+b-2) and recommend the use of

$$I_x(a,b) = 1 - I_{1-x}(b,a) \tag{5}$$

for larger x. The author's experience with this algorithm indicates that often 10 or more decimal place accuracy is achieved using significantly less than 80 continued fraction divisions, even above the range for x given by Kennedy and Gentle. Also a study by a former student, J. Leitz, over 91 beta distributions with $a \geq b$ and $0.1 \leq a,b \leq 500$ and over 15 x-values in the probability range 0.0001 to 0.9999 for each distribution, supports this conclusion. Leitz's study though suggests that this continued fraction algorithm may not be very effective for the case when a and b are both less than unity, as also suggested by Bargmann and Ghosh. An exception to this is the case $a = b = 0.5$, when the method is quite accurate (10 or more decimals).

A second continued fraction expansion from Aroian (1941) (See also Aroian (1959) and Tretter and Walster (1979)) is also given by Abramowitz and Stegun (1964), Bouver and Bargmann (1975), Boardman and Kopitzke (1975) and Kennedy and Gentle (1980). It has the form

$$I_x(a,b) = \Gamma(a+b)\,\frac{x^a(1-x)^b}{\Gamma(a+1)\,\Gamma(b)}\,CF \tag{6}$$

where

$$CF = \frac{1}{1-}\;\frac{c_1}{1+}\;\frac{d_1}{1-}\;\frac{c_2}{1+}\;\frac{d_2}{1-}\;\cdots \tag{7}$$

with

$$c_k = \frac{(a+k-1)(a+b+k-1)\;x}{(a+2k-2)(a+2k-1)}, \tag{8}$$

$$d_k = \frac{k\;(b-k)\;x}{(a+2k-1)(a+2k)}$$

The algorithm in recursive form is given in Table 2.

Bouver and Bargmann indicate that this continued fraction expansion provides a superior algorithm for evaluating the beta distribution function over an extensive range of values of a and b. They refer to the range $10^{-8} \leq a,b \leq 70{,}000$ and specifically state that the method is efficient for extremely J-shaped and U-shaped distributions. They recommend its use for $x < a/(a+b)$ (i.e., for $x <$ mean), and the use of the complement form (5) for larger x. The author's experience indicates

TABLE 2. Second Continued Fraction Algorithm for the Beta Distribution Function.

STEP 1: Input values of a. b, x and n = the number of recursive steps to be used (effective number of divisions used in the continued fraction expansion is then 2n).
STEP 2: Evaluate $S_n = c_n/(1 + d_n)$ using (8) to evaluate c_n and d_n.
STEP 3: For k = n-1, n-2,... , 1, evaluate $S_k = c_k/[1 + d_k/(1 - S_{k+1})]$. using (8) to evaluate c_k and d_k.
STEP 4: Evaluate $CF = 1/(1 - S_1)$ and $LA = \ln\Gamma(a+b) - \ln\Gamma(a+1) - \ln\Gamma(b)$.
STEP 5: Evaluate $LF = LA + a\ln x + b\ln(1\text{-}x) + \ln CF$ and $I_x(a,b) = \exp(LF)$.

that this algorithm is quite effective often yielding at least 10 decimal accuracy using n = 100 for x values in the probability range up to 0.90.

3. FOURIER SERIES CHEBYSHEV POLYNOMIAL ALGORITHM

Another effective algorithm for evaluating the beta distribution function is based on a Fourier series Chebyshev expansion (Woods and Posten (1977)). The algorithm is effective, computationally, due to a recursive strategy used with the expansion

$$I_x(a, b) = 1 - (\theta/\pi) - S \tag{9}$$

where

$$\theta = \cos^{-1}(2x - 1) \quad \text{and} \quad S = \sum_{j=1}^{\infty} b_j \sin(j\theta) \tag{10}$$

where the b_j coefficients are defined below. The series S is evaluated by

$$S_n = \sum_{j=1}^{n} b_j \sin(j\theta) \tag{11}$$

for some convenient value of n. The coefficients are evaluated recursively by

$$b_1 = (2/\pi)(a - b)(a + b)^{-1} \tag{12}$$

$$b_2 = (1/\pi)(a+b)^{-1}(a+b+1)^{-1}\{2(a\text{-}b)^2 - (a+b)(a+b\text{-}1)\}$$

$$b_{j+2} = (a+b+j+1)^{-1}\left[2(a\text{-}b)\frac{j+1}{j+2}b_{j+1} + (j+1\text{-}a\text{-}b)\frac{j}{j+2}b_j\right]$$

$$j = 1, \ldots, \quad n - 2.$$

S_n is then evaluated recursively and quickly (without use of any trignometric functions) by

$$U_{n+1} = U_{n+2} = 0 \tag{13}$$

$$U_k = b_k + (4x - 2)U_{k+1} - U_{k+2.} \qquad k = n, \ldots, 1 \qquad S_n = 2U_1 (x - x^2)^{1/2}.$$

The algorithm is given in Table 3.

TABLE 3. Fourier Series Chebyshev Polynomial Algorithm the Beta Distribution Function.
STEP 1: Input parameter values a and b and the values of x and n = the number of terms to be used in the series (11).
STEP 2: Evaluate b_k, $k = 1, \ldots, n$, recursively from (12).
STEP 3: Evaluate U_1 recursively from (13) and S_n from (13)
STEP 4 : Evaluate $I_x(a, b) = 1 - (\theta/\pi) - S_n$.

Studies by former students (F. Rockhold, A. Tupek and T. Ashbee) and the author indicate that this algorithm is effective over a broad range of values of a, b and x, achieving eight or more decimal place accuracy if enough terms are used. Usually 200 terms provide at least 6-8 decimal accuracy. Because of the recursive approach these terms are computed rapidly. The method is not as accurate as indicated if the distribution shape is such that the probability is strongly compressed towards either end-point (e.g., a=100, b=2). In this case it appears that one can usually expect four or five decimal accuracy with 200 terms. Additionally, this method is quite weak when either parameter is very near zero (a or b $\leq$ 0.2) and is not very strong when either parameter is small (0.2 $\leq$ a or b $\leq$ 1.4), though in the latter case one can expect several decimal place accuracy. This general weakness for small parameter values, however, does not exist when a or b = 1/2 (unless the other parameter has a different low value, $\leq$ 1.4). The opposite is, in fact, true. When a or b = 1/2, the method has very high accuracy. But, when a or b = 1, the method is weak often providing as little as three decimal accuracy. The latter weakness is important when using this method to evaluate F probabilities (to be discussed). In particular, when the F random variable has one of the degrees of freedom parameters equal to two, the corresponding beta parameter is equal to one. This and the general weakness for small a, b can be overcome by use of an identity that evaluates the beta distribution function by first sufficiently incrementing the low parameter. The beta distribution function can be evaluated by evaluating the corresponding beta distribution function with first parameter incremented, using the identity

$$I_x(a,b) = (1-x)^b \sum_{i=0}^{k-1} C(a+i,b)\, x^{a+i} + I_x(a+k,b) \tag{14}$$

for arbitrary k = 1, 2, ... , with

$$C(m,n) = \Gamma(m+n)/\{\Gamma(m+1)\,\Gamma(n)\}. \tag{15}$$

If b is the small parameter, (14) may be used after reversing the order of the two parameters using (5).

The Fourier series Chebyshev polynomial method is not as fast as the foregoing two continued fraction methods, but it is effective more broadly over the range of x. Also, if probabilities are required for a number of x values for the same values of (a,b), the b_j coefficients need only be evaluated once and the algorithm can be revised to take this into account. Further this algorithm does not require the availability of a log gamma subroutine (unless (14) is being used to increment a parameter) as do the two continued fraction algorithms and the fourth algorithm to be presented. For use as part of a general program requiring the inclusion of a beta distribution subroutine, the author has often preferred the Fourier series Chebyshev algorithm because it seems to avoid the problem of overflow from a division by zero. Infrequently , with the continued fraction algorithms, the author has had such division problems. The latter though have always been handled by repeating the run with use of a larger value of n. Finally, the Fourier series Chebyshev algorithm is the only one presented here with at least an approximate error bound available. For this method, an approximate error bound on the error one makes in truncating the series (11) at N terms is

$$\text{Approx. Error Bound} = N\,|b_N| \,/\, \{2 \min(a,b)\} \tag{16}$$

where b_N is the last coefficient evaluated in (11).

4. SERIES ALGORITHM

A fourth algorithm for the beta distribution function is given by Bosten and Battiste (1974), (See also, Kennedy and Gentle (1980), Pike and Hill (1967) and Pike and SooHoo (1976)). This algorithm is used in the IMSL subroutine for the the beta distribution function and Kennedy and Gentle indicate that this method is among the best available. The author's experience with this algorithm is limited but preliminary evaluations indicate it to be quite effective and that it provides high accuracy even when using only a moderate number of terms. The algorithm is

$$I_x(a,b) = x^a \frac{\Gamma(s+a)}{\Gamma(s)\Gamma(a+1)} SUM1 + x^a (1-x)^b \frac{\Gamma(a+b)}{\Gamma(a)\,\Gamma(b+1)} SUM2, \tag{17}$$

where

$$SUM1 = \sum_{k=0}^{\infty} A_k, \quad SUM2 = \sum_{k=1}^{IB} B_k, \tag{18}$$

IB = integer part of b, s = 1 if b = integer and s= b - IB otherwise and

$$A_k = a\,x^k (1-s)_k / \{k!\,(a+k)\}, \tag{19}$$

$$B_k = \frac{b\,(b-1)\ldots(b-k+1)}{[(a+b-1)(a+b-2)\ldots(a+b-k)(1-x)^k]}$$

Also, $(1-s)_0 = 1$ and $(1-s)_k = (k-s)\,(k-1-s)\ldots(1-s)$, for $k = 1,2,\ldots$. Additionally, SUM1 is defined = 0 (not = 1 as indicated in Bosten and Battiste) if s = 1, and SUM2 = 0 if b < 1. The coefficients in (19) can be evaluated recursively as in the algorithm given in Table 4.

5. ALGORITHM FOR THE NATURAL LOGARITHM OF THE GAMMA FUNCTION

Most of the foregoing algorithms for the beta distribution function require the evaluation of the natural logarithm of the gamma function. If a built-in function for this is not available, a very effective algorithm for this is the Euler-Maclaurin expansion provided by Abramowitz and Stegun (1964).

$$\ln\Gamma(y) = (y - 1/2)\ln y - y + (1/2)\ln(2\pi) + \sum_{i=1}^{4} T_i \tag{20}$$

TABLE 4. Series Algorithm for the Beta Distribution Function
STEP 1: Input values of a, b, x and n = the number of terms to be used in SUM1 in (18).
STEP 2: Evaluate IB = integer part of b. Set S = 1 if b = integer; otherwise set S = b - IB. Set $A_0 = 1$ and $SUM1 = 1$.
STEP 3: If S=1, set SUM1 = 0; otherwise for $k = 1, \dots,$ N, evaluate $A_k = [(k-s)(a+k-1)\,x / \{k\,(a+k)\}]A_{k-1}$, and SUM1 = SUM1 + A_k.
STEP 4: Evaluate $B_1 = b / [(a+b-1)(1-x)]$ and $SUM2 = B_1$.
STEP 5 : If IB = 0, set SUM2 = 0; else, if IB not= 1, for k=2,3, ... ,IB evaluate $B_k = [(b-k+1) / \{(a+b-k)(1-x)\}]B_{k-1}$ and SUM2 = SUM2 + B_k.
STEP 6 : If S = 1, then set F1 = 0; otherwise calculate $F1 = \exp[\,\ln\{\Gamma(s+a) / (\Gamma(s)\Gamma(a+1))\}] + \ln(SUM1) + a\ln(x)$.
STEP 7 : If IB = 0, then set F2 = 0; otherwise evaluate $F2 = \exp[\,\ln\{\Gamma(a+b) / (\Gamma(a)\,\Gamma(b+1))\}] + \ln(SUM2) + a\ln(x) + b\ln(1-x)$.
STEP 8 : Evaluate $I_x(a,b) = F1 + F2$.

with $T_1 = 1/12y, \quad T_2 = -1/360y^3, \quad T_3 = 1/1260y^5, \quad T_4 = -1/1680y^7.$

The expansion (20), as truncated at four terms, provides an extremely accurate method for evaluating $\ln\Gamma(y)$ with the error being less than 10^{-14} if $y \geq 18$ and enough digits are carried in the calculations. When y < 18, one can increment the argument of the gamma function and use

$$\ln\Gamma(y) = \ln\Gamma(y+k) - \sum_{i=0}^{k-1} \ln(y+i) \tag{21}$$

where k is an integer which, when added to y, produces a value between 18 and 19. The algorithm is given in Table 5.

6. EVALUATION OF t AND F DISTRIBUTION FUNCTIONS

The distribution functions for the t- and F-variates are simple to evaluate using any of the foregoing algorithms for evaluating the beta distribution function. for the t-variate with v degrees of freedom, the identity

$$P(t_v < t) = (1/2)[1 + I_x(1/2, v/2)], \quad \text{for } t > 0, \tag{22}$$

TABLE 5. Algorithm for the Natural Logarithm of the Beta Distribution Function
STEP 1: Input y and initialize PROD = 1.
STEP 2: If y is less than 18, then for j=0,1, ... ,k-1, evaluate Z = y + j and PROD = Z (PROD), where k is the smallest integer such that y+k is greater than or equal to 18.
STEP 3: Evaluate W = y + k and $T_1 = 1/12W,\ T_2 = -1/360W^3,\ T_3 = 1/1260W^5,\ T_4 = -1/1680W^7.$
STEP 4: Evaluate $\ln\Gamma(y) = (W - 1/2)\ln W - W + (1/2)\ln(2\pi) - \ln PROD + \sum_{i=1}^{4} T_i.$

with $x = t^2/(v + t^2)$ provides the algorithm. For $F(v_1, v_2)$, the F-variate with v_1 and v_2 degrees of freedom, the identity

$$P(F(v_1, v_2) < F) = I_x(v_1/2, v_2/2)\ , \tag{23}$$

with $x = (v_1/v_2)F/[1 + (v_1/v_2)F]$, provides the algorithm. The identities (22) and (23) are given by Johnson and Kotz (1970).

7. EVALUATION OF THE NONCENTRAL F DISTRIBUTION FUNCTION

The evaluation of power for regression or analysis of variance F tests requires an algorithm for evaluating the noncentral F distribution function. The noncentral F variate is defined by $F(a, b; \Delta^2) = [\chi^2(a; \Delta^2)/a]/[\chi^2(b)/b]$ with the numerator chi-square being noncentral and independent of the denominator chi-square. The noncentrality parameter Δ^2 is defined by

$$\Delta^2 = \sum_{j=1}^{a} \mu_j^2, \text{ with } \chi^2(a; \Delta^2) = \sum_{j=1}^{a} W_j^2 \text{ and } W_j \sim N(\mu_j, 1) \text{ i.i.d.} \tag{24}$$

Any of the forgoing algorithms for evaluating beta distribution functions can be used to evaluate the noncentral F distribution function by using a relationship between the noncentral F and the noncentral beta distribution function. This relationship is given on page 192 of Johnson and Kotz (1970), in terms of central beta probabilities, as

$$P[F(a,b\,;\,\Delta^2)<F_0] = \sum_{j=0}^{\infty} p(j,\Delta^2/2)\; I_t(a/2+j,\; b/2) \tag{25}$$

where

$$p(j\,;\lambda) = e^{-\lambda}\lambda^j/j! \tag{26}$$

and

$$t = (a/b)\,F_0\,/\,[1+(a/b)\,F_0] \tag{27}$$

The form (26) is a Poisson probability and the infinite series is a sum of products of Poisson and central beta probabilities. Clearly, if the series (25) is truncated at N terms, The error made due to this truncation is less than the one minus the sum of the Poisson terms evaluated up to this point. One can require the algorithm to evaluate terms in the series until the truncation error is less than some pre-specified ERROR (e.g., ERROR=.000001). This can be achieved by accumulating the calculated Poisson terms until they exceed 1 - ERROR.

The two expressions (25) and (26) can each be evaluated recursively by using

$$p(0;\lambda) = e^{-\lambda}, \qquad p(j\,;\lambda) = (\lambda/j)\,p(j-1\,;\,\lambda), \quad j=1,2,\ldots \tag{28}$$

and

$$B_j = B_{j-1} - \frac{\Gamma[\{(a+b)/2\}+j-1]}{\Gamma[(a/2)+j]\Gamma(b/2)}\, t^{(a/2)+j-1}(1-t)^{b/2} \tag{29}$$

where $B_0 = I_t(a/2, b/2)$ and $B_j = I_t\,(a/2+j, b/2)$.

A similar approach can be used to evaluate the power of two=tailed t-tests but squaring the t statistic provides an F statistic and the noncentral F distribution function can then be used to evaluate the power of the t-test.

8. REFERENCES:

Abramowitz, M. and Stegun, I.A. (1964) Handbook of Mathematical Functions. National Bureau of Standards, Applied Math. Series, U.S. Govt. Printing Office, Wash. D.C.

Aroian, L.A. (1941) Continued Fractions for the Incomplete Beta Function. Annals Math. Statist., 12, 212-223.

Aroian, L.A. (1959) Corrections to "Continued Fractions for the Incomplete Beta Function." Annals Math. Statist. 30, 1265

Bargmann, R.E.and Ghosh S.P. (1963) Statistical Distribution Programs for a Computer Language. IBM Watson Research Center Report RC-1094.

Boardman, T.J. and Kopitzke, R.W. (1975) Probability and Table Values for Statistical Distributions. Proc. Statist. Comput. Section, Amer. Statist. Assoc., Washington D.C.

Bosten, N.E. and Battiste, E.L. (1974) Remark on Algorithm 179 (S14); Incomplete Beta Ratio. Commun. Assoc. Comput. Machin. 17, 156-157.

Bouver, H. and Bargmann, R.E. (1975) Computer Algorithm and Modules for The Evaluation of Statistical Distribution Functions. Tech. Report 110, Statistics Dept., Univ. of Georgia, Athens Ga. U.S.A.

Johnson, N.L. and Kotz, S. (1970) Continuous Univariate Distributions, vol. 2, Houghton and Mifflin Co. New York.

Kennedy, W.J. and Gentle,J.E. (1980) Statistical Computing. New York: Marcel Dekker, Inc.

Muller, J.H. (1931) On the Application of Continued Fractions to the Evaluation of Certain Integrals, Biometrika 22, 284-297.

Pike, M.C. and Hill, I.D. (1967) Remark on Algorithm 179: Incomplete Beta Ratio, Comm. Assoc. Comp. Machin. 10, 375-376.

Pike, M.C, and SooHoo, J.S. (1976) Remark on Algorithm 179: Incomplete Beta Ratio. Trans. Math. Software 2, 201-208.

Posten, H.O. (1982) Computer Algorithms for the Classical Distribution Functions. Proc. First Internat. Conf. on Teaching Statistics. Univ. of Sheffield, England, 313-330.

Tretter, M.J. and Walster, G.W. (1979) Continued Fractions for the Incomplete Beta Function: Additions and Corrections. Annals Statist. 7, 462-465.

Woods, J.D. and Posten, H.O. (1977) The Use of Fourier Series in the Evaluation of Probability Distribution Functions. Commun. Statistics B6, 201-219.

Interactive User-Friendly Package for Design and Analysis of Experiments

By A. Baines and D. T. Clithero, Leeds

Summary

An interactive package for design and analysis of experiments is being developed for use both by statisticians and by scientists, engineers and managers with a minimal knowledge of statistics. For the non-statistician user the package contains guidance, explanation, validation and interpretation facilities, which aim to represent the assistance of a statistical consultant. The present scope of the package includes aspects of simple comparative, factorial and response surface experiments.

Keywords : statistical package, design of experiments, statistical analysis, expert systems, statistical consultancy.

1. Introduction

Most routine statistical analyses are done using a statistical package such as SAS, SPSS, GENSTAT, GLIM, MINITAB, etc. Such packages, primarily developed for use by statisticians, are also widely used by scientists, engineers and managers with a minimal knowledge of statistics. For the inexpert user, the packages are relatively unfriendly and unsafe because they are chiefly concerned with the computational aspects of statistical analysis rather than the formulative and interpretative aspects of statistical consultancy. Similarly, validity checks are often available in the packages but the user is required to have sufficient statistical expertise to request and understand their use.

This paper describes the underlying principles and structure used to develop a package for design and analysis of experiments, which incorporates statistical consultancy as well

Compstat 1986

as computation, and is particularly aimed at the inexpert user. The topic of experimental design was chosen because of its relative dependence on statistical consultancy rather than computation and, perhaps because of this, it is not covered by any of the commercially available packages. The associated statistical analyses are also included because it is clearly unfriendly for the inexpert user to deal with different packages for experimental design and for statistical analysis.

2. Underlying Principles and Structure of Package

The feasibility of formally representing the consultancy process between statistician and client was demonstrated for design of simple comparative and response surface experiments by Jones (1980) and for statistical advice and guidance within multiple regression analysis by Wetherill (1981). In both cases the program structure was conceived as an essentially open-tree network with all possible outcomes of every possible sequence of nodes or steps predetermined and provided for in the package. It was decided that the present package could be similarly structured, particularly if exceptional combinations of outcomes were dealt with by indicating that they were outside the scope of the package and suggesting that the user consults a statistician.

One consequence of assuming such a program structure was that it facilitated the use of Fortran as programming language rather than a specialist Expert System language, e.g. Lisp, Prolog, which, while perhaps allowing a more flexible formulation of the consultancy process, are relatively unsuitable for statistical computation.

The package starts with an 'Overview' module in which the user's problem is specified in sufficient detail for it to be mapped to an appropriate design and analysis module. Each design and analysis module corresponds to an area of statistical methodology of which simple comparative, factorial and response surface experiments are currently considered.

Each module is a network of steps. Each step requests the user to provide specific information about his problem, to approve a proposed course of action or to choose among alternative courses of action. A step comprises one or several interrelated sub-steps, each requiring the user to respond with a single name, number or keyword. An invalid response results in an error message and a request to try again. Steps may be single pass, e.g. specifying the maximum number of observations, or multi-pass, e.g. describing the variables involved in the problem. After valid completion of a step the program automatically selects the appropriate next step or sub-step.

A set of keywords is available at every step to invoke management and help facilities for

(i) including or excluding a background description associated with each step and sub-step.

(ii) further explanation of the step and guidance on the appropriate response.

(iii) explaining any technical terms used in the background description or statistical analysis displays

(iv) displaying the current problem specification, the proposed design, the experimental data and its statistical analysis, and for obtaining hard copy of such information.

(v) saving, updating and recalling the current problem specification and for terminating a run.

(vi) moving back to a prior step for purposes of amendment and for moving forward again through completed and valid steps.

The first of these facilities allows the practised user to move more quickly through the program by reducing the information display. Another facility enables a series of responses to be input at the same time, thereby eliminating intermediate displays. Moreover a keyword can be input as the minimum number of characters required to discriminate it from other possible keywords. These features should make the package more efficient for practised users and particularly for statisticians.

Logical, arithmetical and statistical validity checks are automatically applied as and when necessary. A validity check which is successfully passed is generally not referenced, whereas failure results in an error message and reference to an appropriate step. In particular, when a problem description is detected as being outside the current scope of the program, a suitable message is displayed and the user is directed towards an appropriate problem revision or encouraged to consult a statistician.

The actual package structure comprises a Fortran program and five data files. The program incorporates the network and consultancy logic, the statistical computation(for which use is made of the NAG library),and the user/program interface. It does not generate most of what is seen on the screen but rather copies, edits and displays material taken from the data files. The five data files contain respectively the background descriptions for each step or sub-step, the supplementary information providing further understanding of a step and guidance on the appropriate response, the explanations of technical terms, the error messages and comments on validity checks, and the summary of the current problem specification, design, data and statistical analysis.

3. The Overview Module

This module is concerned with obtaining a sufficient specification of the user's problem so that it can be allocated to a design and analysis module or identified as being outside the scope of the package.

The initial steps request information on the experimental unit of the problem, the available or feasible maximum number of such units, whether the experimentation is simultaneous or sequential and, for sequential experimentation, whether the design is to be pre-planned or adaptively developed.

The next step requests the identification and specification of the factors and variables involved in the problem, starting with an initial classification as response variable, design factor, block factor or associate variable. An associate variable is not taken into account during design but may be involved in the statistical analysis. Each variable and factor is further specified as quantitative or qualitative, continuous or fixed number of levels, the range of values or number of levels, and the blocksize for block factors. Optionally the variables, factors and factor levels can be named.

On completion of this step the program attempts to allocate the problem to a design and analysis module. If successful the user is advised as to which statistical area his problem matches and the program moves to the appropriate module. If unsuccessful the user is so advised and the program moves to a negotiation step. This invites the user to consider revising his problem specification so that a successful allocation can be made. A series of sub-steps guides the user on which revisions, if feasible,would result in an acceptable specification. The negotation step is completed when a suitable revised specification is chosen by the user. The alternative advice is to stop the run and consult a statistician.

4. Design and Analysis Modules

Currently three design and analysis modules are being implemented, covering the areas of simple comparative, factorial and response surface experiments.

The simple comparative experiments module deals with the estimation of confidence limits or with tests of significance for means and standard deviations, or for the comparison of two means or standard deviations, with a pre-planned experimental design. The statistical methodology is essentially that described in Davies (1954).

The factorial experiments module covers the pre-planned design and analysis of replicated, unreplicated and incomplete factorial experiments, with or without block factors. The range of incomplete factorial experiments and their methods of generation is described by Zemroch (1986).

The response surface experiments module deals with adaptively planned experiments based on the methodologies of Determination of Optimum Conditions, Box (1951), Evolutionary Operation, Box (1957), and Simplex Evolutionary Operations, Spendley (1962).

The initial steps in these modules request additional information to refine the problem specification so that the precise design methodology can be selected. When relevant, information is also requested so as to determine the minimum number of observations to give the required degree of discrimination. If this minimum number is larger than the previously specified maximum a negotiation step is entered in which the user is invited to revise his problem specification until it is acceptable. Further steps describe the recommended design, handle data input and initiate and describe its statistical analysis. For the response surface experiments module the network of steps starting with the design of a sub-experiment to the interpretation of the statistical analysis is a cyclic procedure.

Acknowledgements

The authors are grateful to Peter Zemroch for many helpful comments on the structure and content of the package. The research was funded by the Cooperative Research Grant GR/C68576, of the Science and Engineering Research Council with Shell Research Ltd. as the cooperating organisation.

References

Box,G.E.P. and Wilson,K.B. (1951),On the experimental attainment of optimum conditions. J.R. Statist.Soc.B, 13, 1-45.
Box, G.E.P. (1957), Evolutionary Operation: a method for increasing industrial productivity. Appl.Statist., 6, 81-101.
Davies,O.L. (1954), Design and analysis of industrial experiments. Oliver and Boyd, Edinburgh.
Jones, B. (1980), The computer as a statistical assistant. BIAS, 7, 168-195.
Spendley, W., Hext, G.R. and Himsworth,F.R.(1962), Sequential application of simplex designs in optimisation and evolutionary operation. Technometrics, 4, 441-461.
Wetheril, G.B. (1981), A regression package for non-statisticians with emphasis on response surface analysis. Internal Report, Statistics Department, University of Kent.
Zemroch, P.R. (1986), The computerised generation of block incomplete factorial designs in the conversational experimental design and analysis package SELINA. Compstat 1986 (to be published).

NL: A Statistical Package for General Nonlinear Regression Problems

By S. Huet and A. Messéan, Jouy-En-Josas

Summary :

NL is a statistical package designed for nonlinear regression problems, taking into account the heteroscedasticity of variance, if any. The algorithm for the estimation of the regression parameters is adapted to the topic but it also presents possibilities for future extensions. NL is composed of independent routines, each of them devoted to a statistical task. In the present version, an host system is used to ensure the interface between the user and the routines.

Keywords : nonlinear regression, statistical package, host system, Gauss-Marquardt algorithm.

1. INTRODUCTION

NL deals with nonlinear regression problems, when the variance of errors are allowed to depend on the unknown parameters.

The observed variable Y is related to one or more independent variables x, through a known function f, depending on unknown parameters denoted by θ. For each of the k values of x, $(x_i,\ i=1,\dots,k)$ we observe n_i independent variables $(Y_{ij},\ j=1,\dots,n_i,\ i=1,\dots,k)$. The variance of the error ϵ_{ij} is either constant or dependent on x_i and/or on θ through a known function v.

Thus NL deals with the following models : $Y_{ij} = f(x_i, \theta) + \epsilon_{ij}$ and $Var(\epsilon_{ij}) = \sigma^2 v(x_i, \theta)$
For simplicity, let us denote F_i for $f(x_i, \theta)$ and V_i for $var(\epsilon_{ij})$. $F_{i.a}$ and $V_{i.a}$ will denote the derivatives of F_i and V_i with respect to θ_a. Some regularity conditions are assumed, as the continuity of $F_{i.a}$ and $V_{i.a}$.

A lot of packages solve standard nonlinear regression problems. Some programs, that maximize a likelihood, allow variances depending on θ , but the likelihood must be introduced by the user (GENSTAT for example).

From a software point of view, we designed NL as a collection of standard algorithms with great portability in order to obtain independence with respect to the operating system, the

Compstat 1986

computer and the user interface.
At the present time, its implementation is carried out by the host system "Consistent System" available under Multics system.

2. STATISTICAL INFERENCE ON θ

2.1 Estimation

Four different estimators of θ are proposed :

(1) If the errors are assumed to have a gaussian distribution, it is natural to look for $\hat{\theta}_{ML}$, the maximum likelihood estimator of θ. $\hat{\theta}_{ML}$ is consistent and efficient and minimizes the function $-2L_n(\theta)$ defined by :

$$-2L_n(\theta) = \Sigma_i\ n_i \log(V_i) + \Sigma_{ij}\ (Y_{ij} - F_i)^2/V_i$$

(2) The Ordinary Least Squares estimator of θ, $\hat{\theta}_{OLS}$, minimizes $S_n(\theta)$, defined by $S_n(\theta) = \Sigma_{ij}\ (Y_{ij} - F_i)^2$. $\hat{\theta}_{OLS}$ is consistent but less efficient than $\hat{\theta}_{ML}$ except if V_i is constant and equal to σ^2.

(3) The Weighted Least Squares estimator of θ, $\hat{\theta}_{WLS}$ minimizes $SW_n(\theta)$, defined by $SW_n(\theta) = \Sigma_{ij}\ (Y_{ij} - F_i)^2/V_i$. $\hat{\theta}_{WLS}$ is not consistent except if V_i does not depend on the unknown parameters θ.

(4) The Modified Least Squares estimator is derived from Weighted Least Squares estimator. $\hat{\theta}_{MLS}$ satisfies $\Sigma_{ij}F_{i,a}(Y_{ij} - F_i)/V_i = 0$. This estimator is consistent and more efficient than $\hat{\theta}_{OLS}$.

2.2 Testing hypotheses

Any test which considers a sub-model of a general model as the null hypothesis is carried out by NL : goodness of fit test when replications, test of affine hypotheses as "$\theta_a = \theta_b$" and "θ_a = fixed value", comparison of c curves are of that type.

If the Maximum Likelihood Estimator has been chosen, the likelihood ratio test is used. Let $L_{n,r}$ be the maximum log likelihood when r parameters are estimated and $L_{n,r-s}$ when r-s parameters are estimated, then $-2(L_{n,r-s}-L_{n,r})$ is asymptotically distributed as χ^2 with s degrees of freedom.

2.3. Model validation

In addition to the tests, NL proposes different statistical tools for model validation. The residuals and the weighted residuals, the sensitivity functions, the fitted values of f are calculated and can be displayed. The sensitivity functions, denoted $S_a(x,\theta)$, are equal to $(\partial f(x,\theta)/\partial\theta_a)/\sqrt{v_i}$. The variations of S_a with x give information about the influence of the experimental design on the estimations.

Moreover if some replications are present, the variation of the sample variances versus the sample means is studied, in order to improve the modelization of the variance.

2.4 Confidence intervals for functions of parameters

Confidence interval, for each of the θ_a's, say the first component θ_1, may be defined in terms of the likelihood ratio statistic. Let $L_{n,p-1}(\theta_1)$ be the maximum log likelihood when θ_1 is fixed.

A confidence interval of level α is defined as the set of those θ_1 which satisfy $-2(L_{n,p-1}(\theta_1) - L_{n,p}) \leq \chi_1^2(\alpha)$ and corresponds to the projection of the likelihood contour $-2L_{n,p}(\theta) = -2L_{n,p} + \chi_p^2(\alpha')$ onto the θ_1 axis, provided α' is such that $\chi_p^2(\alpha') = \chi_1^2(\alpha)$.

Due to numerical problems, we use an approximate confidence interval for a function $\Phi(\theta)$ based on the quadratic approximation of the likelihood contour.

All these intervals should be preferred to the symmetric ones based on the asymptotic Normality of the estimator, because of their robustness with respect to the nonlinearity (Messean, 1982).

3. NUMERICAL ASPECTS

3.1 *Parameter estimation.*

Normal equations

Parameter estimation is carried out by solving the following system of equations : $B^T(\hat{\theta})\ (Z(Y) - \eta(\hat{\theta})) = 0$ where $Z(Y)$ is the vector of sufficient statistics (or a subset of it), $\eta(\theta)$, the expectation or, more generally, the limit of $Z(Y)$ and $B(\theta)$ a h*p matrix where h is the dimension of $Z(Y)$.

Each estimator is defined by setting expressions for B, Z and η. For example, the Maximum Likelihood Estimator of θ is defined by the following formulas

. $Z_i(Y) = (1/n_i)\sum_j Y_{ij}$ and $Z_{k+i}(Y) = (1/n_i)\sum_j Y_{ij}^2$

. $\eta_i = F_i$ and $\eta_{k+i} = F_i^2 + V_i$

. $B_{i,a} = (1/\sigma^2)(F_{i,a}/V_i - V'_{i,a}(F_i/V_i^2))$ and $B_{k+i,a} = (1/\sigma^2)V_{i,a}/V_i^2$

Many estimators can be written in such a form, including all ML estimators in exponential families. Properties of such estimators can be found in Huet & Messean (1986). This kind of generality easily allows future extensions of NL for they only need to add new routines which calculate the values of B, Z and η.

Algorithms

To solve the normal equations, we use a generalization of Gauss-Newton and Gauss-Marquardt algorithms. To implement them, we linearize the vector η, i.e. $\eta(\theta) = \eta(\theta^o) + D(\theta^o)(\theta-\theta^o)$ where $D(\theta)$ is the h*p matrix of the derivatives of η with respect to θ and θ^o the current approximation of θ.

Carrying it in the normal equations, we get the following linear system whose solution gives the new Gauss-Newton approximation :

$B^T(\theta^o)D(\theta^o)(\theta-\theta^o) = B^T(\theta^o)(Z(Y)-\eta(\theta^o))$.

The product matrix $B^T D$ must be inversible : this condition is fulfilled for our estimators under some regularity conditions. The linear system is solved by Cholesky 's decomposition.

As in the least squares problems, we can modify the Gauss-Newton algorithm by adding a diagonal matrix λI to the matrix $B^T D$. By analogy, we call this modified algorithm, the Gauss-Marquardt algorithm.

The properties of these algorithms in the general case are discussed in Huet & Messean (1986).

Constraints on parameters

Constraints can be introduced by the user, in order to take into account some information about the parameters or to define hypotheses to test. As a matter of fact, the likelihood ratio method is carried out by re-estimating parameters under equality constraints.

We thus have to deal with two kinds of constraints $\theta_a^{min} \le \theta_a \le \theta_a^{max}$ and $\theta_a = \theta_b$. The constraints proceeding is based on a reparametrization of the model with reduction of the parameter space dimension. The normal equations are solved in the space of "active" parameters without any constraints.

3.2 Confidence intervals

To compute confidence intervals based on the likelihood

ratio method for each θ_a, we use the projections of likelihood contours upon each parameter axis. We thus have to solve the following problem :
Minimize or maximize θ_a with $-2L_{n,p}(\theta) = K$, where K is a critical value.
Using Lagrange 's method, the problem is reduced to solve the p following equations : $\partial L_{n,p}(\theta)/\partial\theta_b = 0$, $b \neq a$ and $-2L_{n,p}(\theta) = K$
For each confidence interval, we choose as initial values the extreme points of the confidence elipsoid based on asymptotic normality of $\hat{\theta}$.

Maybe due to the particular form of the equations, the iterative process, based on a modified Newton-Raphson method, often fails to converge when the number of parameters is large.

Another characteristic of NL is the use of a symbolic derivator to compute the analytical form of the derivatives of f and v. The user introduces the model through an easy-to-use interface and the adequate subroutines are generated. In that purpose, we adapted a symbolic derivator, written in PASCAL.

4. SOFTWARE ASPECTS

NL has been designed as a package of standard algorithms driven by an external system which ensures the communication with the user.

4.1. Modularity

NL is composed of independent routines, each of them devoted to a statistical, numerical or graphical task. These routines are written in Fortran 77, printing is limited to error messages and intermediate results during the iterative processes, logarithmic and exponential functions are protected from overflow and negative arguments.

The following modules are now available :

- Study of the variability of the data.
- Estimation of the parameters under various hypothesis.
- Study of the residuals
- Calculation of fitted values for f and v
- Study of sensitivity functions.
- Construction of confidence intervals for the parameters θ.
- Construction of likelihood contours.
- Computation of a confidence interval for the F_i, i=1,...k
- Prediction : calculation of a confidence interval for $f(x_0, \theta)$ where x_0 is any value of the range of variation of x.
- Calibration : calculation of a confidence interval for the abscissa x when the ordinate y is observed
- Calculation of a confidence interval for a function $\Phi(\theta)$.

4.2. Driving NL

Of course, the different modules can be called directly from any main program, communication between modules being done through files. But our aim is to use NL through a host system, which involves data management, error management, communication between tasks and which provides an adequate user interface.

Presently, this is done by the Consistent System (Klensin, 1978), a subsystem of the Multics system, which is a general statistical system that allows external routines calls. The user can manage the nonlinear regression analysis from the CS level, through the classical CS program syntaxis. The following command is an example of the use of the estimation module

```
eval Result := estimation (Data, Model_Name, Pinit
                           option estim("mls") print(10)
```

The agent "eval" executes the CS macro "estimation" which decodes the command, checks its validity, allocates the necessary working space, sets the options to the user's values if any, else to default values, calls the appropriate Fortran routine of NL and then, if successful, computes the return arguments. So the macros interface NL and the host system and its data structures.
A user manual of NL under CS describes all its capabilities and is available from the authors (Bouvier et al., 1985).

V. FUTURE EXTENSIONS

Two kinds of extensions are already planned. Firstly, we are extending the capabilities of NL in order to solve the distribution fitting problem (under multinomial distribution), to define the model with differential equations and to use an external integrator for differential equations. Secondly, we intend to interface NL with some other interactive environment.

REFERENCES

Bouvier, A. et al. (1985) : **"CS_NL : manuel d'utilisation"** Technical report. Laboratoire de Biométrie. INRA
Huet, S. & Messean, A. (1986) : **"A generalization of Gauss-Newton and Gauss-Marquardt algorithms for estimation in exponential families"** Technical report. Laboratoire de Biométrie. INRA
Klensin, J. (1980) : "The Consistent System" **The American Statistician, 34**, 3, p 169-176
Messean, A. (1982) : "Régions de confiance dans le modèle non-linéaire" **Journal de la Société de Statistique de Paris**. Vol. 123, 2, p 134,143.
Ross, G.J.S. (1978) : "Exact and approximate confidence regions for functions of parameters in non-linear models" **Proceedings in computational statistics**, p 110,135. Physica-Verlag WIEN.

Statistical Software for Personal Computers

By W. J. Keller, Voorburg

Summary.
In this paper we will discuss 8 selected statistical packages for Personal Computers. These packages are tested and the results are presented in a number of tables. We concentrate on packages for IBM PC's and compatibles.

Keywords: Personal Computers, statistical software, software review.

1. Introduction

In this paper several selected statistical packages for Personal Computers (PC's) will be evaluated. We concentrate on statistical software for the IBM PC and compatibles. After scanning the literature and a large number of software descriptions, we tested more than a dozen packages. A selection of these packages will be described in this paper. We will concentrate in this paper on statistical software and therefore leave out the packages which focus mainly on econometric and time series analysis (we tested, among others, Soritec, Micro TSP, Gauss, Rats, SST, and Shazam). Furthermore, a few other tested statistical packages did not, in our opinion, offer enough value for money and were also excluded. The resulting eight packages, viz. ABC, BMDPC, NCSS (also called Number Cruncher), P-STAT, SPSS/PC+, STATA, STATGRAPHICS and SYSTAT, are the subject of interest in this paper.

Since we believe that actual using packages reveals important characteristics which are not revealed by other methods (e.g. by examining manuals), all tested packages are uses in a battery of tests. Therefore, we are able to include our judgment of characteristics as ease of use, and quality of presentations. The reader should, however, bear in mind that these judgements are, by nature, subjective and therefore should be interpreted not as absolute measures.

The characteristics examined and discussed in the next sections are "Use of resources", "Ease of use", "Data management", "Descriptive statistics and tests", "Multivariate Analysis", "Time series analysis", and "Presentation of tables and graphs". Most evaluations will be measured on a three point scale: "-" , ".", and "+", standing for "bad/fair", "fair/good", and "good/excellent", respectively. In the tables 4, 5 and 6 a "-" score will in general mean that the feature mentioned is not available in the package. Each table will be concluded with some total evaluation.

Compstat 1986

The results are summarised in the last section and table, where also some total scores in terms of "Value for money" will be given.

2. Use of resources

In table 1 we give some general characteristics of the packages tested. Please note the version: sometimes versions differ more than packages so this might be important to know. The price is the non-academic, single unit price in Europe. After the minimal internal memory (RAM) requirement we state if a hard disk is necessary or preferred, and give an indication of the amount of space occupied on your disk by the programs. The optional mathematical coprocessor (Intel 8087) for the PC can speed up computations considerable (3 times on average), and we list if this device is necessary, supported but not necessary, or not supported.

People with local area networks (and others) will be interested to know if a not copy-protected version of the program is available. In the last colums we mention the maximum number of variables in a record (sometimes only limited by the amount of RAM), and the maximum number of data elements (integers, in order to see if space is used efficiently). Note that in particular modules of a package (i.e. the regression module) often heavier restrictions on the number of variables are imposed. The packages which process record by record from disk are only limited by disk space. The last column gives our final judgment of the bearing of the packages on yout account and your PC.

3. Ease of use

This is a rather subjective set of scores, so be warned! In table 2, first column, we indicate whether a package is mainly command or menu driven. The last type is often easier to learn but less powerfull for the more experienced user. Under "Ease of learning" this issue is evaluated, while under "Syntax" we give our impression of the ease of use of the language (or menu's), its simplicity, power, and flexibility. Under the headings "Input" and "Output" various aspects of interactions are evaluated. Under "Input" the use of abbreviations for variables and commands, of ranges (x1-x5), of functionkey's and keyboard macro's are considered, as well as the editing of command lines and the availability of an descent editor (or can we use the nice resident screen editor of Sidekick?), and the batch facilities. Under "Output" we total our view on things as screen management (is the output scrolling continously or screenwise), the speed and appearance of screen output, the use of log-facilities to reroute results or commands or both to files, etc.

Under "Documentation & help" the quality of the user's guide, reference guide, reference card and available help screens are listed. "Robustness" refers to the robustness of the programs with respect to operator's errors etc., and the gentle with which the user is handled. The evaluation under "Module swap" gives an indication about the speed of switching modules (e.g from the regression module to the graph module to see the residuals); some packages really slow down interactions by long pauses for this kind of switches.

4. Data management

Often people only look at the advanced capabilities of statistical packages, but experiences show that the large majority of jobs are concerned with simple statistics and, in particular, in data management. Therefore we have put some emphasize on these often forgotten features. The first column in table 3 gives the score for import/export of data: free format and fixed format are used often, while some packages also provide direct access to spreadsheet and database files (e.g. of Lotus 1-2-3 or dBase II/III). Very important for complex data structures are the possibilities to join and merge data (case wise, and variable wise, preferable on a key variable), and to sort data. In order to select observations, a flexible "if" clause is preferable ("if income <50000"), while subgroup processing ("by sex: tabulate income") allows for the analysis to be repeated over subgroups.

Under "Compute/batch/language" we evaluate the possibilities to generate new variables and to recode variables, as well as true programming features as loops, if-then-else constructs, procedures with parameters, etc. This is especially important in setting up batch programs or other means to extend the package' command language. The last criterium, "Labels, missing weights" focusses on variable and value labels, facilities for missing data, and the use and possiblities with weights. All these last features are especially important for survey's.

5. Statistical methods and presentation

In table 4, 5 and 6 some scores on statistical methods are given. Since most headings speak for them selves (we hope), we will only provide here some specific comments. Under "Test of hypotheses" we evaluate the facilities for simple t-tests, and other parametric tests; we also looked if not only test statistics but also probability levels are supplied. In table 5 the possiblities for more advanced methods are listed. Finally, in table 6, we included a short list of features relevant for time series analysis. The column marked "Lags" shows if the packages could generate lagged observations (e.g. $income_t$=$income_{t-1}$). "Box-Jenkins" refers of course to the toolbox of methods known under that name, while "Smoothing/ seasonal adjustment" provide some information on the use of moving averages, exponential smoothing, filters, etc.

In table 7 we list our impression of the quality of the tables and the facilities in composing your own reports (report writer), on the on hand, and the quality of the graphs on the other. Only high resolution graphics packages obtained a "+".

6. Summary

In table 8 we summarize our evaluations, and add two criteria: "Speed of computation" and "Accuracy". The range of speed going from the slowest program to the fastest is about 1 to 3 (with an average of 2). These figures are based on a test with 560 cases on 6 variables (integer scores), with a weight added, and a mix of excercitions as import data, tabulate, recode, regress, save data, etc. To test the accuracy we did the well known Longley regression test (with very collinear variables), a test of moments (on 7.0000001, 7.0000002, etc) and an ANOVA test with similar tricky data. All these dirty tests also helped us to evaluate the robustness and the "feel" of the programs.

The last column provides our final judgment, "Value for money". ABC is easy to use by novices, can read SPSS9 setups and is not expensive. BMDPC provides the most extended library of statistical programs. P-STAT shines when it comes to datamanagement and survey processing. STATGRAPHICS is somewhat slow but can definitely be recommended at fast PC's (like the IBM AT). Of our favorites, we think that NCSS is a cheap and nice menu driven program with lot's of possibilities. SPSS/PC+ is rather expensive but, in our opinion, worth the price. We also liked STATA very much. This small program provides all the tools you will need for basic statistical problems, and is very flexible and easy to use. SYSTAT, finally, really offers a large value (especially in terms of advanced methods) for a limited amount of money.

Table 1. Use of resources

Package (1)	Version	Price in Europe (US $)	Min. RAM (Kb)	Hdisk necessary/preferred	Min-max size on disk (Mb)	8087 necess./support.	Copy prot. ?	Max. # of variables	Max. # of integers	Score
ABC	2.2	200	256[2)]	no	1	supp.		200[2)]	disk	+
BMDPC (basis)	PC	450	640	nec.	1-3	nec.	prot	300	disk	-
BMDPC (full)	PC	1500	640	nec.	1-20	nec.	prot	300	disk	-
NCSS	4.2	80	196	no	0.7			250	32K	+
P-STAT	8.21	600	640	nec.	2-4	nec.		150	disk	-
SPSS/PC+ (basis)	PC+	1000	384	nec.	1-3	supp.	prot	200	disk	-
SPSS/PC+ (full)	PC+	1700	450	nec.	2-6	supp.	prot	200	disk	-
STATA (basis)	1.3	400	256	no	0.3	supp.		RAM	250K	+
STATA (full)	1.3	600	256	no	0.4	supp.		RAM	250K	+
STATGRAPHICS	1.2	700	384	pref.	1	supp.		RAM	65K	.
SYSTAT	2.2	500	256[2)]	pref.	1-2	supp.		200[2)]	disk	.

(1) SPSS basis does not include the options Advanced Statistics and Tables, the full package does (both options are also available separately). BMDP basis consists of the so-called package I (6 programs), the full package of all 28 programs. STATA basis consists of STATA/Professional + Kit, the full package also of STATA/GRAPHICS.
(2) There is also a 512 Kb version supporting more ariables.

Table 2. Ease of use

Package	Command/ menu	Ease of learning	Syntax	Input	Output	Document. +help	Robustness	Module swap	Score
ABC	menu	+	.	.	+	-	.	-	.
BMDPC (basis)	com	-	-	-	-	.	-	-	-
BMDPC (full)	com	-	-	-	-	.	-	-	-
NCSS	menu	+	+	.	+	.	-	.	+
P-STAT	com	-	.	-	.	.	+	-	.
SPSS/PC+ (basis)	com	.	.	+	+	+	+	+	+
SPSS/PC+ (full)	com	.	.	+	+	+	+	+	+
STATA (basis)	com	+	+	+	+	.	+	+	+
STATA (full)	com	+	+	+	+	.	+	+	+
STATGRAPHICS	menu	+	+	.	.	.	.	-	.
SYSTAT	com	.	+	.	-	+	.	-	.

Table 3. Data management

Package	Import/ export	Join/ merge/ sort	Select if/ select by	Compute/ batch/ language	Labels/ missing/ weights	Score
ABC	.	-	+	.	+	.
BMDPC (basis)	+	-	+	+	.	.
BMDPC (full)	+	-	+	+	.	.
NCSS	.	.	.	.	.	.
P-STAT	+	+	+	+	+	+
SPSS/PC+ (basis)	+	+	+	.	+	+
SPSS/PC+ (full)	+	+	+	.	+	+
STATA (basis)	.	+	+	+	+	+
STATA (full)	.	+	+	+	+	+
STATGRAPHICS	.	.	.	.	.	.
SYSTAT	+	+	+	+	.	+

Table 4. Descriptive statistics and tests

Package	Means/ variances/ percentiles/ etc.	Tables of means/ frequencies	Correlation/ association	Tosts of hypo- theses	Non-para- metric tests	Graphics/ Expl. data analysis (EDA)	Score
ABC	.	.	.	-	-	-	-
BMDPC (basis)	+	.	+	.	.	.	.
BMDPC (full)	+	.	+	+	+	.	+
NCSS	+	.	+	.	+	+	+
P-STAT	+	+	.	+	-	+	.
SPSS/PC (basis)	+	.	+	+	+	.	+
SPSS/PC (full)	+	+	+	+	+	+	+
STATA (basis)	+	.	+	+	+	.	+
STATA (full)	+	.	+	+	+	+	+
STATGRAPHICS	+	.	+	+	+	+	+
SYSTAT	.	.	+	+	+	+	+

Table 5. Multivariate analysis

Package	Regres- sion analysis	Anova	Factor analysis/ princ.comp.	Cluster analysis	Discrim. analysis	Logit/ probit	Log.lin. models	Score
ABC	.	-	-	-	-	-	-	-
BMDPC (basis)	+	.	-	-	-	-	.	.
BMDPC (full)	+	+	+	+	+	+	+	+
NCSS	+	.	.	-	.	-	-	.
P-STAT	+	+	.	-	+	-	-	.
SPSS/PC (basis)	+	.	-	-	-	-	-	-
SPSS/PC (full)	+	+	+	+	+	-	.	+
STATA (basis)	.	.	-	-	-	.	-	-
STATA (full)	.	.	-	-	-	.	-	-
STATGRAPHICS	+	+	.	.	.	-	-	.
SYSTAT	+	+	.	+	.	.	.	+

Table 6. Time series

Package	Lags	Box Jenkins	Smooting/ associated adjusted	Spectral analysis	Score
ABC	-	-	-	-	-
BMDPC (basis)	.	-	-	-	-
BMDPC (full)	.	+	+	+	+
NCSS	.	-	+	-	.
P-STAT	.	-	-	-	-
SPSS/PC (basis)	.	-	-	-	-
SPSS/PC (full)	.	-	-	-	-
STATA (basis)	.	-	-	-	-
STATA (full)	.	-	.	-	-
STATGRAPHICS	.	+	+	+	+
SYSTAT	.	+	+	+	+

Table 7. Presentation

Package	Quality tables/ reportwriter	Quality graphics	Score
ABC	.	-	-
BMDPC (basis)	.	.	.
BMDPC (full)	.	.	.
NCSS	.	.	.
P-STAT	+	.	+
SPSS/PC (basis)	+	.	.
SPSS/PC (full)	+	.	+
STATA (basis)	.	.	.
STATA (full)	.	+	+
STATGRAPHICS	.	+	+
SYSTAT	.	.	.

Table 8. Summary

Package	Use of Resources	Ease of use	Data manage- ment	Descr. statis- tics & tests	Multiv. ana- lysis	Time series	Presen- tation	Speed of comput.	Preci- sion	Value for money
ABC	+	.	.	-	-	-	-	.	.	.
BMDPC (basis)	-	-	.	.	.	-	.	.	.	.
BMDPC (full)	-	-	.	+	+	+	.	.	.	.
NCSS	+	+	.	+	.	.	.	.	.	+
P-STAT	-	.	+	.	.	-	+	-	.	.
SPSS/PC (basis)	-	+	+	+	-	-	.	+	+	+
SPSS/PC (full)	-	+	+	+	+	-	+	+	+	+
STATA (basis)	+	+	+	+	-	-	.	+	+	+
STATA (full)	+	+	+	+	-	-	+	+	+	+
STATGRAPHICS	.	.	.	+	.	+	+	-	+	.
SYSTAT	.	.	+	+	+	+	.	.	+	+

Recursive Techniques in Statistical Programming

By B. P. Murphy, J. S. Rohl, and R. L. Cribb, Nedlands

Synopsis Recursive algorithms in the language Pascal are given to display useful features in statistical computing procedures such as branch and bound searching and indexing multi- factor tables. Concepts of provability and efficiency of such algorithms are also discussed and study of principles strongly recommended.

Keywords RECURSION, PASCAL, INDEXING, BRANCH and BOUNDS, MULTIFACTOR TABLES

0. Introduction

Recursive technique are much studied in computer science courses, as they underlie much practical methodology in compiler writing and allied work. There are only a few situations in statistical computing where recursion as a programming tool is virtually essential - e.g. most branch and bound model searches, where the technique, together with statistically sound exclusion rules can efficiently and transparently pare down the number of models to be considered. However there are many less obvious uses for recursion where great programming simplifications can be obtained, particularly in dealing with the creation and manipulation of explicitly, and implicitly, ordered table structures.

In this paper we examine some recursive algorithms and data structures for searching model fits, and for table storage, indexing, collapsing and expansion such as are used in AOV procedures utilising sweeping methods. Since every recursive algorithm (Rohl, 1977) has a non-recursive analogue (as should be obvious from the fact that machines are intrinsically non-recursive), comparable non-recursive implementations should also be studied to examine the myth that recursion is intrinsically inefficient; space does not here permit, but see Rohl (1984). The conclusion is drawn that inefficiency lies usually in the user's programming, and the visual elegance of recursive techniques possesses the usual concomitants of any other true elegance - power through simplicity, and that there are many places in statistical computation where they can be very profitably employed by those using structured programming languages like Pascal, C and their derivatives.

1. Brief Description of Recursion

Recursive procedures call themselves, and are analogous to mathematical functions, such as the gamma function, which can be defined in terms of themselves. A good introduction to recursive programming is Rohl (1983), in which the relations between recursion, iteration and recurrence are fully discussed, and illustrates that mathematical recursion does not necessarily imply programming recursion. The stacking principle employed by the compiler is also fully discussed, and must be clearly understood by those wishing to use the technique efficiently. In particular, the distinction between what should and should not be stored at each level of the recursion is a central point of efficiency. The examples below illustrate some of these essential points.

Compstat 1986

2. Branch and Bound Searching

The use of Pascal procedures in 'all regressions' searches has been indicated by Clarke (1982), and in log-linear searches by Murphy (1981,1983). The problem is to examine the fit of all loglinear models to a contingency table, and preserve the 'best'. This is an interesting topic in itself (Edwards, 1984; Murphy, 1984), but we are concerned here only with ensuring the enumeration of all models, and the efficient bypassing of obviously fruitless paths. The model terms are often called u-terms (Bishop et al.; 1975), and here they have been set up as integer indices, and the model as a set of integers.

The procedure *DeleteUterm* below for a backward search keeps calling itself to delete another u-term from *currentUmodel* while a fit is deemed to still hold. Testing the fit and deciding whether it holds are matters of statistical decision and are outside the recursive procedure, and irrelevant to it. The recursion simply ensures that all models are tested once only, and only if necessary, i.e. if *fitted* is true at every call, then *deleteUerm* can be shown to visit every one of the 2^n-1possible models.

```
procedure DeleteUterm (nextu : natural);
var  i : natural;

begin
     for i := nextu downto 1 do
     if effect[i] in currentUmodel then
          begin
          currentUmodel := currentUmodel - [effect[i]];
          If not (ResultObvious( currentUmodel )) then
                    If fitted(data, currentUmodel, chisq, probability) then
                              DeleteUterm(i - 1);
          currentUmodel := currentUmodel - [effect[i]];
     end;
end; {of procedure "DeleteUterm"}
```

Use of such a formal procedure relieves the statistical programmer of much hack work in deriving a working central algorithm, and encourages concentration on the severe statistical problems involved. Even the problem of modifying the default search order implied by this procedure , which may be felt necessary for statistical reasons (e.g. fixing or excluding certain terms, or to aid presentation and interpretation) is also external to the recursion, which simply exists to guarantee the complete coverage of potential models consistent with any set of exclusion principles.

The 'provability' of the procedure is here its greatest asset. Further, the readability of the code is in itself an argument for the use of the Pascal language, and even manages to make the procedure look trivial. However, much proctical experience in recursion is often necessary to adequately read such code, for another feature of recursion is that small changes in code usually have quite dramatic, though often intended, effects.

3. Indexing Multifactor tables

Though one can effectively define multidimensional arrays in Pascal at compile time, it has the usual (lack of) generality problem, and we are better advised to store all tables as vectors, as we used to do in FORTRAN days, indexing separate data items by some in-line calculation. The following procedure *ReadValues* reads in a sequence of numbers which constitute an implicitly ordered multidimensionar table, and performs certain functions on the incoming data which require knowledge of the current item's indices held in the array *index* defined statically outside the recursion, but which functions do not otherwise involve the recursion.

```
procedure ReadValues (dimension : dimensionrange);
var levelno : levelrange;

begin
for levelno := 1 to nooflevels[dimension] do
     begin
     index[dimension] := levelno; }
     if dimension < noofdimensions then ReadValues(dimension + 1)
     else begin
          unitno := unitno + 1;
          read(table[unitno]);
          {Check(table[unitno]);}               {do what is wanted}
          end;
     end;
end; {of procedure "ReadValues"}
```

Of course, the indices could be obtained without recursion, by for-loops. With the current algorithm we avoid the messy 'end-effects' problems of that approach. More importantly, the indices are automatically available at each level of the recursion and are handled totally by the compiler, removing a large source of potential human error. Finally, it is a method by which many other table accessing problems can be attacked.

4. Sweeping ANOVA effects

One such extension is the algorithm below, which implements the forward and backward (Wilkinson,1970) sweeps for an *effect* in an ANOVA calculation. The procedure *OneSweep* in direction *up* calculates a marginal table for a prescribed *effect*, while in direction *down* it subtracts out the means obtained from the marginal table from each appropriate original data value. The enclosing procedure *PairSweeps* initialises and controls the order of sweeping. This procedure is a simplified version of the one used in the MASS (Westat Associates; 1983,1986) statistical package. A slightly more extensive variant handles estimation of missing values; iterations to convergence combine usual looping with these recursive procedures to yield central code (exclusive of input/output) for the full single strata ANOVA procedure, including missing values, in under 50 lines of standard Pascal.

Some further features of recursion are apparent here. A function of the enclosing procedure *PairSweeps* is statistically obvious in that it controls the sequence of sweeps. But it also protects the recursive*OneSweep* procedure, in that it can be pared down to the bare essentials of what must be stacked, passed and computed at each level of the recursion. Such enclosures are often necessary for the same reasons of efficiency, and also to simplify initialisation and termination and the human interface. Many examples of inefficient recursion we have seen were results of the failure of the programmer to realise this fact.

```
procedure PairSweeps(termno : effectsrange; startfactor : factorrange
          currenteffect : factorset ; noofmeans : levelrange;
          nooffactors : factorrange);

type   upordown = (up,down);
var     mean  : meanarray ;  unitno : unitrange ;  n,  count, meanno : levelrange;

     procedure OneSweep(thisfactor : factorrange; n : levelrange ; direction :
upordown);
     var    no, levelsinfactor : levelrange;

     begin
     levelsinactor :=  factor[thisfactor].nooflevels;
     if thisfactor in currenteffect then n := n * levelsinfactor;
     count := n;         {establishing index base for this and lower level factors}
     for no := 1 to levelsinfactor do
          begin
          if thisfactor in currenteffect then meanno := count+no;
          if thisfactor< nooffactors then
               begin
               OneSweep(thisfact+1,n,direction);      { down another level}
               if thisfactor in currenteffect then n:=n+1;
               end
               else begin                          { at bottom so now to add/subtract}
                    unitno := unitno + 1;          { item into  correct  cell}
                    if direction = up then
                         mean[meanno] :=  mean[meanno]  + table[unitno]
                    else  table[unitno] :=  table[unitno] - mean[meanno];
                    end;
          end;
     end;      {onesweep}

 begin    {PairSweeps proper}
 for meanno := 0 to noofmeans do  mean [meanno] := 0;
 n := 0; unitno := 0;
 OneSweep(startfactor, n, up);                      {set by model effects control code}
{Getmeans(mean); }                                  {another external procedure}
 n := 0;  unitno := 0;
 OneSweep(startfactor, n, down);
 end ;     {of procedure "Pairsweeps"}
```

5. Computing degrees of freedom

The function *DegreesOfFreedom* computes the degrees of freedom associated with an hierarchical model for a multidimensional contingency table, assuming all cells have non-zero entries. It is one of a set of alternative algorithms for the problem benchmarked by Murphy and Rohl (1978), and for which further 'efficiency' considerations are now being examined. It is simply a forward variant of the branch and bounds technique above, where the bounding principles are the existence of the current term in the model being fitted, and the constraint of marginality inclusion in hierarchical log-linear modelling.

```
function DegreesOfFreedom (cliques : factorsetarray ; noofcliques : natural ;
                    level : naturalarray ; n : natural) : natural;
{This function yields the d.f of an hierarchical loglinear model specified by an
  array of cliques e.g. 123/145/23 . Each clique e.g.123 is a set of factor indices,
  indicating a table to which the iterative scaling must converge (Bishop et al.;
  1975). A clique contains all subeffects within it e.g. clique 123 contains
  (factors/effects) 1,2,3,12,13,23, and 123 . However they are not met in this
  order, but as 1, 12, 123, 145, 2, 23, 3. Level[i] contains the number of levels for
  i'th factor.}

var  currenteffect : factorset;
     i, df, currenteffectdf : natural;

     procedure Extend (first : natural);
     {This procedure seeks to extends currenteffect by one factor by trying all
       possible factors from first to n'th in turn. If the new currenteffect is
       contained  in the model its d.f. are added, and by calling itself recursively,
       extend attempts  to extend currenteffect by the next factor.}

     var  i, j : natural;
          found : Boolean;

     begin
     for i := first to n do
          begin
          currenteffect := currenteffect + [i];
          currenteffectdf := currenteffectdf * (level[i] - 1);
          j := 0;
          repeat
               j := j + 1;
               found := currenteffect <= cliques[j];
          until (j = noofcliques) or found;
          if found then
               begin
               df := df + currenteffectdf;
               if i <> n then extend(i + 1) ;
               end;
          currenteffect := currenteffect - [i];
          currenteffectdf := currenteffectdf div (level[i] - 1);
          end;
     end; { of procedure "Extend" }

begin
df := 0;
currenteffectdf := 1;
currenteffect := [];
Extend(1);
```

```
DegreesOfFreedom := df;
end; {of function "DegreesOfFreedom"}
```

This procedure has many lessons. Firstly it is provable, elegant and transparent, though statisticians would not consider it an obvious approach (computer people without our impedimenta do see it as natural). It also illustrates the use of the enclosing procedure for initialisation and retention of variables which do not need to be multiplied on the recursion stack. It is a template we have used for a number of problems, some on the surface quite removed. The simple control structure of this type of algorithm seems inevitably correct, so we feel very confident of the integrity of any recursive solution which passes our tests.

It is not uniformly efficient - models such as 'all k-factor interactions' are better handled by alternative algorithms on machines where multiplications and divisions are expensive (Murphy and Rohl, 1978). This exposes the dilemma of choice of economies of execute time versus programmer time, to which our response is always to prefer the latter until such (rare) circumstances that the former is seen to be quite imperative .

6. Conclusions

The examples have illustrated many points about recursive procedures in general, and in particular statistical uses. As noted, a full account of recursion theory and practice is given in Rohl (1983). Programming statisticians would profit by examination of at least the first chapters. It is a requirement of Westat Associates that MASS programmers in its employ are competent with recursion, as it has been found demonstrably highly cost-effective in many problems; unfortunately Westat have never had a very successful experience with statistical graduates without formal computer science qualifications as programmers. It is a sad state of affairs, and it is not the students' fault; the statistical teaching fraternity's nonchalent attitude to such study has been inflicted on them. This attitude needs wholesale revision.

7. Acknowledgements

Work on these matters has taken place over many years, and has had external funding at times by many bodies. Most recently, the Commonwealth Government of Australia has supported the ANOVA work through the Australian Industries Research and Development Incentives Board.

8. References

Bishop YMM, Feinberg SE and Holland P (1975) : **Discrete Multivariate Analysis**, MIT Press.

Clarke MRB (1982) : A Givens algorithm for moving from one model to the next without going back to the data, **Appl. Statist.**, 30, 198-203.

Edwards D (1984) : A computer intensive approach to the analysis of sparse multidimensional tables , **Proc. COMPSTAT'84**, 355-359.

Murphy BP (1983) : Production and reproduction of statistical software for Z80 microprocessors, **Contemp. Maths, 9**, 397-400.

Murphy BP (1981) : **The LOLITA Program Manual**, Raine Medical Statistics Unit, University of Western Australia.

Murphy BP (1984) : Similarities and differences between linear, logistic and log-linear models for survival analysis, **Proc. COMPSTAT'84**, 378-382.

Murphy BP and Rohl JS (1978) : Degrees of Freedom Algorithms. Manuscript.

Rohl JS (1977) : Converting a class of recursive procedures into non-recursive ones, **Software - Practice and Experience, 7**, 231-238.

Rohl JS (1984) : **Recursion via Pascal**, Cambridge University Press.

Westat Associates (1983) : The MASS Program as described in : **The MASS Program Manual,** Henstridge JD, Murphy BP, Poulsen JC and Wearne KL (1986), Westat Associates, Western Australia.

Wilkinson G (1970) : A general recursive procedure for the Analysis of Variance, **Biometrika, 57**, 19-46.

Design Criteria for a Flexible Statistical Language

By R. W. Payne and P. W. Lane, Herts

Summary

Statistical packages that constrain their users to choosing from a fixed set of analyses restrict the types of investigation that scientists may undertake, or lead to approximate or inappropriate analyses. We describe methods and criteria for designing a flexible package that avoids such deficiencies, with illustrations from our involvement with Genstat, in particular from the design of the new version called Genstat 5. Use of comprehensive algorithms and the ability to include user-supplied source code are important, but the crucial feature is that the commands of the package should themselves constitute a programming language for defining statistical, and other, computations.

Keywords

Data structure, Genstat, programming language, statistical language, statistical package design.

1. Introduction

The statistical package is now overwhelmingly the most popular tool for statistical analysis. One reason is that the major packages are well maintained and reliable so their results can be trusted; their developers test thoroughly before release and any remaining serious errors are likely to be found quickly by one of the many users. Another important reason is that packages enable people to do analyses very conveniently and with far less demands on their time than would be required, for example, to write a special-purpose program or to use a subroutine library. Thus, once someone has chosen a package, there is a considerable disincentive to go outside either to some other package or to a Fortran or Pascal program, when he is unable to get exactly the right analysis or output.

Consequently, package designers have a duty to avoid unduly restricting the range of available analyses and facilities. Otherwise, those who rely on

Compstat 1986

the package will be tempted to constrain the types of investigation that they undertake: for example, if with the software available, "analysis of variance" means either one-way or two-way, experimenters will not be encouraged to examine more than or two factors at a time. Alternatively, they may adopt an
inappropriate or, at best, an approximate analysis which need not generate the correct scientific conclusions.

One sensible strategy for design is to select algorithms for the package that are as general and comprehensive as possible. For example: the algorithm for analysis of variance in Genstat (Wilkinson 1970, Payne & Wilkinson 1977) handles the whole class of generally-balanced designs, which includes all the commonly-occurring experimental designs; the algorithm for fitting generalized linear models (Nelder & Wedderburn 1972) allows Genstat to fit not just classical linear regression models but also many others, including log-linear models for contingency tables and bioassay models (Lane & Payne 1985); and the clustering algorithm allows groups to be constructed by optimizing one of several different criteria (Banfield & Bassill 1977). In Genstat 5 there are new facilities for the randomization of generally-balanced designs and for fitting a range of standard growth and exponential curves.

However this is not the full solution. No designer can envisage the full range of requirements of the users of a widely-distributed package such as Genstat, nor can they predict what new methods may be suggested after the package has been released. Thus, packages should be designed in order to make it easy for users to extend them. We suggest that there are three ways in which this should be possible: by including new source code supplied by the user, by writing a program in the command language or by using a *procedure* written by some expert in the statistical area concerned.

2. Inclusion of new source code

Many packages provide a dummy command, which can be used to invoke the user's own source code. In Genstat, for example, the command is called OWN; articles in Genstat Newsletter 15 (1985) describe the use of this command to implement a monotone regression algorithm and an interface to a data-base package. With modern operating systems, it is a simple task to re-form the executable program to include the extra code. With Genstat 5 this can be taken much further, since the syntax of the command language is bootstrapped into the system. There are special directives which define the syntax of the others, generating codes that are stored in a binary file for subsequent runs. Hence it is very easy to add new commands or parts of commands into the syntax.

The implementation of extra techniques by this method requires expertise in the source language (Fortran 77 for Genstat 5) and the commitment to master details of the internal workings of the package. The amount of time needed by a programmer working in a low-level language (compared to the command language of the package) makes this option suitable only for frequently recurring tasks, or when a considerable amount of existing code is to be linked to the package - as in the examples described in Genstat Newsletter 15.

3. Programming in the command language

A quicker way of extending the package is for its control language to be usable as a programming language in its own right to link together existing algorithms and facilities to generate new types of analysis. The most important requirement for the control language to be used in this way is the ability to store output from the algorithms of the package in data structures within the package, so that it can be used as input to other algorithms. Of course this presupposes that the package supports a sufficiently wide range of data structures (Payne & Nelder 1976). With Genstat, virtually anything that can be printed as output from an analysis can also be saved in a suitable data structure. There are structures to store single values (i.e. scalars), vectors of numbers or of text, pointers (i.e. vectors whose elements are references to other data structures), factors (or grouping variables), multi-way tables with or without margins and rectangular, symmetric or diagonal matrices.

The pointer allows the range of data structures within the package to be extended by forming new compound structures from collections of standard data structures. Data which is to be passed from one part of a program to another can then be collected together and referenced by the single name of the pointer. In Genstat 5 there are several compound structures defined as standard because of the convenience of their use in some of the statistical commands. For example, the LRV structure used in multivariate analysis has a matrix containing the latent vectors, a diagonal matrix containing the latent roots, and a scalar for the trace. The components of the pointer must also be accessible individually for operations which do not require them all. Genstat 5 has a suffix notation allowing elements of pointers to be referred to by numerical or textual labels, either singly or in a list. Pointers can have other pointers as their elements, so all the facilities for hierarchical and compound data structures suggested by Lamacraft & Payne (1980) are provided.

To enable general programs to be written, the package must allow for selection between alternative sets of commands and for repetition of series of commands. In Genstat 5 there is a powerful looping structure and IF-THEN-ELSE and CASE constructions, together with an EXIT command to allow any such program control structure to be abandoned during execution, dependent on the result of a logical expression.

The Genstat 4 syntax has had a reputation for being rather difficult to learn. This is partly because of its power and generality, analogous to the difference between learning Fortran or Pascal, compared to learning Basic. With Basic, simple programs can be written very quickly and easily, but it becomes very tiresome for anything complicated; conversely, Fortran or Pascal (like Genstat) take longer to learn but are much less restrictive. However, the Genstat 4 language does have its quirks and inconsistencies, and so the syntax has been completely revised for Genstat 5, to give a clear and consistent set of "grammatical" rules that are obeyed by every Genstat 5 directive. Most of the directives and concepts map across from Genstat 4 to Genstat 5 in a fairly obvious way, so that users at the test sites have adapted to the new syntax with very little difficulty. However the opportunity has been taken to impose a consistency on the vocabulary of commands and to standardize conventions. These should all make Genstat 5 easier to learn and remember.

4. Procedures

Once a program has been developed, it is convenient to be able to store it in a procedure structure for subsequent use. If the package also allows libraries of procedures to be defined, this provides a third means of extension by which such programs can be made available to other users who may not have the time nor expertise to master a programming language. For example, many scientists need to use statistics, but lack of computing expertise or infrequency of use may make it difficult for them to use a statistical language. The best way of making flexible statistical methods directly available to them is for the statisticians that they consult to set up a "sub-language" of procedures to do their analyses, effectively customizing the package for their particular requirements.

The sub-language of procedures must be made easily accessible. In Genstat 5, the invocation of procedure looks just like a standard command in the Genstat
language. A standard library of procedures, which is attached automatically when Genstat 5 is run, is supplied with the package and includes such diverse methods as the analysis of censored data, canonical correlation analysis and correspondence analysis. The user may also attach his own library and many Genstat sites will wish to define their own standard library, which will also be attached automatically. Use of a non-standard command name in a program initiates an automatic search for a procedure of that name, first in the user's library, then in the site library and finally in the Genstat procedure library. Genstat can thus be customized for novices or for groups of users with common requirements. Furthermore, new methods can be programmed by experts and quickly made available to interested users, either via a procedure library or through an article in the Genstat Newsletter.

5. Conclusion

Packages have had much influence on the way in which statistical analysis is carried out, but unless they are designed to be flexible they can limit the horizons of their users or mislead them into resorting to inappropriate techniques. The command language of a package, like Genstat, that is designed for programming will inevitably take longer to learn than one that merely selects one of the analyses from a fixed repertoire, but for serious users the extra versatility acquired is well worth the effort; for novices or those with no aptitude for computing the package can be made accessible by writing procedures suitable for their particular needs. Such a package should also encourage the development of new statistical techniques by making it easy for researchers to try out new ideas.

References

Banfield C.F. & Bassill L.C. (1977), A transfer algorithm for non-hierarchical classifications, Applied Statistics, 26, 206-210.

Lamacraft R.R. & Payne, R.W. (1980), A new look at data structures for statistical languages, in "COMPSTAT 1980: Proceedings in Computational Statistics", 463-469, Vienna: Physica-Verlag.

Lane P.W. & Payne R.W. (1985), Genstat 5: a general-purpose interactive statistical package, with facilities for generalized linear models, in "Lecture Notes in Statistics No.32. Generalized Linear Models, Proceedings, Lancaster 1985", 105-111, New York: Springer Verlag.

Nelder J.A. & Wedderburn R.W.M. (1972), Generalized linear models, J. R. Statist. Soc. A, 135, 370-384.

Payne R.W. & Nelder J.A. (1976), Data structures in statistical computing, Proceedings of the 9th International Biometric Conference, Vol.II, 191-208.

Payne R.W. & Wilkinson, G.N. (1977), A general algorithm for analysis of variance, Applied Statistics, 26, 251-260.

Wilkinson G.N. (1970), A general recursive algorithm for analysis of variance, Biometrika, 57, 19-46.

Statistical Software SAM - Sensitivity Analysis in Multivariate Methods -

By T. Tarumi and Y. Tanaka, Okayama

Summary: A statistical software SAM is developed for performing sensitivity analysis in multivariate statistical methods. The main objectives are to detect influential observations and to evaluate the stability or reliability of the results. The outline of SAM is shown with a numerical example.
Key words: sensitivity analysis, influential observations, perturbation theory, multivariate methods.

1. Introduction

In these ten years or so various methods have been proposed to evaluate the influence of a single or multiple observations in regression analysis (Belsley, Kuh and Welsch,1980; Cook and Weisberg,1982). If the conclusion depends heavily upon a few observations, we must be very careful in using the result. Similar problems occur in other multivariate analysis. As a natural expression we can consider sensitivity analysis as follows.

In general we can recognize that a statistical method is a system, a set of data is an input and the result of analysis is an output. Our concern is the sensitivity of this system, that is, how a small change of data (input) affects the result of analysis (output). As a tool to investigate the sensitivity of various multivariate methods we developed a statistical software SAM (Sensitivity Analysis in Multivariate Methods).

2. Functions

From the viewpoint of sensitivity we shall deal with

Compstat 1986

multivariate statistical methods such as

[M01] principal component analysis (PCA),

[M02] biplot,

[M03] canonical correlation analysis (CCA),

[M04] canonical or multiple discriminant analysis (CDA),

[M05] the second method of quantification (= categorical discriminant analysis, QUANT-2),

[M06] the third method of quantification (= equivalent to multiple correspondence analysis, QUANT-3),

[M07] the forth method of quantification (= one of MDS's, QUANT-4),

[M08] linear regression analysis,

[M09] nonlinear or generalized linear regression analysis

[M10] the first method of quantification (= a type of linear model, QUANT-1),

[M11] linear discriminant analysis (LDF),

[M12] factor analysis (FA),

[M13] K-L type quantification method (QUANT-KL),

among others. The methods from [M01] to [M07], which are formulated as singular value decompositions or eigenvalue problems, have already been incorporated in SAM. The others are the methods to be incorporated in the next step.

As small changes of data we treat the following kinds.

[D1] Downweighting a single or multiple individuals

[D2] Deleting a single or multiple individuals

[D3] Deleting a single or multiple categories

[D4] Deleting a single or multiple items

[D5] Deleting a single or multiple variables

[D6] Pooling categories within an item

[D7] Pooling items

The main objectives of sensitivity analysis can be expressed as follows.

[O1] To evaluate the influence of each individual in turn and detect influential observations.

[O2] To evaluate the influence of a specified set of individuals.

[O3] To search for an influential set of individuals.

[04] To evaluate the influence of the most influential, the second most, the third most, ... successively in backward manner.

[05] To obtain jackknife or cross-validatory estimates.

[06] To select variables (items, categories).

In the present paper we shall discuss the objectives from [01] to [03]. To objective [05] is mainly related to regression analysis or discriminant analysis, in which statistical inference plays an important role.

The mathematical or computational techniques used in SAM are:

[T1] Exact method (recomputing slightly changed data),

[T2] exact method using simplified formula,

[T3] perturbation theory of eigenvalue problems (simple/ multiple eigenvalues),

[T4] Taylor expansion (one-step estimation).

Table 1. Relationships among the multivariate methods, the changes of input data and the mathematical techniques

	M01	M02	M03	M04	M05	M06	M07	M08	M09	M10	M11	M12	M13
D1	#	#	#	#	#	#	#	o	o	o	o	o	o
D2	#	#	#	#	#	#		o	o	o	o	o	o
D3						o							
D4					o	o				o			
D5	o	o	o	o				*			*	o	
D6					o	o							
D7					o	o							
T1	#	#	#	#	#	#	#		o			o	o
T2								o		o	o		
T3	#	#	#	#	#	#	#						
T4									o			o	o

#:already implemented *:omitted o:planned to be implemented

As a standard output the amount of influence of each individual is summarized in a table and the influence of a specified individual or a specified set of individuals is shown graphically. In order to detect an influential set of individuals it is useful to gather individuals which have similar influence patterns. For that purpose we can use principal component analysis, biplot and cluster analysis.

The relationships among the multivariate methods to be investigated, the changes of input data and the mathematical techniques are summarized in Table 1.

3.Example

We shall show an example of sensitivity analysis in the third method of quantification (multiple correspondence analysis) applied to the following data.

	Category									
Individual	1	2	3	4	5	6	7	8	9	10
1	1	1	1	0	0	1	1	1	1	0
2	0	0	0	0	1	1	0	0	0	1
3	1	1	1	1	0	1	1	1	1	0
4	0	1	1	1	1	1	1	1	1	1
5	1	1	1	1	1	1	1	1	1	1
6	1	1	0	0	0	1	1	0	0	0
7	1	1	1	1	0	1	1	0	1	0
8	0	1	1	1	1	0	1	1	1	1
9	1	1	1	1	1	0	0	0	0	1
10	1	1	0	1	1	1	0	1	1	1
11	0	0	1	1	0	0	1	1	1	1
12	0	1	1	1	1	0	1	1	1	1

<u>Sensitivity analysis procedure by using SAM</u>

```
*****************************************************
*                                                   *
*       SAM (Sensitivity Analysis System)           *
*                                                   *
*****************************************************

Which do you use for the output device LP(1) or CRT(C) ? L
Number of characters per line (80 or 132) ? 80

Statistical Method
  (01) Principal component analysis
  (02) Biplot
  (03) Canonical correlation analysis
  (04) Canonical discriminant analysis
```

```
  (05) Quantification 2
  (06) Quantification 3 (multiple correspondence analysis)
  (07) Quantification 4
  (08) Linear regression analysis
  (09) Nonlinear or generalized linear regression analysis
  (10) Linear model (Quantification 1)
  (11) Linear discriminant analysis
  (12) Factor analysis
  (13) K-L type quantification
which ? 6

Number of dimensions ? 3

      (unperturbed result. omitted)

Display scattergram ( Y or n ) ? N
Kind of changes of data
  (1) Downweighting individuals
  (2) Deleting individuals
  (3) Deleting categories
  (4) Deleting items
  (6) Pooling categories
  (7) Pooling items
which ? 1

Epsilon (rate of down weighting) for the perturbation ? .3

Check each individual in turn(E)
or specify a set of individuals(s) ? E

Specify a computational method ( y or N ) ? N

The amount of influence of each individual.
```

No.	theta1	theta2	G(1)(1)	G(1)(2)	C1	C2
1	-0.015088	0.012334	0.346307	0.408670	0.360786	0.208218
2	-0.016770	-0.043615	1.152240	1.578490	1.169300	0.405120
3	-0.007500	0.011458	0.410044	0.508710	0.420899	0.177818
4	0.013231	0.014263	0.212241	0.254768	0.219675	0.155791
5	0.023069	0.015970	0.104038	0.129349	0.147766	0.141972
6	-0.034850	-0.011489	0.733119	1.019190	0.754010	0.267726
7	-0.012632	0.011648	0.277822	0.320839	0.291771	0.191020
8	0.006041	0.001614	0.501304	0.638261	0.508787	0.182017
9	0.014055	0.002784	0.270303	0.351653	0.290578	0.263448
10	0.019929	0.006413	0.202394	0.299917	0.230652	0.195265
11	0.004472	-0.022994	0.414009	0.597378	0.419883	0.216479
12	0.006041	0.001614	0.501304	0.638261	0.508787	0.182017

```
Individual number ? 2
Individual number ? _

Specify a computational method ( y or N ) ? N

      ( perturbed result, omitted )
```

Plot the movement of row or column profile (Y or n) ? Y
Which plot, 1)row 2)column 3)both 0)end ? 3
Component assigned to the horizontal axis ? 2
Component assigned to the vertical axis ? 3

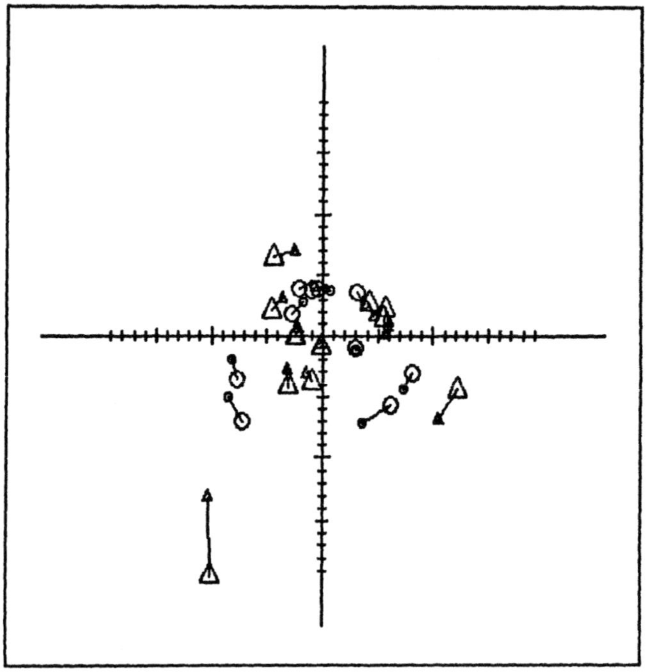

Which plot, 1)row 2)column 3)both 0)end ? 0

Another set of individuals (Y or n) ? N

References.

Belsley D.A., Kuh E. and Welsch R.E.(1980). Regression Diagnostics, John Wiley and Sons.

Cook R.D. and Weisberg S.(1982). Residuals and Influence in Regression, Chapman and Hall.

Tanaka Y.(1983). Sensitivity analysis in Hayashi's third method of quantification, Behaviormetrika Vol. 16, 31-44.

Tanaka Y. and Tarumi T.(1985). Computational aspect of sensitivity in multivariate methods, Technical Report No.12, Okayama Statisticians Group.

Tanaka Y. and Tarumi T.(1986). Sensitivity analysis in Hayashi's second method of quantification, Journal of the Japan Statistical Society, Vol. 16, No. 1.

Tarumi T. (1986). Sensitivity analysis of descriptive multivariate methods formulated by the generalized singular value decomposition, Math. Japon. Vol. 31, No. 6.

Data Handling and Computational Methods of Clinical Trials

Data Handling and Computational Methods in Clinical Trials

By E. Marubini, Milano

SUMMARY: After outlining the complexity of the chronic diseases and commenting briefly upon the levels of strength of evicence of different studies commonly adopted in bio-medical research, the paper focuses on the problem of using data-bases fcr the assessment of therapies and some tipical subjects arising in analyzing data collected in cancer randomized clinical trials and in combining findings from trials carried out in different situations.

Chronic diseases are, nowadays, the main concern of medicine in developed countries. Chronic disease has several features which lead to its complexity and make it difficult to trace its "natural history" and to develop therapeutical strategies. Among these features are a multifactorial ætiology, nearly non-specific and often endogeneous, and a long time course with continuous interaction between the host biological system and the surrounding environment. Several important medical questions arising in such a complex framework can only be faced by means of observational studies. In contrast to controlled experiments, in observational studies the investigator lacks the power to create the groups to be compared but is restricted to take selected observations or measurements which seem appropriate for the objectives.
As far as the evaluation of the effectiveness of a therapy is concerned, the hierarchy of strength of evidence pertinent to different types of study is shown in Table 1. If one thinks of the table as a pyramid and proceedes from the top to the bottom, the base on which he/she can build up his/her conclusion is becoming larger and larger.
The selection of the control group is a basic requirement for a clinical study aiming at evaluating the effectiveness of treatments. As pointed out by Chalmers and Schroeder (1979) and by Sacks et al. (1982), some authors use retrospective responses to a standard treatment as an alternative to randomization for selecting control groups. Gehan and Frierich (1974), Budzar et al. (1978) and Gehan (1982) advocate the use of historical control groups in cancer clinical trials. The risks one can take by adopting this approach have been widely debated in literature (Byar et al., 1976; Simon,

Compstat 1986

1982, Armitage, 1982; Sacks et al., 1982; Green and Byar, 1984; Zelen, 1985; Micciolo et al., 1985).

TABLE 1. Hierarchy of strength of evidence concerning efficacy of treatment

1. Anecdotal case reports
2. Case series without controls
3. Series with literature controls
4. Analyses using computer data-bases
5. 'Case-control' observational studies
6. Series based on historical control groups
7. Single randomized controlled clinical trials
8. Confirmed randomized controlled clinical trials

(from:S.B.Green,D.P.Byar: Stat.Med., 3: 361-370, 1984)

In the present paper after commenting upon the 4th level of the pyramid, attention will be focused on some topical subjects arising in analysing data collected in a randomized clinical trial (RCT) and in combining findings from several RCT's. Since my activity concerns mainly the oncological research, the above issues will be discussed by emphasizing problems concerning this field.

MEDICAL DATA BASE

The impressive advances in computer storage and retrieval has prompted a growing fascination with medical data base since the middle 1970's (Starmer et al.,1974; Rosati et al.,1975; Fries,1976).These data base were defined as "... a shared repository of data elements describing attributes of patients, interventions,and outcome... .Data are acquired in line with routine medical care by members of the medical staff." (Starmer and Lee, 1982).
Data banks enable the clinical researcher to pursue various aims: /i/ To define Disease States. Collecting time-oriented data in a large number of cases the data-base can be used to develop useful definitions of disease states by symptoms,signs and progression with confirmation by laboratory and pathology findings. As an example think of the clinical staging system

developed by Lamberg et al.(1984) for cutaneous T-cell lymphoma; /ii/ To evaluate Prognostic Factors. To relate the initial characteristics of the patients to their eventual outcome and to identify homogeneous risk groups is one of the most important tasks of a clinical researcher and medical data bases have been shown to be particularly useful to this end. As an example think of the revision of the role of prognostic factors for Stage I Melanoma of the skin made feasible by the register of the WHO Collaborating Centre for Evaluation of Methods and Diagnosis and Treatment of Melanoma (Cascinelli et al.,1985); /iii/ To plan Clinical Trials. The goal here is to perform exploratory investigations to search for subgroups of patients showing good response to some therapy and to try to get rough estimates of effectiveness of given treatments. These estimates enable the investigator to give insight to the sample size needed to find out the clinically relevant differences between treatments in the clinical trial to be planned.
Dambrosia and Ellenberg (1980), Byar (1980), Simon (1982), Mantel (1983) and Green and Byar (1984) have raised several methodological criticisms on the use of data banks to evaluate treatments efficacy. This use of a medical data base would be a poor substitute for a controlled RCT.
Being aware that the credibility of any research rests mainly on its adherence to high scientific standards in collecting data and analyzing them, Hlatky et al., (1984) adopted very stringent criteria to build up the Cardiovascular Data bank at Duke Medical Center. In discussing the main characteristics of a modern database they say: "Despite the difference in objectives, the methods used in the modern observational database and in a well-conducted clinical trial are similar, except for the means of selecting therapy. Both take care in defining terms, collecting data prospectively and completely, managing data in computerized system, following all patients and employing multidisciplinary research teams."
With this structure the observational database can play a complementary role to the RCT in assessing effectiveness of therapy.

DATA HANDLING IN A SINGLE RCT

The object of a RCT is usually formalized as comparison of two or more treatments, interpreting such a comparison in a very broad sense. On the other hand the complexity of the chronic disease yields a setting in which patients with the same illness form groups remarkably heterogeneous in terms of prognosis. Think, for instance, of stage I melanoma of the skin; three factors clearly influence prognosis: sex, maximum tumor thickness and ulceration (Cascinelli et al., 1985). In analyzing results of a possible RCT carried out to test the efficacy of a new treatment of stage I melanoma, the investigator builds up a regression model both to increase the accuracy of the comparison, by making allowance for the three above cited prognostic factors and to detect their interaction with treatment effects and thus to possibly find out which patients need which treatments.

Testing for a treatment-prognostic factor interaction deserves some comments because misinterpretation of the implications of test for interaction is widespread in RCT. Peto (1982) has distinguished quantitative from qualitative interactions. A quantitative interaction involves variation in the magnitude, but not in the direction, of the true treatment differences according to patients features. On the contrary a qualitative interaction implies a change in the direction of treatment. An interesting example of both types of interaction is given by Micciolo et al., (1985); by means of Cox's (1972) model they investigated the efficacy of chemotherapy (CMF) in breast cancer. Considering the disease-free interval as response, they found a quantitative interaction CMF x menopause, whilst, considering the survival time as response, the same interaction appeared to be of qualitative type suggesting that CMF benefited patients in premenopause but not those in menopause.

Recognizing that significant results in all standard statistical tests for interaction do not give any kind of evidence to the presence of a qualitative interaction, some new approaches have been recently published.

Shuster and van Eys (1983) present a technique to divide a case series into three subsets: one region of superiority for each of the two treatments and one region of uncertainity. This method appears to be particularly useful in the case of a single continuous prognostic factor.

Azzalini and Cox (1984) devise a test of qualitative interaction in a two-way analysis of variance and give approximate distributional results assuming that there are no treatment effects. Gail and Simon (1985) suggest a likelihood ratio procedure suitable for assessing the statistical significance of

qualitative treatment x prognostic factors interactions with specification in advance of the set of disjoint subgroups of patients. Owing to its theoretical properties this test compares favorably with its competitors; in addition it is easy to compute on all the response variables commonly used in RCT and can be applied to published data.
Availability of great amounts of data collected in large multicentric RCTs (as an example,see Moon et al.,(1984) for breast cancer) and of powerful regression computer programmes has stimulated a great deal of studies aiming at investigating the role of putative prognostic factors. These studies are likely to be of an explanatory nature,the analyses of data should therefore be accomplished by different methods and their results should be interpreted with caution.
It is known that stepwise addition of variables tends to examine only a few combinations of the original variables and the models finally selected may not be the best one; nevertheless stepwise regression is the procedure of selecting prognostic factors more widely used. With normally distributed responses Armitage and Gehan (1974) adopted an alternative approach which identifies,among all possible subdivisions, of discrete (or made discrete by paritioning,if continuous)prognostic factors the one which maximizes the "between subgroups sum of squares" of the analysis of variance. By means of Cox's model (1972) this approach was extended to the analysis of censored survival times by Marubini et al.(1983).The final remarks of their paper are pertinent here: "We are aware that the forward stepwise approach of selecting variables in multiple regression may be less effective than the backward method which has many satisfactory properties... .The algorithm presented here appears more similar to the forward approach. However it has proved to be of value in carrying out exploratory analysis of large sets of data...; we have found that the information given by the fashion-tree graph not only reveals the role of the more important prognostic factors but also identifies subsets of patients at different levels of risk and helps to untangle the essential features of a complex network of relationships."
A modified version of the step-down procedure was adopted by the WHO group (1980) to investigate risk factors of stage I melanoma; they started with three main groups of variables: host characteristics, clinical aspects and histological features of the primary tumor. For each of these groups a

"minimal" set of significant factors was selected and in a second stage of analysis these three "minimal" sets were analysed together. A similar strategy was adopted later by Drzewiechi and Andersen (1982).
A recent exhaustive review of methods suitable for modelling response in terms of prognostic factors and the relative problems has been published by Harrell et al.,(1985). In my opinion two issues of this paper are particularly relevant. The first one concerns a five-steps strategy to build up a prognostic model. By arguing that the more clinical insight one inserts into the analysis, the better is the final result, this strategy takes advantage of statistical methodology and clinical experience through a strict interaction of these two components in consecutive steps and enables the researchers to compact the information given by the original set of variables into a few clinical indexes easily interpretable by clincians. However capacities of this strategy need to be deeply tested in different branches of clinical research. The second issue regards the strong suggestion of splitting the dataset into a "training" or model development sample and a "test" sample. This latter makes feasible the validation of the derived model by assessing its predictive accuracy in an unbiased way.

COMBINING FINDINGS FROM RCTs

Conflicting reports of RCTs purporting to evaluate the effectiveness of the same treatments are often found in literature. Such inconsistencies involve variability in admission-exclusion criteria, treatments administration, outcome evaluation and the statistical properties of RCTs with modest numbers of subjects. On the other hand the experience of the last decade tends to suggest that in the near future it is unlikely to devise treatments for any major kind of carcinomas which have large effects, such as preventing or retarding in a substantial way as many as half of the deaths now expected. This implies facing the problem of assembling conclusions from different sources and thus resorting to statistical methodology of meta-analysis (Hedges and Olkin, 1985). Although it has a long history in many research fields, for instance in agriculture, it has only recently concerned medicine (Louis, et al., 1985). Laird and Dersimonian (1982) suggest a simple random effects

model for combining the results of RCTs comparing two treatments which are appropriate with dichotomous responses. The model allows them to estimate the overall difference of the treatments effect as well as the degree to which this difference varies among the RCTs. It is similar to the model proposed by Cochran (1954) for normally distributed variables.
The Bayesian formalism is likely to be valuable here, but, as far as I know, its applications in this context are only marginal. Furthermore methods for handling covariate information particularly when it may be lacking in some studies are needed to increase our skillfullness to assemble findings from different RCTs.

REFERENCES

Armitage P. (1982), The role of randomization in clinical trials, Stat.Med., 1, 345-352.

Armitage P., Gehan E.A. (1974), Statistical method for the identification and use of prognostic factors, Inter.J. Cancer, 13, 16-36.

Azzalini A., Cox D.R. (1984), Two new tests associated with analysis of variance, J.Roy.Stat.Soc. B, 46, 335-343.

Budzar A.M., Gutterman J.U., Blumenscein G.R., Hortobagji G.N., Toshima C.K., Smith T.L., Hersh E.M., Freireich E.J., Gehan E.A. (1978), Intensive postoperative chemoimmunotherapy for patients with stage II and stage III breast cancer. Cancer, 41, 1064-1075.

Byar D.P. (1980), Why data bases should not replace randomized clinical trials?, Biometrics, 36, 337-342.

Byar D.P., Simon R.M., Friedewald W.T., Schlesselman J.J., De Mets D.L., Ellenberg J.H., Gail M.H., Ware J.H. (1976), Randomized clinical trials: prospectives on some recent ideas, New Engl.J.Med., 295, 74-80.

Cascinelli N., Marubini E., Morabito A., Bufalino R. (1985), Prognostic factors for stage I melanoma of the skin: a review, Stat.Med., 4, 265-278.

Chalmers T.C., Schroeder B. (1979), Controls in journal articles, New Engl.J.Med., 301, 1293.

Cochran W.G. (1954), The combination of estimates from different experiments, Biometrics, 10, 101-129.

Cox D.R. (1972), Regression models and life tables (with discussion), J. Roy.Stat.Soc. B, 34, 187-220.

Dambrosia J.M., Ellenberg J. II (1980), Statistical considerations for a medical data-base, Biometrics, 36, 323-332.

Drzewiecki K.T., Andersen K.P. (1982), Survival with malignant melanoma. A regression analysis of prognostic factors, Cancer, 49, 2414-2419.

Fries J.F. (1976), A data bank for the clinician? New Engl.J.Med., 294, 1400-1402.

Gail M., Simon R. (1985), Testing for qualitative interactions between treatment effects and patient subsets, Biometrics, 41, 361-372.

Gehan E.A. (1982), Design of controlled clinical trials: use of historical controls. Canc. Treat. Rep., 66, 1089-1093.

Gehan E.A., Freireich E.J. (1974), Non-randomized controls in cancer clinical trials, New Eng.J.Med., 290, 198-203.

Green S.B., Byar D.P. (1984), Using observational data from registries to compare treatments: the fallacy of omnimetrics, Stat.Med., 3, 361-370.

Harrell F.E., Lee K.L., Matchar D.B., Reichert T.A. (1985), Regression models for prognostic prediction: advantages, problems, and suggested solutions, Canc. Treat.Rep., 69, 1071-1077.

Hedges L.V., Olkin I. (1985), Statistical methods for meta-analysis, Academic Press,, New York.

Hlatky M.A., Lee K.L., Harrell F.E., Califf R.M., Pryor D.B., Mark D.B., Rosati R.A. (1982), Tying clinical research to patient care by use of an observational data-base, Stat.Med., 3, 375-384.

Laird N.N., Dersimonian R. (1982), Issues in combining evidence from several comparative trials of clinical therapy, Proceedings of the XIth International Biometric Conference, Toulouse, 6-11 September.

Lamberg S.I., Green S.B., Byar D.P., Block J.B., Clendenning W.E., Douglass M.C., Epstein E.H. Jr., Fuks, Z.Y., Golitz L.E., Lorinez A.L., McBurney E.I., Michel B., Roenigk H.H. Jr., Van Scott E.J., Yonderheid E.C. (1984), Clinical staging for cutaneous T-cell lymphoma, Ann.Inter. Med., 100, 187-192.

Louis T.A., Fineberg H.V., Mosteller F. (1985), Findings for public health from meta-analyses, Am.Rev.Pub.Hlth., 6, 1-20.

Mantel N. (1983), Cautions on the use of medical data-bases, Stat.Med., 2, 355-362.

Marubini E., Morabito A., Valsecchi M.G. (1983), Prognostic factors and risk groups: some results given by using an algorithm suitable for censored survival data, Stat.Med., 2, 295-303.

Micciolo R., Valagussa P., Marubini E. (1985), The use of historical controls in breast cancer. An assessment in three consecutive trials, Con. Clin.Trials, 6, 259-270.

Moon T.E., Jones S.E., Bonadonna G., Powles T.J., Rivkin S., Budzar A., Montagne E. (1984), Using a data-base of protocol studies to evaluate therapy: a breast cancer example, Biometrics, 3, 333-339.

Peto R. (1982), Statistical aspects of cancer trials. In "Treatment of cancer", Chapman and Hall, London.

Rosati R.A., McNeer J.F., Starmer C.F., Mittler B.S., Morris J.J., Wallace A.G (1975), A new information system for medical practice, Arch. Intern.Med., 135, 1017-1024.

Sacks H., Chalmers T.C., Smith H. (1982), Randomized versus historical controls for clinical trials, Am.J.Med., 72, 233-240.

Shuster J., van Eys J. (1983), Interaction between prognostic factors and treatment, Contr.Clin.Trials, 4, 209-214.

Simon R. (1982), Randomized clinical trials and research strategy, Canc. Treat.Rep., 66, 1083-1087.

Starmer C.F., Lee K.L. (1982), A data-based approach to assessing clinical interventions in the setting of chronic disease, Canc.Treat.Rep., 66, 1077-1082.

Starmer C.F., Rosati R.A., McNeer J.F. (1974), Data-bank use in management of chronic disease, Comp.Biom.Res., 7, 11-116.

WHO Collaborating Centres for the evaluation of methods of diagnosis and treatment of melanoma (1980), Prognosis of stage I melanoma of the skin, Int.J.Canc., 26, 733-739.

Zelen M. (1985), The role of statistics in the design and evaluation of trials in cancer medicine. In "Clinical trials in cancer medicine", Academic Press, New York.

Database Assisted Conduct of Clinical Trials

By N. Victor and R. Holle, Heidelberg

1. Introduction

In the last few years increasing use of the computer and of software in planning, conducting and analyzing clinical trials has been made. With regard to large multicenter studies it is now no longer conceivable to dispense with these tools. The practical importance attached to them by the trialists is shown by the agendas of recent conferences on clinical trials. During the 85-meeting of the Society for Clinical Trials, two whole sessions were spent on discussing this subject alone (v. SOC. CLIN. TRIALS, 1985). In books on clinical trials, however, this aspect is often omitted, and where it is treated at all the presentation lags behind the rapid development (see e.g. POCOCK (1983), LEE (1984)). A description of the tasks which can be facilitated by the computer and an explanation of how to make use of it are given in the papers of GILLESPIE & FISHER (1985) and VICTOR (1986).

This article is not intended to be a review and an evaluation of existing software products, but deals with tasks which are still performed without using - or without sufficiently using - the potential offered by the computer, as these tasks cannot be solved with the ordinary software systems (unless they are subject to further improvements). For in spite of recent improvements many requirements on software for clinical trials have not been fulfilled yet. One of these requirements is in the centre of this essay: a <u>continuous data structure</u> for all software components. The importance of adequate data models and data designs in this context has been pointed out by HOLLE & LEIBBRAND (1985) and HAUX (1981).

Our presentation is based on the experience we have made in the last six years in the 'Zentrum zur Methodischen Betreuung von Therapiestudien' of the University of Heidelberg (ZMBT), which is conducting more than a dozen of large multicenter studies (mainly in the fields of cardiology and oncology).

Compstat 1986

2. The actual situation in the use of the computer

A first step towards using the computer in the field of clinical studies was made by using statistical program packages (e.g. BMDP, SPSS) for analysis. The use of specialized database systems (e.g. SIR) to manage the extensive data was the next important step. Recently, comfortable data entry systems and flexible graphical systems have been added. However, the parallel use of several software tools within one study is uneconomic and ineffective as a result of unnecessary additional work which has to be performed (data transfer, repeated data definition, learning of several command languages). The integration of the systems by adequate interfaces (v. chapter 4) would rationalize this work considerably.

There are a large number of tasks that can be facilitated by using the computer; the list in Tab. 1 does not claim to be complete.

Table 1 Computer assisted tasks during a clinical trial

Study phase \ Type	Statistical	Organizational
Planning	sample size estimation	**design of data forms**
Conduct	**randomization** **study monitoring**	**registration of forms** data entry and update **organizational support**
Analysis	final analysis **presentation of results**	**information archiving**

The tasks which are not printed in boldface letters are largely supported by existing software systems, those printed in boldface have been only insufficiently supported by software tools until now; these will be dealt with in chapter 3. In addition to the usual differentiation by study phases the table is also grouped with regard to the fields of statistics and organization. This shows that the amount of work performed in the field of organization largely exceeds the work necessary to solve statistical problems. It also shows that there is a considerable lack of software tools for these organizational tasks. But exactly here it would be worthwhile to rationalize work by using the computer, so that the biometrician could concentrate on his essential work which is statistics.

The estimation of the necessary sample size is a field where a lot of statistical research has been done. As a result of this there are a large number of programs. Here we only want to mention the system STPLAN (BROWN & HERSON, 1980), because it covers many aspects in the field of study design; the other single programs are usually limited to particular tests and situations and are not integrated (or cannot be integrated) in software systems for clinical studies. If one wishes to estimate statistical power while taking into account balanced randomization, covariate adjustment and group sequential analysis the estimation is feasible only with the aid of simulation techniques. Therefore the integration of a random number generator and of the statistical procedures in a way which allows direct execution and storage of the results in the general system (demanded later in this text) will improve the situation.

Until now the use of the computer in clinical studies has been concentrating on statistical analysis. Most of the former gaps in the method spectrum of the usual packages (e.g. survival analysis) have now been filled. However, as a result of the continuing rapid progress being made in this field, the software producers must make constant efforts to keep the time lag between the development of new methods and their implementation as small as possible. In the practical realization of the propositions valid in theory, statistical and computational problems are strongly connected and Computational Statistics will be very important for further progress. Examples of open problems are the nominal α's for group sequential analyses, closed sequential plans for main and side effects, confidence bounds for empirical survival curves and tests for hypotheses of the type $H_0: \mu_A = \mu_B + \delta$. We cannot deal with any of these problems in this survey. Most of them can only be solved by simulations and to each of them a separate paper under the topic of this session could very well be dedicated.

3. Areas requiring extended computerization

The **design** and testing **of forms** is a time-consuming process, and software tools are urgently needed. Usually the database scheme definition is not started before the finalization of the documentation forms which makes it unusable for this purpose. However, some data base systems (DBS) offer data entry features by allowing the creation of (graphical insufficient) screen images of forms according to the database scheme. Requirement: It should be possible to design forms on the screen and then to print them while storing the design information as part of the data description in the database (DB).

In large studies **randomization** should be carried out centrally so that the procedure can be implemented on a microcomputer with a direct interface to the database (LANGE & MacINTYRE, 1985). This would allow the use of complicated designs for balanced randomization (e.g. data dependent biased coin procedures) as well as immediate checking of inclusion criteria and automatic patient registration. Requirement: Provision of interfaces to randomization PCs.

While remote data entry at clinical sites, as described by GILLESPIE & FISHER (1985) is and will remain an exception, data entry is commonly performed centrally and therefore every week hundreds of data forms from the clinical sites and reference centers arrive at the data center. In case of erroneous or missing forms, requests are sent back and have to be answered. Thus, the clerical task of **forms registration** is a very time-consuming one, but it is most important in order to maintain a high data quality. By using extended DB software instead of paper and pencil, reporting on and ordering of missing forms could be facilitated. For maintaining an overview on the data flow, it is not sufficient to state whether a form is stored or not, but one needs detailed information about the form and its items (completeness, correctness, availability). Requirement: Extension of the DBS by appropriate procedures.

The regular **monitoring** of the study progress requires many (to a great extent monthly) activities by the biometrician: The control of patient recruitment and drop out rates of each clinical centre, the control of the quality of collected data, the sequential control of severe side effects and the control of the protocol compliance. The last point is of special importance since too many protocol deviations may jeopardize the analysis of the whole study unless they are noticed immediately. Requirement: Appropriate routines stored as procedures within the data and method base.

Another set of procedures is needed for the **organizational support** of the participating clinicians, e.g. by means of appointment lists for follow-up visits, requests for certain diagnostic procedures or laboratory tests. Forms supply and drug supply are organizational tasks which might also well be supported by using a computer. Such support seems important especially with regard to the motivation of the clinicians and in order to avoid the impression that computerization serves only to supervise the study participants by the data center; on the contrary, the clinical routine work has to be facililitated for the study patients. Requirement: Extension of the method base by routines for organizational support.

With statistical packages there is still an obvious discrepancy between their range of computational methods and their lack of adequate **presentation of results.** POCOCK (1983, p. 173) is right in stating that "line-printer output is usually not a very effective means of communicating results", but his opinion that "graphical display of results is often best achieved by hand" neglects recent improvements on the graphic software sector, mainly for PCs. To use this software data has to be transferred from the printer output into the PC which is cumbersome and prone to errors. Requirement: An interface between statistical and graphical systems and possibilities to store elaborate graphs, tables and reports as templates, which can be used again in later reports based on updated results.

For the **archiving** of the **data** material without loss of information in a retrievable but condensed form it is not sufficient to merely copy all data on magnetic tapes. Requirement: DB-procedures for creation of self describing data files and printing of directly understandable 'case reports' (condensed description of every case by the most important items).

Nota bene: Most of these requirements are included in the more general requests of section 4 (inclusion of organizational data and metainformation, continuity of data structure, linkability, extensibility).

4. The Central Role of the Database

In our opinion, it is the DBS which has the central role in the methodical and organizational conduct of a clinical study. This is because the DB is the most appropriate tool for steering all organizational and data-processing actions and because it seems therefore reasonable to use the DBS as the central software component to which all other systems can be linked.

An important condition for the use of the DBS as steering tool is that the conceptual study model can be adequately described in the data definition language (DDL). In view of an efficient performance of the organizational task, however, the data structure must include more than the definition of the patient data record. The organizational data and the metainformation contained in the study protocol must also be described in the DB in order to largely automate the conduct of the study and to relieve the responsible personnel from extensive work with paper and pencil.

The organizational data include details about clinics and reference centres (addresses, competences, responsibilities etc.), about the flow of information (listing of all forms, distribution channels), about drug supply and the dispatch of clinical samples. In order to manage these additional details and the data records which are generally no longer hierarchical, the DBS must be based on a particularly flexible data-model. Therefore, as HAUX (1981) and HOLLE & LEIBBRAND (1985) have already observed, those DBS which are based on the relational data model are especially useful in clinical studies.
As opposed to the organizational data the metainformation in the study protocol remains valid for the entire course of the study. The metainformation comprises the inclusion and exclusion criteria, the randomization method and the patient-specific schedule and dosage scheme for therapy and follow-up (with modification rules depending on success or side effects if necessary); it also comprises the scheme of the study as a whole with the intended sample-size, the planned recruitment time, dates of controls, study meetings, interim analyses etc. The schemes in Fig. 1 drawn up for the individual patient and the study as a whole are to give an idea about the size and the importance of this metainformation. Part of it can be directly stored in the data description or in form of appropriate procedures, another part can be formulated as semantical integrity conditions (cf. SCHLAGETER & STUCKY (1983), p. 292).

We will now consider the request that all software components should be linkable to the DB as the central system. By linkability we mean the possibility of simply changing from one system to another in the course of one work session and the easy transmission of all necessary information (also metainformation). By making this request we are not advocating the unmanageable integrated supersystem, but we aim at making work more effective by using several specialized systems at the same time. As for statistical analysis, one needs for the efficient conduct of a clinical trial "a set of software tools between which one can move easily" (NELDER, 1985). Even now, there are appropriate systems for many tasks available. What is missing or insufficient, however, are the interfaces which are necessary for the simple changeover from one system to another. Efforts are being made in order to extend and link individual systems (e.g. SAS data step/procedure step/SAS-GRAPH, SIR/SIR-FORMS), but the commercial software producers are guided by marketing considerations. Thus they extend their own systems with components demanded by the users even if specialized software already exists. The user, however, should be free to compose his own software tools from the components which are the most adequate for his needs in these particular areas.

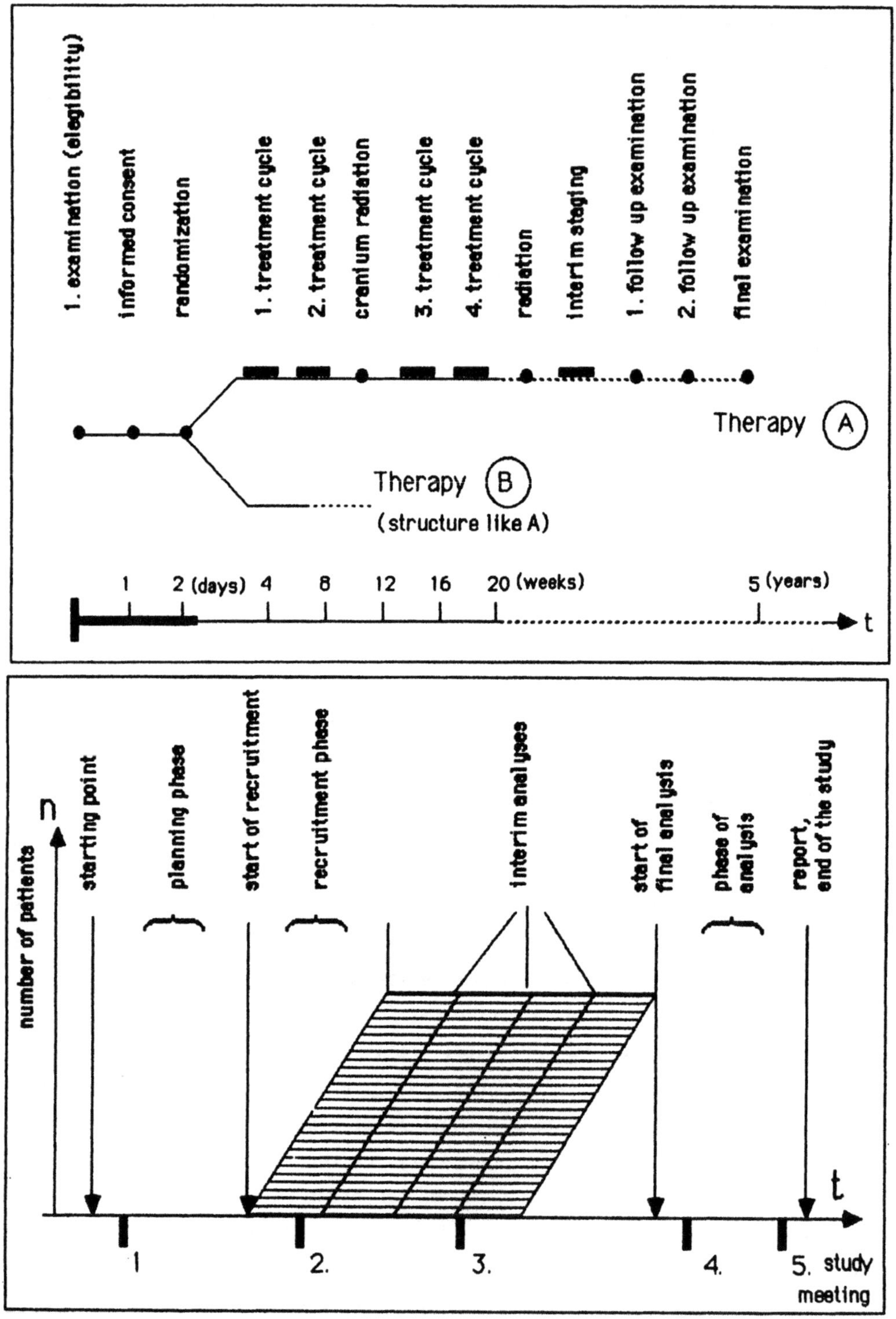

Fig 1: Time schedules for the individual patient (above) and the whole study (below)

We want to consider more closely two demands on the interfaces:

a) continuity of the data structure and

b) extensibility by user-specific routines.

a) Data description in the DDL is the enumeration of all items including their range of legal values and time and logical dependencies; it is correspondingly work-intensive; it is fixed when the forms are designed. When using independent software components this data definition must be done three times, that is when designing the forms, when working out the data entry programs and when setting up the database. It should be possible to define the data structure only once in the DDL of the central DBS, where it would then be accessible to all other work phases and their respective software. For example it must be accessible to the data entry system in order to support the generation of data entry masks and accessible to the form design system in order to generate screen images which are - after having been modified by the user if necessary - ready for print. By using adequate software architecture (as presented in Fig. 2 and also similarly in RASSMANN (1985)) such efficient use of DBS

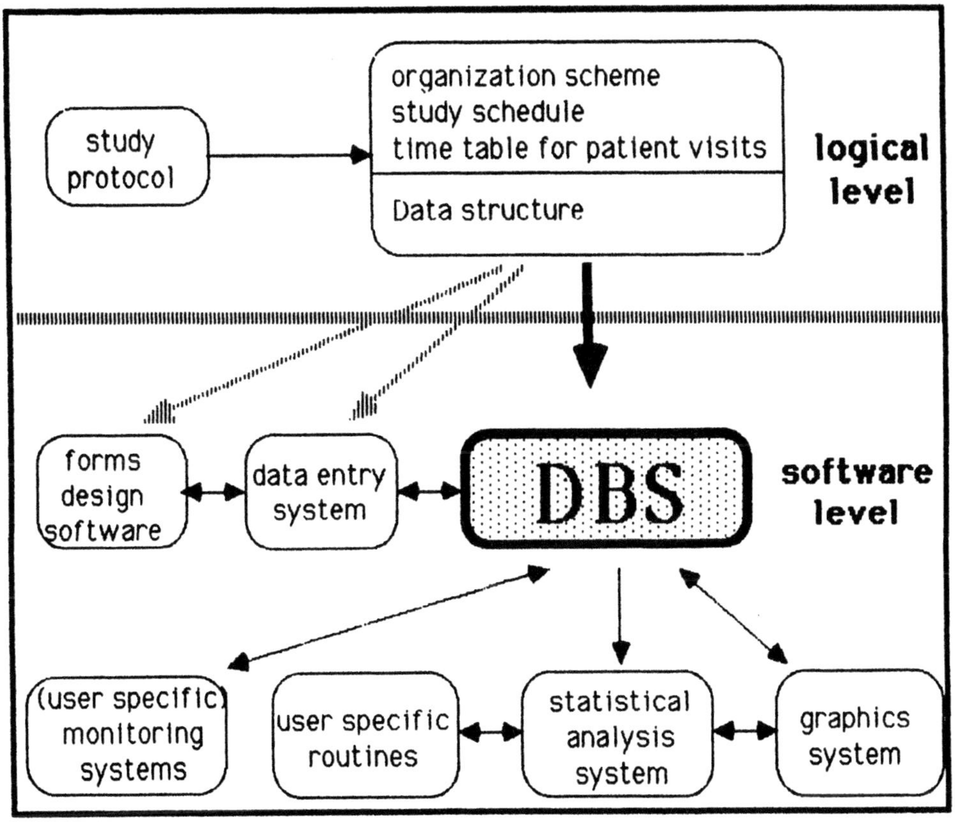

Fig. 2: Linked software systems for clinical trials

data description in other software components is possible, although the sequence demanded 'data description → DBS → data entry screen → printed forms' is opposed to the chronological order 'form design/data entry/data storage and manipulation'. This continuity puts high demands on the flexibility of the individual software components, which must offer several input modes, e.g. with regard to the design of the data entry mask, by the keyboard, by the DDL of the DB and by direct screen manipulations (mouse). It is also important to integrate high-quality graphic systems which are offered by many PCs, but which are not used by present systems for form design, data entry and presentation of results.

b) The DBS used in the study and the analysis system - or the Data- and Methodbase System as defined by HAUX (1983), which contains both - must allow external programs (command sequences of the other systems in Fig. 2 as well as user-specific routines) to be called up in the system as procedures. To avoid that software construction gets unmanageable by comprising too many components, the methodbase for example should be extended by language elements in order to carry out sample-size estimation, randomization, form registration and the writing of appointment letters. The possibility to execute BMDP commands under SAS is a good example for this kind of extensibility. However, the extensibility by user-specific routines on which VICTOR (1984) and MURPHY (1985) have insisted, has been neglected up to now. Nevertheless this is important, because as a result of the variety of analysis problems it will never be possible to solve all problems with standard software and because data transfer to user programmed software and its access outside the general command language are particularly work-intensive.

5. Final remarks and summary

The conduct of large clinical studies without adequate software tools is not conceivable, therefore intensive use of the computer is usual. The computer facilitates the work of the trialist, reduces the number of errors, improves the quality of data and thus the quality of the study results. At the moment the trialists have the choice between numerous individual products for particular tasks of the study, some areas being covered almost completely, others less completely. But more important than filling the gaps in the method and task spectrum is the development of possibilities for the coordinative and simultaneous use of all the necessary software tools. Since the individual products are developped and offered by different firms, one cannot expect an improvement in this direction, if the users do not urgently request compatibility. In this article we want to give the producers some ideas for improvements. However, it is necessary that large centres for the conduct of clinical trials, through their own interface developments, assume a leading role in showing the way for innovation.

References:

BROWN, B.W. & HERSON, J. (1980): STPLAN. A Computer System for Performing Statistical Calculations Needed to Plan a Study. Tumor Institute, University of Texas, Houston.
GILLESPIE, M.J. & FISHER, L.D. (1985): Software for Clinical Trial Data Management. Stat. Softw. Newsl. **11**, 3-8.
HAUX, R. (1981): Einige Bemerkungen zur Frage: Soll man zur Speicherung hierarchischer Datenstrukturen ein Datenbanksystem mit hierarchischem Datenmodell verwenden? Stat. Softw. Newsl. **7**, 58-63.
HAUX, R. (1983): Statistical Analysis Systems - Constructions and Aspects of Method Desig. Stat. Softw. Newsl. **9**, 106-115.
HOLLE, R. & LEIBBRAND, D. (1985): Data Design in Clinical Trials. Stat. Softw. Newsl. **11**, 9-14.
LANGE, N. & MacINTYRE, J. (1985): A Computerized Patient Registration and Treatment Radomization System for Multi-Institutional Clinical Trials. Contr. Clin. Trials **6**, 38-50.
LEE, E.T. (1984): Computerization of Clinical Trials. In: Buyse, M.E., Staquet, M.J., Sylvester, R.J. (Eds.): Cancer Clinical Trials. Oxford Univ. Press, Oxford.
MURPHY, B. (1985): User Extensibility of Statistical Computing Packages. In: Proc. Pacific Statist. Congress, North-Holland, Amsterdam.
NELDER, J.A. (1984): Discussion of the Paper 'Computational Statistics' by N. Victor. Stat. Softw. Newsl. **10**, 122-123.
POCOCK, S.J. (1983): Clinical Trials (Chapter 11). Wiley, New York.
RASSMANN, B. (1985): The Use of Database Systems for Clinical Studies. Stat. Softw. Newsl. **11**, 63-68.
SCHLAGETER, R. & STUCKY, W. (1983): Datenbanksysteme: Konzepte und Modelle. 2. Aufl. Teubner, Stuttgart.
SOC. CLIN. TRIALS (1985): Final Program of the 6th Annual Meeting of the dsoc. Clin. Trials. Contr. Clin. Trials **6**, 209-219.
VICTOR, N. (1984): Extensibility. Contribution in the Panel Discussion 'Standard Packages versus Tailor made Software' at COMPSTAT'84. Stat. Softw. Newsl. **10**, 63-64.
VICTOR, N. (1986): Planning, Execution and Analysis of Therapeutic Studies. Bull. Pol. Academ. Sc. (in press).

An APL-System for Analysing Multivariate Data

By E. P. Liski and T. Nummi, Tampere

Summary: This paper introduces an interactive computer system based on the generalized multivariate analysis of the variance model (GMANOVA). This approach provides comprehensive possibilities for processing multivariate data. Handling of missing data is an integrated part of the system and it is done by using the EM algorithm. Missing data analyses can be easily carried out by using this system. This system also contains flexible possibilities for diagnostic checking of the model. The problem of influential measurements is considered. The influence analysis envisaged also serves as a means for comparing the robustness of various models to missing measurements and to different study designs. In addition to the usual tests of hypotheses within the GMANOVA model, also the single curves and their residuals can be analysed. Comprehensive summaries over individual curves can be performed. Graphical options for looking at the data and the family of growth curves, for example, are available. These programs are written in APL. The users familiar with APL can easily carry out further calculations, prepare further programs and intervene any sequence computations, and check the results immediately.

Keywords: APL, Generalized multivariate analysis of variance, missing data, influential observations, residual analysis.

1. INTRODUCTION

In this paper we consider a computer system which is useful especially for analysing longitudinal data. The basic frame underlying the system is the generalized analysis of the variance model (GMANOVA)

$$Y = XBT' + E, \tag{1.1}$$

where Y is an $n \times q$ matrix of responces, X is an $n \times m$ matrix of known constants, B is an $m \times p$ parameter matrix and T is an $q \times p$ regression matrix. We suppose that both X and T are of full rank. The rows of E are independently distributed with mean vector 0 and covariance matrix Σ. The model (1.1) was put forward by Potthoff and Roy (1964), and it is usually called a growth curves model. This model provides a general format for a variety of multivariate situations. For example medical researchers may be interested in following individuals over a sequence of doses, or epidemiologists may wish to develop pediatric growth curves for blood pressure and other anthropometric variates.

Compstat 1986

In the applications of the growth curve model a time factor is involved. As an example we use data on growth traits of 168 Ayrshire and 40 Finncattle bulls tested at station. The testing covers the interval of 30-365 days of age. Bulls were weighted every 30th day and they were born in 1968 (Liski 1986). In Figure 1 we see the growth of 40 Finncattle bulls plotted using our system.

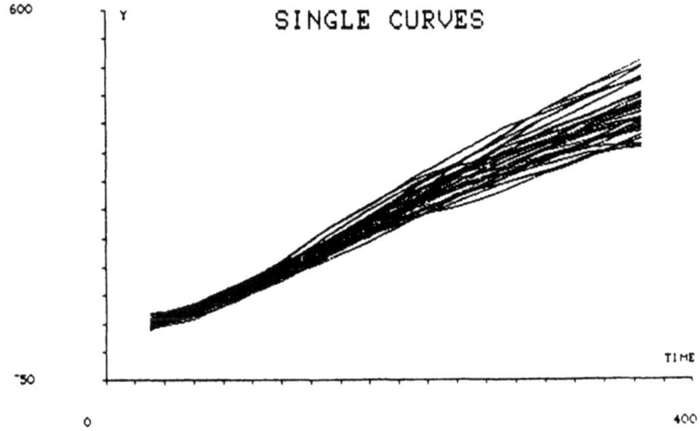

Figure 1. 40 Finncattle bulls weighted every 30th day.

In this kind of longitudinal study each subject is measured at ages which are the same for each subject. The model (1.1) cannot be used in the applications where each sample member can have measurements on arbitrary occasions. For such cases the theory developed by Lindby and Smith can be utilized. Fearn (1975) and Strenio, Weisberg and Bryk have presented such approaches. Also this kind of options will be included in this system, but here we concentrate on the applications of the model (1.1). It should be noted that for example the MANOVA model, the multivariate regression model and the corresponding models of one dependent variable are special cases of the GMANOVA model. However, we do not pay any attention to these special cases. The system represented here is implemented on the IBM Personal Computer AT. The system is easily established also in other computer systems using APL.

2. THE BASIC ANALYSES PROVIDED BY THE SYSTEM

The matrix X in the model (1.1) is called an across-individual design matrix, consisting of indicator variables specifying treatment groups and possible covariates. The matrix T is a within-individual design matrix whose rows, especially in the growth curve applications, are powers of time t_i; $i=1, 2,..., q$. Within the system the across-individual and the within-individual design matrices can be easily specified according to the needs of the problem under study. For the jth subject we could have the univariate linear model

$$\mathbf{y}_{(j)} = \mathbf{T}\boldsymbol{\beta} + \boldsymbol{\varepsilon}_{(j)};\ j=1,2,...,n, \tag{2.1}$$

where $var(\boldsymbol{\varepsilon}_{(j)}) = \Sigma$ for every j and $\boldsymbol{\beta}' = (\beta_0, \beta_1,..., \beta_{p\text{-}1})$. The curves associated with individual can be analysed also separately. Also summaries over specified subgroups of individual curves can be obtained.

The system provides a user with estimates of parameters of the model (1.1). There are several options for estimation, but the maximum likelihood estimate

$$\hat{\mathbf{B}} = (\mathbf{X}'\mathbf{X})^{-1}\mathbf{X}'\mathbf{Y}\mathbf{S}^{-1}\mathbf{T}(\mathbf{T}'\mathbf{S}^{-1}\mathbf{T})^{-1}, \tag{2.2}$$

where $\mathbf{S} = \mathbf{Y}'[\mathbf{I} - \mathbf{X}(\mathbf{X}'\mathbf{X})^{-1}\mathbf{X}']\mathbf{Y}$, is the first choice as an estimate of $\mathbf{B}$. Note that $\mathbf{S}/(n-m)$ is an unbiased estimate of Σ.

According to the Figure 1 we decided to fit a polynomials of third degree on the bulls born in 1966. It is easily done by using GCURVE-function in our system. As the right argument we have given the previously made regression matrix T. It should be noted that we have used centralized time points. The result of the function is the estimated parameter matrix $\hat{\mathbf{B}}$.

Example 1. The use of GCURVE-function on the bulls born in 1968.

```
      B←GCURVE T

**********************************************************************
                                                    12.27  15.5  1986

                        - ESTIMATION OF GROWTH CURVES -



IT IS POSSIBLE TO CHOOSE 3 TYPES OF ESTIMATORS

1. WEIGHTED LEAST SQUARES ESTIMATOR
2. USUAL LEAST SQUARES ESTIMATOR
3. MAXIMUM LIKELIHOOD ESTIMATOR

GIVE YOUR CHOICE: 3

------------------------------
| ESTIMATED PARAMETER MATRIX |
------------------------------

  1  2.201E002  1.232E000  3.423E-04 -4.206E-06
  2  2.385E002  1.268E000 -7.494E-05 -5.104E-06
```

The hypotheses of the form

$$H_0: \mathbf{CBD} = \mathbf{M} \tag{2.3}$$

can be tested under the model (1.1). In the hypothesis formulation (2.3) **C**, **D** and **M** are given matrices of order $g \times m$, $p \times v$ and $g \times v$ respectively. The most typical tests in GMANOVA applications are the equality of group mean curves, the parallelism of the mean curves and the significance of individual parameters. These three basic alternatives are offered to the user as standard tests. The user is also provided with several test statistics and with the probability connected with the value of the statistic. The basic alternative is the likelihood-ratio approach.

3. HANDLING OF MISSING DATA

It is often the case in longitudinal analyses that over time there will be an attrition of the original cohort. In case of missing data the consequence usually is that the original problem turn into a more complex one and the standard approaches are no longer applicable.

The EM algorithm (Expectation and Maximization algorithm) for the model (1.1) is presented in Liski (1985). The algorithm is iterative and each iteration consists of two steps. First, the missing values are replaced by their expectations under the assumption that the parameters are fixed and the observations are given. Thereafter the log-likelihood function is maximized under this completed data. Here we have assumed the multivariate normal distribution. Dempster, Rubin and Laird (1977) have studied the general properties of the EM algorithm. It has been exhibited also conditions under which convergence to maximum likelihood is obtained. The strict application of the EM algorithm provided only linear convergence to maximum likelihood. Our programs apply this approach to the estimation of parameters in the growth curve model (1.1).

The procedure described above provide us with the estimates of the parameters although a part of data is missing. Given the estimated parameters of the model, we often wish to impute missing measurements by assigning suitable estimates in place of missing values and to use this completed data as inputs to further statistical analyses. We have applied this approach also to our data on the bulls born in 1968.

Occasionally, also computational aspects support the completing of data. In many longitudinal studies the basic data sets are very large, and the estimation process under missing data is iterative. However, we might repeteadly analyse the varying subsets of this same data using iterative techniques. Therefore, completing first the basic data may greatly shorten the overall computation time, reduce costs and increase flexibility in data analysis. Of course, this strategy must be applied with great care, and by considering approximation errors.

4. CERTAIN DIAGNOSTIC CHECKINGS

The problem of influential measurements both at the design level and at the inference level can easily be studied within the system. Several types of curves can also be fitted on the data in order to find the most appropriate model.

The system affords flexible possibilities for assessing the influence of different subsets of data on the estimate of the parametric functions $\mathbf{CBD}$. Let $\hat{\mathbf{B}}_{(IJ)}$ denote the estimate of $\mathbf{B}$ based on the subset of the data obtained by deleting the measurements whose indeces contain in the index set IJ. We define the measure of influence as follows:

$$\mathbf{D}_{(IJ)}(\mathbf{CBD}) = (n-m)\,\mathrm{tr}(\mathbf{S}_H\mathbf{S}_E^{-1}), \tag{4.1}$$

where

$$\mathbf{S}_H = [\mathbf{C}(\hat{\mathbf{B}}-\hat{\mathbf{B}}_{(IJ)})\mathbf{D}]'[\mathbf{C}(\mathbf{X}'\mathbf{X})^{-1}\mathbf{C}']^{-1}[\mathbf{C}(\hat{\mathbf{B}}-\hat{\mathbf{B}}_{(IJ)})\mathbf{D}] \quad \text{and} \quad \mathbf{S}_E = \mathbf{D}'(\mathbf{T}'\mathbf{S}^{-1}\mathbf{T})^{-1}\mathbf{D}.$$

In fact $\mathbf{D}_{(IJ)}(\mathbf{CBD})$ is like the Lawley-Hotelling trace statistic. Therefore $\mathbf{D}_{(IJ)}(\mathbf{CBD})$ provides a measure of the distance between $\hat{\mathbf{B}}$ and $\hat{\mathbf{B}}_{(IJ)}$ in terms of descriptive levels of significance.

The influence analysis described in the foregoing can be carried out when all the data to be studied are available. However, knowing the importance of an observation at the design stage would be highly useful. It may be a good idea not to collect observations which are least influential. Of course, detecting influential observations at the inference stage is also useful for the future planning of similar experiments. We can also use a measure of influence that can be applied at the design stage (Liski 1986).

In order to study the adequacy of the models, we can calculate summary statistics from the residual matrix, a summary from the standard deviations corresponding to single curves and also a summary from the root mean square of different time points. Note that the user can print standard error for every single curve and root mean squares for every time point.

5. DISCUSSION

This system provides good possibilities to study a large family of curves simultaneously. For individual curves different test statistics and residual analyses can be carried out and the curves can be displayed on screen, for example. Further analysis and summaries based on these individual curves can be easily accomplished.

AKNOWLEDGEMENT

This paper was prepared while the first author was a senior Research Fellow in the Academy of Finland.

REFERENCES

Beale, E.M.L. and Little, R.J.A. (1975). Missing values in multivariate analysis. *J. R. Statist. Soc.*, B, 37, 129-145.

Dempster, A.P., Laird, N.M. and Rubin, D.B. (1977). Maximum likelihood from incomplete data via the EM algorithm. *J. R. Statis.Soc.*, B, 39, 1-22.

Fearn, T. (1975). A Bayesian approach to growth curves. *Biometrika* 62, 1, 89-100.

Greenlees, J.S., Reece, W.S. and Zieschang, K.D. (1982). Imputation of missing values when the probability of response depends on the variable being imputed. *J. American Statist. Ass.*, 77, 251-261.

Lindley, D.V. and Smith, A.F.M. (1972). Bayes estimates for the linear model (with discussion). *J. R. Statist. Soc.*, B, 34, 1-41.

Liski, E. (1985). Estimation from incomplete data in growth curves models. *Commun. Statist. - Simul. and Computation*, B 14(1).

Liski, E. (1986). A growth curve analysis for bulls tested at station. To appear in *Biometrical Journal*.

Orchard, T. and Woodbury, M.A. (1972). A missing information principle: theory and applications. *Proc. 6th Berkeley Symposium on Math. Statist. and Prob.* 1, 697-715.

Potthofff, R.F. and Roy, S.N. (1964). A generalized multivariate analysis of variance model useful especially for growth curve problems. *Biometrika* 51, 313-326.

Sternio, J.F., Weisberg, H.I. and Bryk, A.S. (1983). Empirical Bayes estimation of individual growth-curve parameters and their relationship to covariates. *Biometrics* 39, 71-86.

AMS - A Computer Program for Survival Analysis, General Linear Models and a Stochastic Survival Model

By A. Morabito, Milano

ABSTRACT: AMS is a computer program of general use in the field of statistics which resort to generalized linear models theory. It permits a discrete handling of given data and estimated vectors making use of a simple language, documented on line.

The reason for presentation of AMS is not the will of competing with excellent statistical packages (PSTAT, BMDP, SAS, GLIM..), but simply the fact that an old unsophisticated product GLIMOD with good informatic structure has been improved to obtain the actual performances. The main characteristics are:

1) interactive computing with video, print and graphic frames
2) high number of statistical models available
3) definition of ad-hoc models adding few FORTRAN lines
4) data transformation , sorting ... , selecting on condition
5) search for maxima, minima any particular values ......
6) design matrix with Wilkinson and Rogers structure formulae
7) examination of all vectors and matrices
8) multiple plots of both estimated and given vectors
9) statistics for goodness of fit and residuals analysis
10) analysis of selected subgroups of cases
11) saving and reloading intermediate elaborations

AMS deals with two statistical types of models:

a) generalized linear models, GLM. (Nelder 1972)
b) survival models. (Kalbfleisch and Prentice 1980)

GENERALIZED LINEAR MODELS

Given n independent observations y(i), i=1..n, of a random variable Y, drawn from a distribution f[Y(i),θ(i),Φ] belonging to the family of exponential functions with one parameter, Φ is a nuisance parameter, θ(i) is function of a linear combination of p independent variables X(i)={x(i,1),.,x(i,p)} with weights β :

$$g[\theta(i)] = \sum \beta(j)\cdot x(i,j) = \beta' X(i)$$

β's are estimated with maximum likelihood approach and good-

Compstat 1986

ness of fit is tested with likelihood ratio test. GLM have a flexible structure so they can describe different error distributions, imposing only two constrains on $y(i)$ and pdf equation:

a) $\{y(i)\}$, $i=1..n$ are independent observations and

b) $f[y(i)] = \exp([y(i)\cdot a(\mu) - b(\mu)]/\phi + c[y(i),\phi])$

A lot of pdf fit the constrains a) and b) , provided a suitable choice of functions $a(\mu), b(\mu), c[y(i),\phi]$. All the usual pdf utilized in statistics, and with rank transformation, some non parametric pdf are included. The asymptotic properties of maximum likelihood estimates are valid when :

$$d \ln(L)/d\mu = [\, y(i)\cdot d\, a(\mu)/d\mu - d\, b(\mu)/d\mu \,] \,/\, \phi = 0$$

so that $\mu = E[Y] = db(\mu)/da(\mu)$.

Covariance matrix is computed from Cramer-Rao theorem (Kendall and Stuart vol.2). ϕ, when unknown, is independently estimable.

LINK FUNCTION

The expected values μ are function of a linear form $\mu(i) = \beta' X(i)$ and of the pdf through a link function $g[\cdot]$. The correct choice of link function is the core of the theory and use of GLM:

$$\theta(i) = g^{\sim}[\mu(i)] = g^{\sim}[\beta' X(i)].$$

Moreover, since the parameter $\mu(i)$ span the whole real axis and θ can vary in a limited range, as for Poisson variates, the link function assumes the meaning of transformation of variable.

Examples of link functions are $g[\theta(i)] =$

a) μ Identity that gives rise to linear regression with gaussian error.

b) $\ln(\mu)$ Logarithm that gives rise to analysis of contingency tables with Poisson error.

c) $\ln(p) - \ln(1-p)$ Log-logit that gives rise to analysis of proportions with binomial error.

Not all pdf are described as GLM, as the negative binomial, and survival pdf. In these cases it is necessary to use algorithms capable of estimating all parameters of likelihood function and its derivatives.

ANALYSIS OF SURVIVAL TIMES

Two reasons suggest the introduction of particular models for survival analysis: the first is the need of estimating a scale parameter, as ϕ in GLM context, the second is the task of treating with censored data. Censoring mechanism is often unknown and contrasts with unbiased estimates of parameters. These dificulties suggest sometime to give up the chosen model in favour of good estimates and asymptotic properties.

In survival studies a single piece of information is the type : $\{d(i),t(i),c(i)\}$, where $d(i)$ indicates the status, $[d(i)=0 \rightarrow$ alive, $d(i)=1 \rightarrow$ dead$]$, $t(i)$ is survival and $c(i)$ is censoring time. A linear combination of covariates of i-th subject with β weights is linked to parameter $\theta(i)$:

$\theta(i) = g^{\sim}[\beta' X(i)]$ is the inverse of link function

$F[t(i),\theta(i)]$ is the cdf of death

$f[t(i),\theta(i)]$ is the pdf of death

When the survival times are known these formulae are exaustive, otherwise the censoring distribution has to be considered. $c(i)$ for $i=1..m$ are the observed values of rv $C(i)$ representing censored times with cdf$= H[C(i)]$ and pdf$= h[c(i)]$, independently distributed from true death times $T^{\circ}(1),..,T^{\circ}(m)$:

$$Pr\ \{T \in [t(i),t(i)+dt),d(i)=1;\theta(i)\} =$$

$$Pr\ \{T^{\circ} \in [t(i),t(i)+dt),C(i) \geq t(i);\theta(i)\} = H(t(i)+0\} \cdot f\{t(i);\theta(i)\}dt$$

$$Pr\ \{T \in [t(i),t(i)+dt),d(i)=0;\theta(i)= h\{t(i)\} \cdot F\{t(i)+0;\theta(i)\}dt$$

where $t(i)+0$ is the right limit computed at $t(i)$ when $F(\cdot),G(\cdot)$ are continuous functions. The hypothesis of complete randomness of censoring mechanism generates the likelihood function:

$$f[t(i);\theta(i)]^{d(i)} \cdot F[t(i)+0;\theta(i)]^{1-d(i)}$$

The pair of variables $[t(i),d(i)]$ for $i=1..n$ given $X(i)=\{x(i,1) ...,x(i,p)\}$ are conditionally independent. The likelihood function factorizes into two parts: the first being the contribute of subjects dead at i-th time and the second of subjects still alive at i-th time. Usually, one wants to know whether individuals, with particular values of covariates, are the ones lost to follow up, that is they have an high risk of censoring just before they eventually die. This fact makes impossible to have sensible results from survival studies, since it happens that the censoring mechanism interferes with failure process.

GENERAL MODELS FOR SURVIVAL ANALYSIS

Generally, a researcher prefers models containing linear combination of variables to set up a so called prognostic index linked to location parameter of pdf. A second parameter, a scale parameter, is intrinsically not estimable in a separate context. Let $y(i)= \ln[t(i)]$, $\mu(i)= \beta' X(i)$, $w(i)= [y(i)-\mu(i)]/\sigma$. The log-likelihood function in terms of w is:

$$\ln\{L(\beta)\} = \Sigma\ d(i) \cdot \ln[f\{w(i)\}/\sigma] + [1-d(i)] \cdot \ln\{F[w(i)]\}$$

where: $f(w) = pdf(w)$ $\qquad \lambda(w) = pdf(w)$ conditional

$F(w) = cdf(w) = \int_0^t f(u)\,du \qquad = f(w)\,/\,[1-F(w)]$

From this log-likelihood function it is possible to obtain the estimates of β weights and scale parameter σ for many models, equating to zero the derivatives respect β and σ.

$$0 = U(\beta,\sigma,j,i) = d\ \ln(L)/d\beta(j) = \Sigma\ x(i,j)\cdot a(i)\ /\ \sigma$$

$$0 = U(\beta,\sigma,p+1,i) = d\ \ln(L)/d\sigma = \Sigma\ [w\cdot a(i) - d(i)]\ /\ \sigma$$

Examples of survival models in terms of hazard function are:

a) EXPONENTIAL Hazard is constant : $\lambda(w) = \lambda^{\bullet}$

$pdf(w) = \exp[w-\exp(w)]$

$cdf(w) = \exp[-\exp(w)]$

b) WEIBULL Hazard is time dependent: $\lambda(t) = \lambda^{\bullet}\cdot p\cdot(\lambda^{\bullet}\cdot t)^{p-1}$

$pdf(w) = \exp[w-\exp(w)]$ where $p = 1/\sigma$

$cdf(w) = \exp[-\exp(w)]$

c) LOG-LOGIT Hazard is time dependent: $\lambda(w) = \exp(w)/[1+\exp(w)]$

$pdf(w) = \exp(w)/[1+\exp(w)]^2$

$cdf(w) = \exp(w)/[1+\exp(w)]$

d) LOG-NORMAL Hazard is time dependent: $\lambda(w) = gau(w)/\Phi(w)$

$pdf(w) = \exp(-w^2/2)/\sqrt{2\pi\sigma^2}$

$= gau(w) \qquad \Phi(w) = \int_w^{\infty} gau(u)\,du$

$cdf(w) = 1-\Phi(w)$

e) GAMMA Hazard is time dependent: $\lambda(w) = pdf(w)/[1-cdf(w)]$

$pdf(w) = [\exp(w)-\exp(w/p)]/[p\cdot\Gamma(p)]$

$cdf(w) = \int_0^w u^{p-1}\exp(-u)/\Gamma(p)\,du$ where $p = 1/\sigma$

f) PROPORTIONAL HAZARD When the pdf is unknown, it is useful the Cox's regression model, that assumes that any event gives information on the value of β weights of subjects who die and it doesn't depend on previous history. The proportional hf is:

$$\lambda(\mu,t) = \lambda^{\bullet}(t)\cdot\exp[\beta' X(i,t)]$$

The likelihood function is made by contributions conditioned to subjects at risk $R(i)$ at the time of i-th death:

$$Pr\ \{j,J(i);t(1),\ldots,t(i-1),J(1),\ldots,J(i-1)\} =$$

$$= \exp[\beta' X\{j,t(i)\}]\ /\ \sum_{s\in R(i)} \exp[\beta' X\{s,t(i)\}]$$

$X(i,t)$ = vector of covariates of i-th subject at time t

β = vector of unknown parameters that describes log-linear relation of the risk with covariates.

$J(i)$ = sequence of death and censoring up to i-th time

When it is possible to imagine censored times fixed in advance, asymptotic properties hold for $U(j)$:

$$E\ [U(j)] = E\ [d\ \ln P\{j,J(j)\}/d\beta\ ] = 0$$

$U(j)$ have mean zero, are uncorrelated and have properties similar to those of maximum likelihood.

AVAILABLE MODELS

AMS includes till now some pdf: Gauss, Poisson, binomial, gamma and negative binomial and, for survival, exponential, weibull, log-logit, log-normal, gamma , Cox's model, and a sthochastic survival model (Serio , Morabito 1985).
If a model is not present in AMS, it is possible to include it by writing few program lines into a subroutine FUNCTP.

BX = μ(i) = linear form link(BX) = link function
clik(BX,Y) = logarithm contribute to likelihood function
grad(BX,Y) = first derivative of clik in respect to BX
hess(BX,Y) = second derivative of clik respect to BX

DEVIANCE AND RESIDUALS ANALYSIS

Minus twice the log of likelihood function in a point of maximum has an asymptotic chi-square distribution. The hypothesis testing contrasts the hypothesis H° with alternative H1:

$$H^\circ : \beta(i) \in \Omega \quad \text{and} \quad H1 : \beta(i) \in \Omega - \omega$$

where Ω represents the full model ω represents the reduced one.

$\Omega=\{\beta(1),\ldots,\beta(p)\}$ $\quad\omega=\{\beta(1),\ldots,\beta(s),0(s+1)..,0(p)\}$ $\quad s \le p$

Kendall and Stuart show that the log of the ratio

$$W= \max \{L(\beta);\beta\in\omega\} \;/\; \max \{L(\beta);\beta\in\Omega\}$$

under regularity conditions and when H° is true $-2*\ln(W)$ has a chi-square distribution.
The default residuals given by AMS are: $U[\beta(i)]= d \ln(L)/d\, \beta(i)$
Crude residuals are also available.
In survival analysis, residuals are defined as product of observed times by the expected hazard function, that is shown to have unitary exponential distribution. Graphical capabilities support residuals analysis.

GOODNESS OF FIT

Goodness of fit test is based on weighted differences among observed and expected frequencies on a partition of space of covariates {X(i)}, into a fixed number K of distinct regions.
Incidence function I(m) m=1..K is defined over R(m) regions.
I(m)=1 if $\{x(1,m),\ldots,x(p,m)\} \in R(m)$, I(m)=0 otherwise.
Considering the enlarged model: $g''[\beta(i)]= \beta'X(i) + \Gamma'I^\circ(i)$,
where: $I^\circ(i)=\{i^\circ(i,1),\ldots,i^\circ(i,k)\}$ and $\Gamma=\{\tau(1),\ldots,\tau(k)\}$,
H° assumes that $\tau(1)=\ldots=\tau(k)=0$ and is tested by the Rao efficient-score-statistics (Linear models 1973,pag 418): $T= \delta'V^{-}\delta$.

δ= {d ln(L)/dτ(1),...,d ln(L)/dτ(k)} is vector of derivatives and V~ is the generalized inverse of covariance matrix V.
All terms are computed at τ(m)=0 and β(j)=β~(j) [maximum likelihood estimates]. T has an asymptotic chi-square distribution with degrees of freedom equal to the rank of matrix V.
In the case of dichotomous logistic model one obtain simply a vector {d(i),x(i,1),..,x(i,p)} for each subject where d(i) = 1 indicates a positive response and d(i)=0 indicates no response:

$$\ln(L) = \Sigma\, d(i)\cdot[\beta' X(i)+\Gamma' I^{\bullet}(i)]-\ln(1+\exp(\beta' X(i)+\Gamma' I^{\bullet}(i))]$$

So when Γ'=0 and β=β~, in the m-th region of the partition the difference among observed and expected subjects is: O(j)-E(j)=

$$d\ln(L)/d\tau(j)=\Sigma\, d(i)\cdot i^{\bullet}(i,j)-I^{\bullet}(i)\cdot\exp(\beta' X(i))/[1+\exp(\beta' X(i))]$$

In survival analysis one has to add to the partition the time elapsed from the beginning of experiment. Called f(j) the observed number of deaths into a generic element of the partition [t(j),t(j+1)) one can compute the expected number of deaths:

$$E[f(j)] = \sum_{i\in R(j)} E[I(i);R(i)] = \delta(j)$$

Schoenfeld shows that the statistics W= Σ [f(j)-δ(j)]²/V(j) has asymptotic chi-square distribution with k-1 d.o.f and variance approximated by $V(j) = \sum_{i\in D(j),\, i\in I(j)} \delta(i)\cdot[1-\delta(i)]$

were D(j) is the set of subjects died at j-th time.

REFERENCES

Cox D.R. (1972) Regression models and life tables (with discussion). J.R.Statist.Soc. B, 34, 187-220

Kalbfleisch J.D.,Prentice R.L. (1972) The statistical analysis of failure time data. John Wiley and Sons,New York

Morabito A.,Marubini E. (1976) A computer program suitable for fitting linear models when the dependent variable is dichothomous, polychotomous or censored survival,and when the dependent variable is quantitative. Comp.Prog.in Biomed., 5, 283-295

Nelder J.A.,Wedderburn R.W.M. (1972) Generalized linear models. J.R.Statis.Soc. A, 135, 370-384

Rao C.R. (1973) Linear statistical inference and its applications. John Wiley, New York

Schoenfeld D. (1980) Chi-squared goodness of fit tests for proportional hazard regression model. Biometrika, 67, 145-153

Serio G.,Morabito A. (1985) Stochastic survival model with covariates: an application to breast cancer. 11th IMACS,2,251-253

Wilkinson G.N.,Rogers C.E. (1973) Symbolic description of factorial models for analysis of variance. Appl.Statis.,22,392-399

Econometric Computing

A Multilingual Specification System for Statistical Software

By V. J. de Jong, Groningen

Summary

An important development to improve the maintainability of software is the use of specification systems. In these systems a specification language is defined in which the design of software can be expressed. The specification is automatically checked for consistency and completeness. In this paper CONDUCTOR, a multilingual specification system for statistical software is discussed. The implementation of a bootstrap estimator is used as an example.

Keywords

multilingual environment, maintainability, statistical software, specification languages, formal specification

Goals

- improve maintainability of statistical software
- involve non-computer scientist in software development
- improve communication between experts that cooperate in the use and development of statistical software

1. Introduction

Existing statistical software can only be maintained by professional programmers. The program source of these packages is written in higher level programming languages. The implementation of statistical techniques in these packages is tedious; the program source of a package may contain over 100,000 lines of code. Even in a well organized modular design, of the package modifications can only be made by experts with knowledge of the architecture of the package. It goes without saying, that experts in statistics not having the required programming skills and knowledge of the architecture, can not adapt these software packages to their needs. This reduces the possibility to add valuable knowledge from experts in statistics to statistical software. New ideas first have to be explained to programmers. This takes time, costs money and causes errors.
The problem of maintainability is not unique for statistical software. In computer science this problem recieves a lot of attention. A survey by Lientz and Swantson [1980] revealed that approximately 65% of maintenance

Compstat 1986

was perfective (i.e.: was the result of changes demanded by the user after the first version of the software product was finished). Perfective maintainance is also a common feature in statistical software where users want to apply the most up-to-date statistical techniques.
An important development to improve the maintainability of software is the use of specification systems. In these systems a language is defined in which the design of the software can be expressed. This language can be in predicate calculus form, algebraic form, relational form, program form or in declaration form. The specification is automatically checked for consistency and completeness. Then it is transformed into an (efficient) executable program. The processor doing this transformation does not accept just some variant of a conventional high level programming language. Rather, it accepts a language which is much closer to the problem domain, enabling domain experts to add their knowlegde to the system. An example of a specification system is given in Cheng et al [1984]. This system allows managers to create software for financial accounting.
In this paper a specification environment for statistical software is discussed: CONDUCTOR. CONDUCTOR is multilingual, it enables experts in different specialized statistical areas to cooperate in the use and development of software. CONDUCTOR is multilingual because communication between highly specialized experts is most conviently done in more than one language. What is a clear and understandable language to one expert is tedious and unreadable to another.
To develop statistical software one needs expertise in technical statistics, data production, statistical computing, database systems, user interfaces, graphics and compiler/interpreter construction. In CONDUCTOR each expert is able to contribute his knowledge. Unnecessary detailed information is hidden from an expert through the introduction of separate specification languages (levels): vertical information hiding. In CONDUCTOR languages (levels) are distinguished for:

- technical statistics: the development of appropriate statistical techniques.
- data collection: the development of techniques to collect data for applied research.
- empirical statistics: the actual statistical inference from data.
- higher level programming: the virtual machine level.

The technical statistical level in the current version of CONDUCTOR focusses on econometric parameter estimation techniques (CONDUCTOR can easily be adapted for other applications). In CONDUCTOR a technical statistician can implement statistical techniques and a set of tests on the underlying assumptions of this technique. Using exception handlers the technical statistician can specify how the resulting statistical software reacts if the assumptions are violated or if exceptions are raised on other levels of the software. Technical statisticians can implement and modify estimation techniques, written in the specification language, without involvement of professional programmers. CONDUCTOR performs all the compile-time checking of formal type- and dimension specifications of matrix references and function arguments.

The data collector is able to describe in CONDUCTOR the way the data is collected. The data is not only stored as a collection of numbers. David [1985] pointed out that also sample design, instrument and context should be added. All this information is stored in a relational database system. If the data is loaded in an statistical technique this database is interrogated to check the consistency of the retrieved data. If an inconsistency occurs an exception is raised. CONDUCTOR reacts on this exception using an exception handler written by the technical statistician. If no such handler exists it will use an exception handler written by the data collector.

Information hiding on the virtual machine level is realized through modularization: horizontal information hiding. To enforce modularization on this level of CONDUCTOR we used initial algebraic semantics for algebraic specification with conditional equations (note that we talk about the specification of our specification environment) as discussed in Klaeren [1983] and Bergstra et al [1985]. The formal specification contains definitions of: the data types on each level, the abstract syntax trees for the languages on each level, the virtual machine and the translation of abstract syntax trees including the type checking.

The statistical software created in CONDUCTOR can be used by applied statisticians who want to apply statistical techniques and available databases to their problems. For the applied statistician this software will have two big advantages compared with existing software. The first advantage lies in the maintainability of the resulting software. New developments can be implemented without the need of interference with other experts. The second advantage is that CONDUCTOR produces more than just statistics. CONDUCTOR can also give advice to use alternative statistical methods in case a test indicates that the used technique is not adequate for the problem at hand or in case problems occur at other levels of CONDUCTOR. The technical statisticians and data-experts can write exception handlers to control CONDUCTOR's knowlegde at this point.

2. Example: the implementation of a bootstrap estimator

In this example a bootstrap estimator is implemented in CONDUCTOR, using the specification language for technical statisticians.

2.1. The bootstrap method

The bootstrap is a statistical method for non-parametric estimation of statistical error. In our example we have observations of a dependent variable (Y) and some independent variables ($X_1,\ldots,X_k$). Suppose the distribution (F) of our dependent variable Y is unknown. Therefore standard statistical methods can not be applied. The bootstrap method solves this problem by approximating the distribution F by the empirical distribution of the n observations of Y.

$\hat{F}$ = mass (1/n) on each observed data point of Y.

The bootstrap method now proceeds as follows:

(1) Draw a bootstrap sample from the observations.

(2) Calculate the ordinary least squares estimator for the sample.

(3) Repeat the procedure under (1) and (2) 'n_of_draws' time and calculate the average and variance of the estimators using for example the Welford algorithm.

These calculated averages and variances are called the bootstrap estimators of the population parameter and its variance. For a further theoretical explanation of bootstraps see for example Efron and Gong [1983].

2.2. The statistical specification program

The bootstrap estimator, described in section 2.1, is implemented with the use of the use of the program in fig. 2.1. In this program an exception handler is specified for an interrupt from the virtual machine level; near-singularity. This shows the power of the exception handling mechanism in CONDUCTOR. Near-singularity is an abstract notion for the technical statistician. The check on near-singularity is implemented on the virtual machine level of CONDUCTOR. Still the abstract notion of near-singularity has a clear meaning on the technical statistical level: multicollinearity. The exception handler of the technical statistician communicates this meaning in a understandable message for the application level user in CONDUCTOR. An other important feature of the specification language is also demonstrated in this example; the matrices may all be declared using formal index expressions. The application level user however may not be confronted with run-time type checking errors. Therefore extensive compile-time checking exists in CONDUCTOR to ensure no type checking errors will occur during run-time. CONDUCTOR can check for example that the expression X-temp'* Y-temp in fig. 2.1 will never generate problems during run-time.

Figure 2.1. The specification program for the bootstrap estimator.

```
NAME Bootstrap
%
INPUT
     INDEX number_of_independents, number_of_observations,
           sample_size, number_of_draws;
     MATRIX [1 to number_of_observations,
             1 to number_of_independents]        X
             MESSAGE: "The independent variables are:"    ;
     MATRIX [1 to number_of_observations]        Y
             MESSAGE: "The dependent variable is:"  ;
OUTPUT
     MATRIX [1 to number_of_independents]        beta
             MESSAGE: "The bootstrap estimators is:"  ;
     MATRIX [1 to number_of_independents]        var_beta
             MESSAGE: "The variance of the bootstrap estimator:"
%
IMPLEMENTATION
     VARIABLES
          INDEX      I,J ;
          MATRIX [1 to number_of_observations,
                  1 to number_of_independents]   X-temp;
          MATRIX [1 to number_of_observations]   Y-temp;
          MATRIX [1 TO number_of_independents]   beta_temp;
     EQUATIONS
          beta = 0 ;
          FOR i := 1 TO number_of_traps DO
               {
               /* draw sample */
                FOR j := 1 TO sample_size DO
                    {
                    sample := draw_random_index
                         (1,number_of_observations);
                    X-temp[j] := X[sample];
                    Y-temp[j] := Y[sample];
                    };
               /* calculate the OLS estimator */
               b-temp := inv(X_temp'*X_temp)*(X_temp'*Y_temp);
               /* update beta and beta_var */
               beta := ((i-1)/i)*beta + (1/i)*b_temp;
               FOR k := 1 TO number_of_independents DO
                    {
                    beta_var[k] := beta_var[k] -
                         (k-1)/k * (beta_var[k]-beta[k])^2
                    };
               }
%
     WHEN near_singular:
     MESSAGE:
     "THE INDEPENDENT VARIABLES IN SOME BOOTSTRAP SAMPLES ARE STRONGLY
     CORRELATED (MULTICOLLINEARITY). DUE TO THIS THE NUMERICAL RESULTS
     OF THE TECHNIQUE MAY BE INACCURATE."
```

2.3. Using the bootstrap level on the application level

The implementation of the bootstrap technique does not effect the knowledge implemented on other levels of CONDUCTOR. Exception handlers written by the data_expert, the application user interface, etc. are not effected by this newly implemented technique. As a result the technique can be used on the application level instantaneously. The technique is used in the following application level session

```
NAME TECHNIQUE: Bootstrap
SAMPLE PERIOD:  1945-1985

number_of_independents =?   3
number_of_observations =?  31
sample_size =?             20
number_of_traps =?        100

The independent variables are: income, indirect_tax, credit
The dependent variable is: consumption
```

Possible warnings and suggestions from experts in other scientific areas may accompany the output.

```
WARNING {statistical expert}

"THE INDEPENDENT VARIABLES IN SOME BOOTSTRAP SAMPLES ARE STRONGLY
CORRELATED (MULTICOLLINEARITY). DUE TO THIS THE NUMERICAL RESULTS OF
THE TECHNIQUE MAY BE INACCURATE."

WARNING {data expert}

YOU USED LONGITUDINAL DATA FOR THE VARIABLE consumption BASED ON
MEASUREMENTS WITH DIFFERENT INSTRUMENTS.
```

3. Current status conductor

- complete formal specification using initial algebraic semantics with conditional equations.
- development of a prototype under UNIX.

References

Bergstra, J.A., Heering, J., & Klint, P. (1985), Algebraic Definition of a Simple Programming Language, Amsterdam: Centre for Mathematics and Computer Science, Report CS-R8504.

Cheng, T.T., Lock, E.D., & Prywes, N.S. (1984), Use of Very High Level Languages and Program Generation by Management Proffessionals, IEEE Transactions on Software Engineering, 552-563.

David, M.H. (1985), The Language of Panel Data and Lancune in Communication About Panel Data, Madison, Wisc.: Draft for Discussion, Institute for Research on Poverty.

Efron, B., & Gong, G. (1983), A Leisurely Look at the Bootstrap, the Jack-knife and Cross-Validation, The American Statistician, 36-48.

Klaeren, H.A. (1983), Algebraische Spezifikantionen: Eine Einfuhrung, Berlin: Springer Verlag.

Lientz, B.P., & Swanson, E.B. (1980), Software Maintenance Management, Reading: Addison-Wesley.

An Efficient Algorithm for Time Series Decomposition

By C. Scarani, Bologna

Summary

The aim of this work is to present an efficient algorithm for time series decomposition. Following Akaike (1980), who uses a linear bayesian model, it is possible to take advantage of the model structure to reduce the computational effort.
Corradi and Scarani (1984) used regularization methods, whereas Ishiguro (1984) trasformed the coefficient matrix to a banded matrix by suitable permutations. Both methods reduce the computational cost with respect to the original code (Akaike and Ishiguro, 1980).
In this work we propose a new approach based on Cholesky factorization of a banded system. Following George (1981) we observed that it is possible to set up an iterative algorithm to find the solution. In typical cases it converges in few steps. The proposed algorithm is very cheap with respect to previous methods. It requires a storage of order N, where N is the observation number of the time series. The operation number is of the same order.
Some applications to economic series are presented.

Key words: Seasonal adjustment; band Cholesky factorization.
Classification: 14-Econometric computing.

1. Introduction

Seasonal adjustment is necessary for a lot of economic time series and there are many types of procedures to do it. Akaike (1980) proposed an interesting method for the Bayesian Decomposition of a time series which is very expensive to run (Akaike and Ishiguro, 1980). The computational aspects have been studied and futher numerical algorithms, see Corradi and Scarani (1984) and Ishiguro (1984), have achieved good results in time and storage. A novel approach based on Cholesky factorization of a banded system is presented here.
In section 2 the problem is described and the solution is given. In section 3 the computational procedure is outlined, The Fortran implementation of this is the program SEA.

Compstat 1986

Its theoretical performances are compared with those of the original code BAYSEA (Akaike and Ishiguro, 1980) and of SEAGIV (Scarani and Sermasi, 1984) in sect. 4 and the real execution time is shown in sect. 5, where the codes are applied to a economic series.

2.The problem

The problem can be summarized as follows. It is supposed that a seasonal time series of period p (e.g. 4, 12,..) $[y_i]$, i=1,2,..N, can be represented as $y_i = t_i + s_i + i_i$ where t_i denotes the trend, s_i the seasonal component and i_i the irregular component. The component i_i is assumed $NID(0,\sigma^2)$. A linear model can be chosen for the following tasks: to model the gradual change of the seasonal component through seasonal differences of order m, to constrain the sum of s_i's within a period be close to zero and to model the trend through differences of order k. These informations are included in a matrix D, partitioned into 3 blocks, which corresponds to the previous costraints. The blocks related with the seasonal component are multiplied by the parameters f and g. See Corradi and Scarani (1984) for detail on the matrix D. The following least square problem is solved:

$$\min_{a} f(a) = \min \| D(a-\mathring{a}) \|^2 + q^2 \| Xa - y \|^2 \qquad (1)$$

where a is the vector of parameters $a=(s_N,t_N,\ldots,s_1,t_1)'$ $\mathring{a}$ is an initial guess for a, D is the 3N*2N matrix of the model, X is a N*2N matrix so made: $x_{i,j}=1$ for j=2*i-1 or j=2*i, =0 elsewhere, and q is a further parameter.The parameter q permits a balance between fidelity to the model and to the data. It is chosen in wise to minimize the criterion ABIC suggested by Akaike (1980)

$$ABIC(q)=-N.\ln(N.q^2)+N.\ln[f(a_*)]+\ln|D'D+q^2X'X|-\ln|D'D| \qquad (2)$$

where a_* is the solution of (1). Consequentely it is necessary to solve eq. (1) for several q and to choose the solution a_* associated with q that minimizes ABIC(q).
The normal equation system associated with eq. (1) is Pa=b, where $P=D'D+q^2X'X$ is a 2N*2N matrix and $b=D'D\mathring{a}+q^2X'y$.
The matrix P turns out to be block tridiagonal and involves only the matrices A, B, C where A and C are symmetric positive definite and B is sparse lower triangular of dimension l*l, where l=max(2pm,2k). A, B, C structure is fixed once defined a particular model. The vectors $b' = (b'_1,b'_2,\ldots,b'_n)$ and $a'=(a'_1,a'_2,\ldots,a'_n)$, n=2N/l, are partitioned in blocks of dimension l. Following George (1981) it is possible to perform the factorization P = L L'

$$P=\begin{bmatrix} A & B & & & \\ B' & C & B & & \\ & B' & C & B & \\ & \cdots & \cdots & \cdots & \\ & & & B' & C \end{bmatrix} = \begin{bmatrix} L_1 & & & \\ W_1' & L_2 & & \\ & W_2' & L_3 & \\ & \cdots & \cdots & \\ & & W_{n-1}' & L_n \end{bmatrix} \begin{bmatrix} L_1' & W_1 & & & \\ & L_2' & W_2 & & \\ & & L_3' & W_3 & \\ & & & \cdots & \cdots \\ & & & & L_n' \end{bmatrix}$$

where L_i , $i=1,2,..n$ e W_i, $i=1,2,....n-1$, are lower triangular matrices of dimension lxl. It is possible to obtain ABIC observing that

$$\ln |P| = \ln|D'D + q^2X'X| = 2 \; \Sigma_{i=1}^{n} \; \Sigma_{j=1}^{l-1} \ln[L_i(j,j)]$$

and

$$\ln \|f(a_*)\|^2 = \ln[(D\mathring{a})'(D\mathring{a})+q^2y'y - z'z]$$

(see Lawson and Hanson,1974, page 122), where $z'=[z_1',...,z_n']$ and z is the solution of the system $Lz = b$.

3.Structure of the algorithm

The algorithm is organized as follows:
assign å and the specific model;
for q=0:

a0) build up A,B,C, $u^2=(D\mathring{a})'(D\mathring{a})$ and $v^2=y'y$. Observe that to compute u^2 it is not necessary to provide the whole matrix D. For the usual choice of å, everywhere zero except the block $\mathring{a}_n$, u^2 can be evaluated easily.

b0) obtain L_1 by factorization of $A = L_1L_1'$, calculate $det=\ln|L_1|$ and W_1 from $L_1W_1 = B$,

repeat the step c0) for i=2,3,..n-1

c0) obtain L_i by factorization of $C-W_{i-1}'W_{i-1}$, calculate $det=det+\ln|L_i|$ and W_i from $L_iW_i=B$,

d0) obtain L_n by factorization of $C-W_{n-1}'W_{n-1}$ and $det=det+\ln|L_n|$,

now $\ln|D'D|=2*det$;

then for every q:

a1) build up A,B,C,b_i, $i=1,2,..n$, and $c^2=u^2+q^2v^2$,

b1) obtain L_1 by factorization of $A = L_1L_1'$, compute z_1 from $L_1z_1 = b_1$ and W_1 from $L_1W_1=B$, calculate $det=\ln|L_1|$ and $d^2=z_1'z_1$

repeat the step c1) for i=2,3,..n-1

c1) obtain L_i by factorization of $C-W_{i-1}'W_{i-1}$, compute z_i from $L_iz_i = b_i-W_{i-1}'z_{i-1}$ and W_i from $L_1W_i = B$, calculate $det=det+\ln|L_i|$ and $d^2=d^2+z_i'z_i$,

d1) obtain L_n by factorization of $C-W_{n-1}'W_{n-1}$, compute z_n from $L_nz_n = b_n-W_{n-1}'z_{n-1}$, calculate $det=det+\ln|L_n|$ and $d^2=d^2+z_n'z_n$, now $\ln|D'D+q^2X'X|=2*\det$ and $\|f(a_*)\|^2=c^2-d^2$

e1) calculate ABIC(q);

choose q_* such that the value of ABIC(q_*) is minimum; save L_i and z_i for i=1,2,...n; then restore the solution for $q=q_*$:

a2) compute a_n from $L_n'a_n=z_n$,

repeat the step b2) for i=n-1,n-2,..,1

b2) compute W_i from $L_iW_i=B$ and a_i from $L_i'a_i=z_i-W_ia_{i+1}$.

4.Comparison among different codes

Two FORTRAN programs are available: BAYSEA and SEAGIV. The procedure described above is implemented in the program SEA. The three codes implement different algorithms, but the results are the same.

BAYSEA realizes a two stage orthogonal decomposition by means of Householder trasformations. The cost is proportional to $(2N)^3$ flops for each value of the parameter d, where d=1/q; (the flop is more or less the amount of work associated with a floating point add, a floating point multiply and a little subscripting, Golub and Van Loan (1983) page 32). The storage is $6N^2+O(N)$ memory locations.

SEAGIV performs a reduction of the least squares problem into standard form, adapting a method developed by Elden (1977). A first step gives a cost proportional to $7/3.N^3+O(N^2)$ flops, but subsequent iterations are proportional to 19.N for each value of q. The storage is $3.N^2+O(N)$ memory locations.

In SEA, all the steps, including the initial and final one, have the same cost. Therefore we will give only the costs for the steps from a1) to e1). There are n Cholesky factorizations, n-1 evaluations of W_i and n-1 products $W_i'W_i$ ($l^3/6+l^2/2+l/3$ flops), n solution of lower triangular system and n-1 products of type $W_i'z_i$ ($l^2/2+l/2$ flops), n det and d^2 computations (l flops). The total amount for every q is $N.l^2+5N.l+8N$ flops. The storage is N(l+6) memory locations. The real costs are even less due to the sparsity of matrix B.Numerical experiments have shown also that the difference between L_i and L_{i-1} rapidly vanishes after few steps of c1) in section 3. This could be utilized to obtain a further reduction in computational cost.

5.Application to an economic series.

The programs are been applied to an economic time series: XG70 (Export of Italian Goods, Thousand Millions Lire, Costant Prices 1970). The series cosists of 88 quarterly observations collected from 1960.1 to 1981.4. The order of differencing of the trend is k=2 and of the seasonal component m=1.

Three samples of the first 24, 44, 88 observations are chosen for comparison of execution times on realistic cases, running on a CDC CYBER 180 model 835. A table of results is reported.

	N=88	N=44	N=24
BAYSEA:			
no. ABIC evaluations	3	3	4
minimum ABIC	871.01	385.80	168.45
for q	0.7	0.7	0.5
CPU time (sec)	34.98	5.27	1.33
SEAGIV:			
no. ABIC evaluations	10	10	10
minimum ABIC	872.00	383.55	168.06
for q	.46	.06	.46
CPU time (sec)	19.25	2.74	.59
SEA:			
no. ABIC evaluations	20	20	20
minimum ABIC	872.00	383.90	168.06
for q	.51	.12	.51
CPU time (sec)	2.82	1.38	.73

The number of ABIC evaluations changes because different minimization routines are used in the codes and usually the function ABIC is almost flat. A simple tabulation is used in SEA.

6.Conclusion

The procedure outlined here has shown to be much more efficient then the previous ones. In this way it appears to be of pratical applicability to deal with operational problem of seasonal adjustment.
The structure of the method looks like a filter approach. The seasonal adjustment problem has been faced by Kitagawa (1981) by means of the Kalman filter: he used the criterion AIC instead of (2), leading to slightly different results. Future research should clarify if the two approches can be unified.

7.Bibliography

Akaike, H. (1980), Seasonal Adjustment by a Bayesian Modelling, J. Time Ser. Anal. 1,1-13.
Akaike, H. and M. Ishiguro (1980), A Bayesian Seasonal Adjustment Program, Computer Science Monographs, no. 13, The Institute of Statistical Mathematics, Tokyo.
Corradi, C. and C. Scarani (1984), A Note on the Computation of the Bayesian Decomposition of a Time Series, J. Time Ser. Anal. 5, 205-212.
Elden,L. (1977), Algorithms for the Regularization of Ill Conditioned Least Squares Problems, BIT 17, 134-145.

George,A. (1981), Direct Solution of Sparse Positive Definite System; Some Basic Ideas and Open Problems, in "Sparse Matrices and their Uses", Duff ed.,Academic Press, London, 283-306.
Golub, G.H. and C. F. Van Loan (1983), Matrix Computation, the Johns Hopkins Univ. Press, Baltimore, Maryland.
Ishiguro, M. (1984) Computationally Efficient Implementation of a Bayesian Seasonal Adjustment Procedure, J. Time Ser. Anal. 5, 245-253.
Kitagawa, G. (1981), A Nonstationary Time Series Model and its Fitting by a Recursive Filter, J. Time Ser. Anal. 2, 103-116.
Lawson, C. L. and R. J. Hanson (1974), Solving Least Squares Problems, Prentice-Hall Inc., Englewood Cliffs, New Jersey.
Scarani, C. and D. Sermasi (1984), SEAGIV: un algoritmo per la decomposizione bayesiana di una serie storica univariata, Rapporto Scientifico no. 24 del Progetto DESEC, Centro di Specializzazione e ricerche economico-agrarie per il Mezzogiorno, Univ. di Napoli, Portici.

Statistical Data Base Management Systems

Database Management and Statistical Data Analysis: The Need for Integration and for Becoming More Intelligent

By R. Haux and K.-H. Jöckel, Bremen

SUMMARY: Database management and statistical data analysis are often considered as two rather independent parts. In this paper we want to point out that database management should be integrated into statistical analysis systems and that difficulties arise when database management is done within a database system whereas statistical data analysis is done separately in a statistical analysis system. Within such a system it is also possible to obtain a more "intelligent" system which is able to support a user more efficiently.

KEYWORDS: Database management; statistical analysis systems; statistical expert systems.

1 INTRODUCTION

Having in mind the existing concepts and realizations of statistical analysis systems and database management systems, we would like to promote the following ideas:

1. Database management should be integrated into statistical analysis systems, because it is not possible to separate data management and data analysis.
2. Such "integrated" svstems should be made more "intelligent" in order to support a user more efficientlv This can be done by adding knowledge (statistical knowledge and knowledge from the subject matter field) into such systems.

It is intended to make some proposals for future

Compstat 1986

developments of statistical analysis systems (or statistical expert systems, respectively). It is not intended to point out possibilities how to work with existing database management and statistical analysis systems. The type of statistical data analysis, we have in mind, is that of a prospective study (randomized clinical trial, cohort study, observational study, ...).

2 THE NEED FOR INTEGRATION

Possibilities for combining database management and data analysis have been discussed rather extensively (cf. Havranek, Sidak and Novak, 1984, pp. 255-290; Haux, 1983/84, p. 112; Rassmann, 1985; Victor, 1984, p. 113 and the references cited in these papers). The need for simultaneously working with both features in obvious. However there are different views and some controversies how to do so, either by

1. separating database management and data analysis on the one hand or by

2. integrating these two parts, on the other hand,

where 1. and 2. may be regarded as the two extremes. The separation can be realized by having two different systems - a database management system (e.g. SIR) and a statistical analysis system (e.g. BMDP). The integration can be realized by imbedding both parts within one system, without having the parts separated (e.g. like SAS). Usually this integration can be achieved by adding a database management component to the data- and methodbase management system of a statistical analysis system (see Haux, 1983/84 for the terminology). Another possibility would be, e.g., the addition of a statistical data analysis component into a database management system (cf. the summary-table-by-example approach from Ozsoyoglu and Ozsoyoglu, 1984).

The arguments for separation and against integration principially follow two lines:

1. During the design phase and during the carry-out phase of a study we need data management support (checks for the data acquisition, sorting data, combining files, ...); we do not need features for statistical data analysis.
2. During the analysis phase of a study, we need statistical data analysis features (for confirmatory, descriptive or exploratory data analysis); we do not need database management support. All data analysis is done by having a case-oriented sequential file (the data "matrix") and by computing some statistical measures on the basis of this file.

Let us now try to point out that this functional separation does not represent adequately the work of a statistician.

With respect to the first point: At least during the carry-out phase a statistician usually does not only check the data by semantic integrity constraints like "if sex is 'male' and if pregnancy is 'yes' then print some error message". In order to obtain a high data quality a statistican will look at univariate distributions (are there extreme values? How does the shape of the distribution look like?), (s)he will possibly look at bivariate, or at bivariate projections of multivariate distributions. In addition it is often required that intermediate (statistical) analyses have to be done. In both cases we need statistical data analysis features besides database management features.

With respect to the second point: During the analysis phase we can have the following type of problem. E.g., for confirmatory data analysis it is intended to compute n tests. In order to take the multiple test problematic into consideration a multiple level of significance is defined. The type one error for the single test problems will be calculated, e.g., by applying the Bonferroni-Holm procedure. This can be easily done if the p-values from the n tests can be stored in the database of a statistical analysis system, combined, eventually sorted, and if the test decisions can be computed on the basis of these p-values. Similar situations occur, if we have measures of location and dispersion for several (sub-)groups of data, if we want to sort them, and print or plot the combined results. In both cases we need database management features besides statistical data analysis features.

If we want to support a statistician (or a user with sufficient statistical knowledge for the problem under investigation) appropriately we should offer tools that have both kinds of features!

3 THE NEED FOR BECOMING MORE INTELLIGENT

Up to now it has been argued that the integration of database management and statistical data analysis features into a statistical analysis system would essentially support the work of a statistician.

We now try to outline that such integrated systems can be improved further by inbedding more a-priori knowledge for the analysis. One way to achieve this is an elaborated type concept, as outlined in Haux (1986). Another way is to represent knowledge in the knowledge base of a statistical expert system (cf. Gale and Pregibon, 1984; Hahn, 1985; Wittkowski, 1986; Wolstenholme and Nelder, 1986 and the references cited in these papers).

It must be admitted that the existing statistical expert systems up to now are far from being saturated. They are now at a point where statistical analysis systems and data base management systems have been many years ago. Some people even doubt or express their reservations against a future wider use and development of these systems, see for example Zelen (1984) for some critical remarks. We do not want to add some new positive or negative statements to this debate. Instead we want to show that the need for more intelligent systems (as which we consider statistical expert systems) releasing the statistician from routine work is and will become a major point. To see this let us start by considering the prinicipal components of statistical work. The (applied) statistician has to combine

1. statistical methodological knowledge,
2. knowledge about the subject matter from which his data are,
3. the ability to formulate and formalize the problem under study and to apply the knowledge in order to solve the problem,
4. knowledge from computer science if the data are of a relevant magnitude,

cf. Jöckel (1986) for a more thorough discussion on these points.

In general a lot of work arises to process the data which are to be analysed. If we accept what has been argued by Victor (1984) that the statistician should have access to the knowledge from the subject matter, this would imply the additional ability to retrieve a knowledge base. If we look at the work of statisticians today we have to admit that a relevant proportion of their time is consumed by these activities, activities that are only of indirect use to the end of a good and adequate statistical analysis. The question arises whether it is possible to release the statistician at least partially from this work without a loss of quality. To our opinion the answer to this question is positive: as the user of a modern, high-level programming language no longer needs to know the internal representation of his statements in machine code there seems to be no need that the user of statistics knows all the details of how the data are stored, retrieved etc.. What (s)he needs to know are the possibilities how to communicate with a system that allows her or him to enter, manipulate and analyse the data in a way statisticians are used to think.

What is needed is a software tool with the following characteristics:

- it offers a communication tool that is orientated towards statistical thinking,
- it handles the data structures adequate to the analysis,
- it gives additional information on the subject matter and on statistical methods if wanted.

With respect to the last point: What is then needed would be to represent knowledge (statistical and from the subject matter field) in a knowledge base and offer a data- and knowledge base management system that processes the different kinds of knowledge and manages the statistical analysis under the direction of the user. As to this point the reader is referred to Jöckel (1986) where an example is given how a data- and knowledge base management system has to operate jointly on several knowledge and data bases in the context of planning a clinical study.

To summarize: by offering software tools that integrate database management and statistical data analysis, we can obtain systems which, in our opinion, are more orientated towards statistical work and towards statistical thinking. However, in order to support a user more efficiently, we can still make a further step and add statistical and subject matter knowledge for better describing data and knowledge. In this sense the integration of database management and data analysis features may be regarded as a first step towards more intelligent systems.

4 ADDITIONAL REMARKS

Another (more technical) aspect concerning integration is, that data structures and algorithms strongly depend on each other. Many of the statistical algorithms are originally based on non-case-oriented data structures (cf. Haux 1983/84 pp. 14-20). Therefore the interfacing of systems by using a case-oriented data structure (e.g. of type data "matrix" or of type relation) will always be problematic. Systems, which support (on the user's (command) language level) only the data "matrix", will remain restricted for a statistician as well as for a more thorough design of a statistical knowledge base. This restriction will not, as sometimes is argued, be improved by storing not only the "usual" set of cases with its variables in the data matrix, but by furthermore allowing the storage of matrices and of two-dimensional contingency tables. It will even become worse: Obviously, all three kinds of data can be stored in such a data matrix. However, they need different sets of operations for managing the different kinds of data. Here the design of analysis systems can be improved, if we use imperative or even functional languages with suitable type concepts and with the possibility of adding (abstract) types.

Of course, in addition, by interfacing systems there may also be a loss of information.

We also want to mention that the demand for needing more intelligent analysis (or expert) systems is not new, and has e.g. already been stated by Nelder (1977). Concerning knowledge management: The maintainance of the represented knowledge will be as difficult as that of statistical method bases. Here we should also use concepts that suitably support creation and management of knowledge bases (cf. Blum, 1982, for possibilities of representing knowledge).

The topics mentioned in our paper could only be discussed in a rough and somewhat simplified way. Problems like language interfaces for different kinds of users and others could not be treated.

We feel that there are lots of possibilities for a further integration of concepts from computer science and from statistics in order to construct better - e.g. more integrated and more intelligent - tools for a statistician.

REFERENCES

Blum, R.L. (1982). Discovery and representation of causal relationships from a large time-oriented clinical database: the RX project. Berlin: Springer.

Gale, W.A. and Pregibon, D. (1984). Constructing an expert system for data analysis by working examples. In: Havranek, T., Sidak, Z., Novak, M. (eds.). COMPSTAT 1984, 227-236. Wien: Physika.

Hahn, G.J. (1985). More intelligent statistical software and statistical expert systems: future directions. Am. Statistician 39, 1-8.

Haux, R. (1983/84). Statistical analysis systems - construction and aspects of method design. Statistical Software Newsletter 9, 106-115 and 10, 14-27.

Haux, R. (1986). Database management and statistical data analysis: the need for appropriate data types and data structure types. To appear in: Proc. XIIIth Int. Biometric Conference, Seattle, USA.

Havranek, T., Sidak, Z. and Novak, M. (eds.) (1984). COMPSTAT 1984. Wien: Physika.

Jöckel, K.-H. (1986). Statistical expert systems and the statistical consultant - considerations about the planning stage of clinical studies. To appear in: Haux, R. (ed.). Expert systems in statistics. Stuttgart: Fischer.

Nelder, J.A. (1977). Intelligent programs, the next stage in statistical computing. In: Barra, R. et al. (eds.). Recent developments in statistics, 79-86. Amsterdam: North Holland.

Ozsoyoglu, Z.M. and Ozsoyoglu, G. (1984). STBE - a database query language for manipulating summary data. Proc. IEEE COMPDEC Conf..

Rassmann, B. (1985). The use of database systems for clinical studies. Stat. Software Newsl. 11, 63-68.

Victor, N. (1984), Computational statistics - tool or science? Stat. Software Newsl. 10, 105-116.

Wittkowski, K.M. (1986). Generating and testing statistical hypotheses: strategies for knowledge engeneering. To appear in: Haux, R. (ed.). Expert systems in statistics. Stuttgart: Fischer.

Wolstenholme, D.E. and Nelder, J.A. (1986). A front end for GLIM. To appear in: Haux, R. (ed.). Expert systems in statistics. Stuttgart: Fischer.

Zelen, M. (1983), Biostatistical sciences as a discipline - a look into the future. Biometrics 39, 827-837.

Privacy, Informatics and Statistics

By G. Marbach, Roma

1. Statisticians have always been quite sensitive to the need for protection of privacy of information regarding both people and legal bodies (institutions, associations and commercial business). This attitude is, as a matter of fact, one of the key conditions to achieve cooperation and reliable data. Information collected through a statistical sample, a census, fiscal registrations, social security sources, etc. has a high risk of different kinds of improper use. Therefore the national statistical offices make every possible effort to render inaccessible individual data by protecting the original files and avoiding publication of statistical tables in which extreme values with a very low frequency appear.

The same attitude marks banking statistics. In Italy, for example, data on bank loans and deposits are only made available for those areas where at least three branches operate and where the amount of transactions goes beyond an established level.

In the fields of information with which the computer deals, the right to privacy is no longer equitable with an aristocratic right to isolation, "to be let alone" (1). On the

(1) S.D. Warren and L.D. Brandeis, *The Right to Privacy*, Harvard Law Review, 4, 1890-91, 193-220.

Compstat 1986

other hand, today everyone must have the right to decide if, when and to whom, information related to his/her person, firm, organization, etc. may be divulged ("right to share and to withhold"). In synthesis, all this deals with the right to preserve one's own identity, even in facets considered by computer files.

2. The fear of extraneous, unauthorized access to information concerning persons as well as legal bodies, dates back to the end of the 1960s. In the U.S. the legislative intervention stems from the 1974 Privacy Act (Public Law 93-579, 31.12. 1974). In Europe the first data protection laws were approved in Germany in two of the States or Länder (Hesse, Oct. 7, 1970 and Rhineland-Pfälz in 1974) and later on, in 1977, the Federal Act on protection against misuses of personal data in data processing appeared. The Swedish Parlament had passed an organic Data Act which came into being on July 1, 1973. The Data Processing and Freedom Act appeared in France in January 1978. In June of the same year laws on this subject were enacted in Denmark and Norway; in Austria the law on protection of personal data appeared in October. In July, 1984, Great Britain enacted the Data Protection Act, which gives British citizens the right to see computerized information about themselves from November 1987 onwards (2).

It should be noted that the Danish law was the first in Europe to extend the right of privacy to legal bodies.

The right of a person to know what data concerning him or her are held in a data bank (3) and the uses for which

(2) *The Economist*, April 9, 1986, p. 26.

(3) *Data bank* is intended to mean a set of informatized data in which technical problems are neglected and socio-economic

./.

that information is employed has even been introduced in the new Portuguese and Spanish constitutions (1977 Act, n. 35 and 1978 Act, n. 18, respectively).

Finally, the Council of Europe approved in September 1980 a convention for the protection of the individual with respect to personal files.

3. As society enters the informatic era and data banks increase in numbers, the need for protection of the individual develops at the same time as the consciousness of the increasing risk of losing control of the information's uses. The "Big Brother" complex is today associated with the syndrome of the goldfish in a transparent bowl - an analogy of the fear of being exposed without escape from scrutinies that register even small and apparently insignificant acts of normal life. As a matter of fact, events which are adaptable to elaboration devices are: telephonic alarms, telephone calls, bank operations, railways and airline reservations, information requests for vacations in specified areas, hotel bookings, car repairs, medical analyses and other aspects related to health, electrical hook-ups, use of water, mail orders, etc. The widespread use of home-tech systems which combine and integrate telephone, television and computer, raises the danger of unauthorized intrusion, interception and abuse in general connected with the information handling.

A large part of our everyday life could, therefore, be

./. aspects primarly considered. This set of data is then suitable for any utilizer, even if not exactly specified in advance. See: A. Rizzi, *Alcune considerazioni sulle problematiche delle banche dei dati*, Univ. of Rome "La Sapienza", Dep. of statistics, probability and applied stat., 1984, n. 5, 3.

reconstructed by registrations contained in electronic files through, fortunately, difficult and expensive, but at the same time possible, operations. It is therefore essential that access to personal data be reserved exclusively to the interested person and this information be given to others only on the explicit and binding consent of the interested party, unless the information is of a general character (such as demographics). This is the same as the protection the law gives to one's reputation and to the secrecy of correspondence and telephone calls (4). People, essential components of social life, would not be in agreement with the changing of their home-fortress into a golfisch bowl (5). Furthermore, the right to privacy entails full freedom for everyone to consider if, when, how and under what conditions personal information may be communicated to a third party and to which third parties. The *right to access* must then guarantee to everyone the ability to discover which information, related to their own person, is inserted into data banks. In addition to this, the right to control the accuracy and appropriate use of this information must be safeguarded, together with the rights to make corrections and additions. Finally, the *right to "oblio"* has to ensure cancelation of those data which are no longer valid or were related to an objective which has already been reached. Such a right must coexist with the rights to privacy and access.

---

(4) R. Borruso, *Per una disciplina delle banche dei dati personali*, see bibl., Autori vari, *Banche di dati e diritti della persona*.

(5) V. Frosini, *Diritto alla riservatezza e calcolatori elettronici*, see bibl., G. Alpa e M. Bessone.

It is easy, therefore, to identify in the society of informatics the need for a new fundamental Charta or "habeas data" (6) or "habeas scriptus" (7), to defend the citizen from the danger of the tyranny of information.

4. Risks of illicit access to and potentially dangerous use of individual information appear in the different phases of elaboration and data dissemination.

First of all, control of input involves the identification of the subjects to which the information refers. Then, when information of different kinds and origins has to be merged, it is again necessary to individually identify the subjects, so as to ensure that the operations carried out are correct.

When data of different kinds supplied by multiple sources are merged, the risk of improper intrusion into the privacy of individuals could increase, regardless of the statistician's intention to submerge individual records into grouped data (Alexander, 1985).

For example, it is useful to combine TV subscriptions files with those of the population census, so as to evaluate the number of TV sets per household or per capita according to geographical area or other characteristics. But this can also be used to identify possible evasion of TV tax payments.

Through statistical analysis of utility billings it is possible to check, for tax purposes, houses which are left vacant.

---

(6) G. Alpa, *Privacy e statuto dell'informazione*, in G. Alpa e M. Bessone, op. cit.

(7) S. Cassese, *Tutela della privacy e banche dei dati della pubblica amministrazione*, in "Privacy e banche dei dati", a cura di N. Matteucci, Il Mulino ed., Bologna, 1981.

The record of publications consulted in a public library could be used to ascertain the preferences of users from a marketing point of view, but also as an indication of political orientation.

To estimate income and expenditures in very restricted territorial areas, statisticians may adopt indicators like - the declarations of income for fiscal ends, domestic electrical energy consumption, utilization of private telephones, automobile possession, etc. But again these same elements can be considered similar to a mechanism like cross-reference control, if used by fiscal administration for the control of fiscal evasion. In addition, by merging these kinds of indicators we are in danger of misinterpretation because statistical incongruancies between income and expenditure are not a reliable sign of a bad fiscal conscience.

5. On the other side of the coin we find the risk of being controlled and constricted in the management of information.

Recently in Italy we had a sensational example of this. Data on the audience rating of different TV programs by the RAI (radiotelevisione italiana), collected by meter and utilizing a national sample of 1.400 households, were contested by a private network. In July 1985, the court of Rome forbade not only the dissemination of these data, but also restricted their collection to public transmiters because "the use of market research data executed by one of the firms involved in the market must be considered an incorrect competitive practice and an unfounded means to ascertain market shares". This is, in addition, an evaluation of the inadequacy of the system utilized and the technique chosen. But this evaluation was made by the Court without the consultation of experts in statistics and market research.

As can be seen, protection of privacy is of primary importance in computerized information. But any attack on the freedom to collect and utilize, for managerial purposes, data about market situations or related subjects should be resisted with great energy. Although it is essential to avoid any inconvenience to the subject to whom the information is related, it is equally necessary to protect the right to collect, elaborate and properly utilize information that has no dangerous implications.

6. Laws concerning privacy in informatics are exposed to the different, not always completely avoidable, risks of generic formulation, insufficient technical skills, impracticable prescriptions or requests for conditions which imply enormous costs and confused procedures, protection of individual information that presumes a not yet existing widespread technical preparation in informatics. On the other hand, the same laws imply protection of the individual, while at the same time exposing the management of data banks to interruptions and supplementary expenses of an unknown amount, as the guaranty is still too general.

Criteria on information privacy regarding people and legal bodies are sometimes inadequate and defective. For example, it is not sufficient simply to forbid queries that ask for information pertaining to a single record. Those interested in subject $k+1$ could correctly request and receive mean data referring to the first k elements. The same procedure could also be applied to $k+1$ subjects, reobtaining the new mean. But in this phase it is easy to break any privacy and pick up the individual information concerning unit $k+1$

(Ullman, 1981).

In general, the problem is to avoid access to data banks through illicit interrogations or through questions that could allow the identification of individual figures (Rafanelli, 1984). It should first be assured that any set of data originated by segmentation and/or intersection identifies at least n_0 elementary units - with the value n_0 sufficiently high - and only at this stage should inquiries be accepted for elaboration.

Protection of individual rights must also consider the generally low level of knowledge about informatics, as well as the high costs of access to information in data banks. Besides this, legislative actions have completely neglected, up to now, the existence and possibilities of the Personal Computer.

Taking all these aspects into account, in a preliminary stage a "code of behaviour" should be elaborated for operative purposes. The legislative interventions - as in the 1978 French law - often contain general principles of the following type: "informatics must not infringe upon the identity and rights of any person, either in private life or individual and public freedom". These general principles should be expressed in operative terms.

Once guidelines are operating, people and legal bodies involved will identify and clearly focus on problematic points. Workable solutions should then be elaborated by a task force with representatives of private data banks, experts in public administration, representatives of information suppliers. Their results will be accepted and became operative in the area of informatics by "self-regulation". In this way it would be easy to build a steady and realistic basis for subsequent legislative interventions. At this stage, such interventions would be specific, up-to-date and therefore valuable.

BIBLIOGRAPHY

Alexander L., *Privacy and Confidentiality Issues in Statistical Data,* Proceedings of the 45th Session, Intern. Stat. Inst., Amsterdam, 1985, Invited Paper 9.1.

Alpa G. e Bessone M. (a cura di), *Banche dati, telematica e diritti della persona,* Quaderni di diritto comparato, Ediz. Cedam, Padova, 1984.

Autori vari, *Banche dati e diritti della persona,* I libri dell'Istituto giuridico italiano, Vol. 2, Cedam, Padova, 1985.

Camera dei Deputati, n. 1144, Proposta di legge "Norme per la tutela del diritto alla riservatezza delle persone fisiche nel trattamento automatizzato dei dati e delle informazioni personali", presentato il 18 gennaio 1984.

Camera dei Deputati, n. 1210, Proposta di legge "Discipline dell'uso dei sistemi informativi personali", presentata il 27 gennaio 1984.

Camera dei Deputati, n. 1636, Proposta di legge "Norme per la promozione e il coordinamento delle attività informatiche nell'amministrazione statale", presentata il 30 aprile 1984.

Camera dei Deputati, n. 1657, Disegno di legge "Costituzione ed esercizio delle banche di dati personali ad elaborazione informatica", presentato il 5 maggio 1984.

Camera dei Deputati, Servizio per la documentazione automatica, Quaderni di documentazione, n. 4, *Banche dati e tutela della persona,* 2^ ediz. ampliata, maggio 1983.

Catania N., *Dossier privacy,* ediz. Sarin, Pomezia (Roma) 1983.

Centro di iniziativa giuridica Piero Calamandrei, *Le banche dati in Italia - realtà normativa e progetti di regolamentazione,* Jovene edit., Napoli, 1985.

Cox L.H., *Emerging Policy Issues in Record Linkage and Privacy,* Proceed ings of the 45th Session, Intern. Stat. Inst., Amsterdam, 1985, Invited Paper, 9.2.

ISI, *Declaration on Professional Ethics,* Amsterdam, 1985.

Marbach G., *Un'agenzia nazionale della informazione statistica: opportunità e problemi,* Istituto centrale di statistica, Annali di statistica, Serie VIII, vol. 26, Roma 1971, pp. 269-282.

Muhsam H.V., *On Two Human Rights Involved in the ISI Declaration on Professional Ethics,* ISI, 45th Session, Amsterdam 1985, Contributed Papers, Book 1.

Pagano R., *Tutela dei dati personali: evoluzione della legislazione europea e stato del dibattito,* Bollettino Informazioni Costituzionali e Parlamentari, Camera dei Deputati, 1985, n. 2.

Rafanelli M., *La sicurezza nelle basi di dati statistici,* Congresso A.I.C.A. 1984 Roma.

Rodotà S., *Protezione dei dati e circolazione delle informazioni,* Rivista critica del diritto privato, dic. 1984.

Tiberio P., (a cura di), *Basi di dati - stato dell'arte e prospettive,* Collana AICA di informatica, Masson Italia editori, Milano, 1985.

Ullman J.D., *Principles of Database System,* Computer Science Press, N.Y. 1981.

Italian Statistical Data Dictionary System Prototype Developed with a 4th Generation Language

By P. Costa, F. Marozza, and F. Vinciguerra, Roma

SUMMARY: Statistical Information Systems are systems conceived to collect, structure and make available the statistical data needed for the correct and prompt investigation and the management of an organism, reducing its entropy level. Data Dictionaries handle data about the data of information systems; they represent a fundamental environment for their development and management; in statistical applications they play an even more important role and assume some peculiarities originated by the special nature and complexity of data. 4th Generation Languages are interactive systems for developing and exploiting automated solution methods and information systems; they increase productivity by 1000%, asking a learning period of a few days and offering the direct visibility of data and procedures; this enhanced performance makes feasible prototypal analyses, incremental developments, case simulations. A prototype of the Statistical Data Dictionary System for the Italian National Institute of Statistics has been realized exploiting the efficiency and the power of the MAPPER 4th Generation Language.
KEY WORDS: data dictionary, data base, statistical data base, statistical information system, 4th generation language

1. STATISTICAL INFORMATION SYSTEMS

Information Systems. An information system is the organized set of information that determines the reduction of the entropy level in a system; lacking an information system, an organism cannot achieve its purposes or can achieve them with difficulty. It may be viewed as the combination of living and material resources and procedures set up to collect, register, manipulate and exchange the proper information for planning, evaluation and control, as well as

Compstat 1986

operational activities. Informatic systems are the automated implementation of portions of information systems, by means of which the latter are enhanced; of course the informatic system includes not only the hardware, but also the software that makes it work, human resources to develop and manage it, communication capabilities and all that is needed to make the whole system operative.

Statistical Information Systems.

The Statistical Information System (SIS) is a subset of the information system: while the latter also contains elementary data -microdata- and supplies individual information, mainly oriented to current activities, the former is conceived for research, planning and auditing; it commonly consists of summary data -macrodata- derived from individual data and sometimes made available to outside users too, and of data on the data, as described below, essential for their exploitation (structure in the figure).

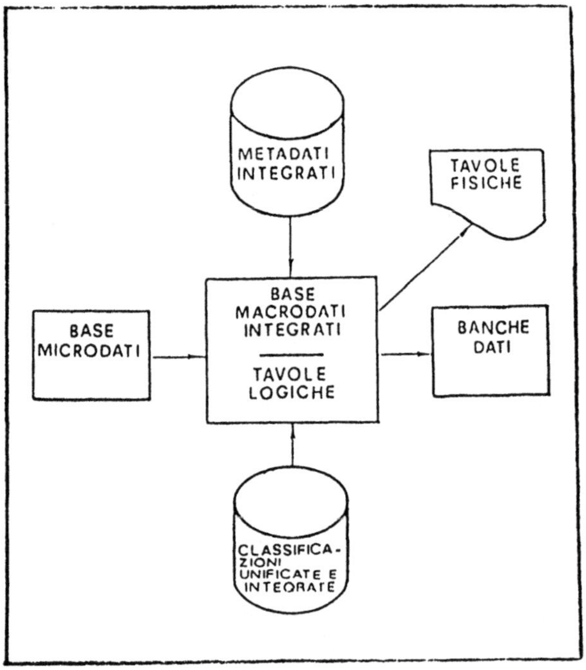

For some years Statistical Information Systems have been partly automated, to acquire, register, maintain, manipulate and distribute the statistical information more efficiently.

Statistical Subsystems at ISTAT. The data banks of the Italian National Institute of Statistics have been developed in two main areas:

a) time series, covering national economic accounts, industrial production, prices, foreign trade, jobs and salaries, press communiques

b) territorial series, covering socio-demographic indicators and census data for population, dwellings, industry and agriculture.

2. DATA DICTIONARIES

Data Dictionaries. A dictionary is an ordered collection of references to terms relating to a given field or a given activity, defining their meanings and describing their uses. A Data Dictionary, naturally, contains «data about data» or, as is commonly said, «metadata»: it is an information system managing data about the data of another information system, intended to optimize information management. Since the mid-70's Data Dictionaries have been fundamental to the efficient development and exploitation of automated information systems. They aim at covering all life cycle phases, from initial analysis to day-by-day production, and tend to deal with all interacting components: users, administrators, EDP specialists, basic software, applications.

Data Dictionaries for Statistical Information Systems. The typical problems raised for information technology by statistical applications offer new challenges and new targets to Data Dictionaries and make them even more relevant, for at least two reasons [Vinciguerra(1986)].

First, an SIS often collects data from various external information systems, each one handling data with different names, definitions and domains, different space and time availability extensions, analytic levels, integrity constraints, measuring and calculating criteria and so on.

Second, in an SIS the classification problem is crucial. No piece of information, either an observed variable or a controlled variable, may be handled unless it is defined and inserted in a classification; but the same phenomenon may be described in terms of different classifications under the same aspect or by different classification for different aspects -e.g. for goods, economic origin, or economic destination-. The same classification may also vary with the time -e.g. new/old goods, new/old economic activities, different grouping-; it is then necessary to establish and offer correspondence and compatibility between classifications.

3. 4TH GENERATION LANGUAGES

Software development and 4th Generation Languages. Traditionally, software is built using the classical programming languages, such as COBOL and FORTRAN, together with the packages for recording data and their relationships, as the various Data Base Management Systems.

However these languages and packages have several limitation that directly affect the final products. This mainly arises from their inability to match the human cognitive apparatus.

Nowadays, the ceaseless technological and methodological evolution of information science has produced new tools that, improving the quality of the correspondence between men and machines, raises individual productivity by better exploiting human perceptive and cognitive faculties.

Although they are not just languages, these tools were termed «fourth generation languages»; their innovative capabilities were evidenced by their ability to allow working by conjectures and refutations [Popper(1963)], through hypotheses and experiments [Laborit(1983)].

Prototyping with a 4th Generation Language. Compared with traditional tools, a 4th generation language has 1000% greater productivity, a learning period of a just few days and direct visibility of data and procedures. This radically improved performance makes feasible prototypal analyses, incremental development and case simulations.

Among the 4th generation languages, the MAPPER stands out because in addition to its complete conformity to the general principles of 4th generation languages, it offers this same technology on mainframes [Martin(1982)]. This implies that a prototype developed in a few days could naturally become a production data dictionary because of the intrinsic MAPPER capability for handling huge Giga-byte data inputs.

4. THE ISTAT STATISTICAL DATA DICTIONARY SYSTEM

The ISTAT project. The Italian National Institute of Statistics decided to set up a Statistical Data Dictionary System (SDDS) to handle all metadata concerning the overall statistical activity of the Institute. The SDDS is intended to serve the needs of managers, designers and planners of surveys and information subsystems at operative, tactical and strategic levels, as well as the requirements of the external user community.

It describes and relates each other surveys, statistical units, micro and macro data at the various aggregation levels, tables and publications. It was also focused on the integration between micro and macro data and classifications, that are meant as the heart of the statistical work.

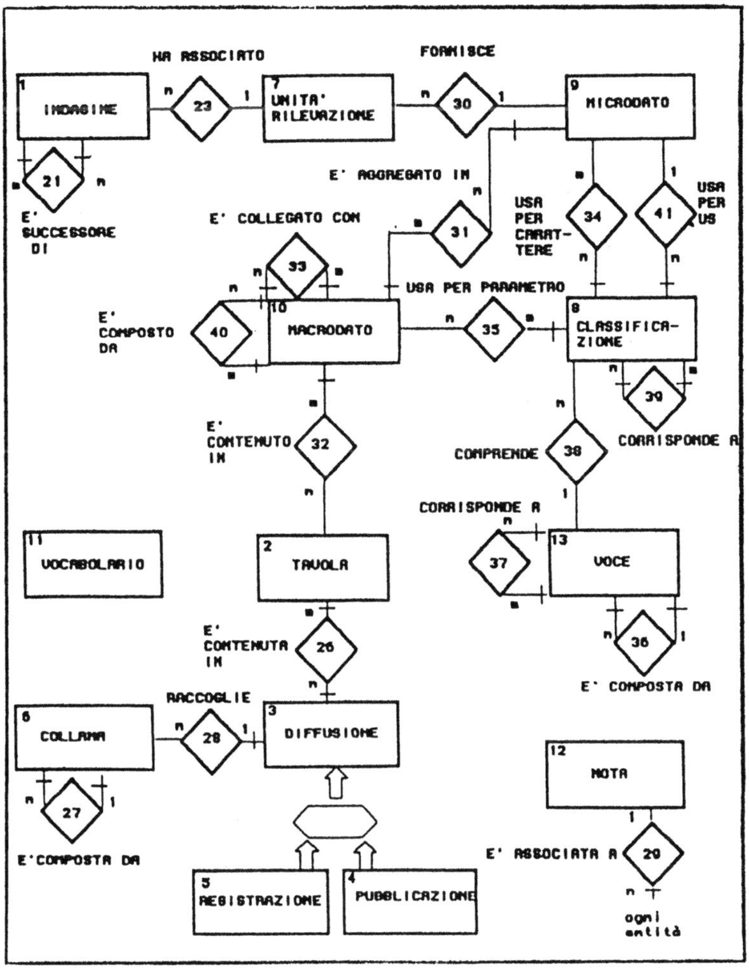

United Nations [ECE/UNDP (1984)] and European Community [CADOS(1985)] reports were also referenced, as well as several ISTAT works, such as Business System Planning for the Institute, Business System Planning for demographic statistics, proposals for official Italian statistics [Moser(1983)], Prices subsystem Design.

So far the Requirement Analysis and Conceptual Design (Entity-Relationship skeleton in the figure) have been performed, correcting and detailing from initial kernels, following the DATAID method produced by the Italian National Reserch Council [Ceri(1983)].

The MAPPER prototype. Since a Data Dictionary system for a statistical institute is a basically strategic and highly costing investment, before going on to the final implementation a quick prototype has been costructed in MAPPER to check the identified requirements, the designed data and functions, and overall operations. To get a prototype of SDDS that is significant for our purposes, it was decided to implement the near totality of data and function types for a subset of ISTAT statistical activity, namely for Price and Labour Force subsystems, as opposed to alternative prototyping criteria. The prototype itself was subjected to an immense amount of unmaking and remaking as the data and relationship concepts, function operations and data values became well cleared.

Hence the near totality of entity, attribute and relationship types were represented; several thousand records were loaded, including attribute domains and codes, samples of synonyms, portions of classifications and their correspondences. Functions for data acquisition, schema navigation and selection by values were made operative. Both functions and data are stored in the MAPPER data base that is structured as a set of file cabinets (in the figure) and the architecture of our prototype comes out straightforward to users.

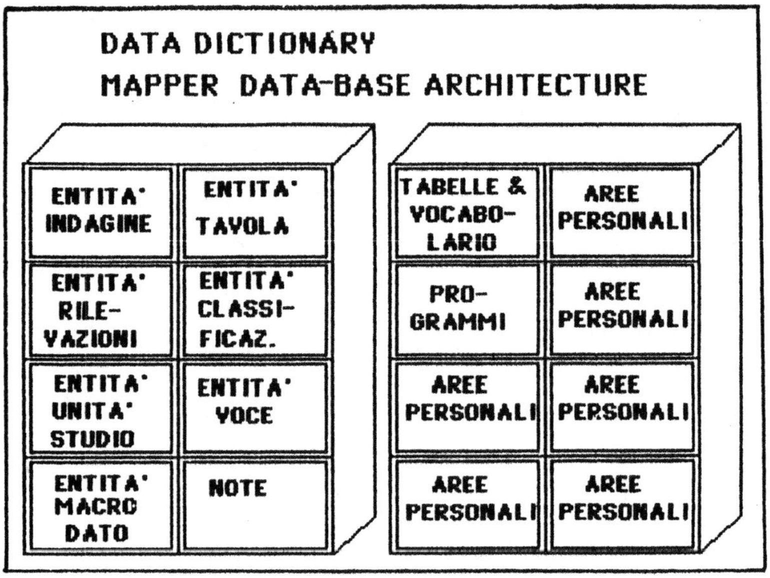

This prototyping work will hopefully offer enough information to proceed to a solid final implementation.

MAIN REFERENCES:

Albano A. Orsini R. (1985), Basi di dati, Boringhieri, Torino

CADOS (1985), Documentary system for CRONOS, EUROSTAT

Ceri S. (1983), Metodology and tools for data base design, North Holland

ECE/UNDP (1984), User guide to metainformation systems in statistical offices, United Nations Development Program, ECE/UNDP/SCP/H.4

Laborit H. (1983), La colombe assassine', Edition Grusse & Fasquelle, Paris

Martin J. (1982), Application development w/o programmers, Prentice Hall, NJ

Moser C. & a. (1983), Aspetti delle statistiche ufficali italiane, ISTAT, Roma

Popper K. (1963) Conjectures and refutations, Routledge & Kegan Paul, London

Rizzi A. (1984), Statistica e Informatica, Atti XXXII riunione SIS, Sorrento

Ruberti A., Isidori A. (1982), Teoria dei sistemi, Boringhieri, Torino

Vinciguerra & a. (1986), Dizionari Dati per S.I. statistici, Sistemi&Automazione

A Methodology for Conceptual Design of Statistical Databases

By G. Di Battista, G. Ferranti, and C. Batini, Roma

Keywords: methodologies, conceptual database design.

SUMMARY

In statistical applications, data are described at different levels of aggregation, from elementary facts of the reality to complex aggregations such as classifications, time series, indexes. This paper describes a methodology for conceptual design of statistical databases that provides the designer suitable strategies for defining such different levels of aggregation starting from user requirements, and checking the completness, coherence and minimality of the conceptual schema at the different levels.

1. INTRODUCTION

In methodologies for data base design, a conceptual design step is usually proposed, whose goal is to obtain a formal description of the application, called conceptual schema. The advantages of performing a conceptual step are:

1. enhanced interaction between user and designer and active role of the user in design decisions;
3. integrated view of user requirements;
4. project's stability;
5. easier maintenance of design products.

In this paper we propose a methodology for conceptual design of statistical databases. The main difference between statistical applications and traditional business data bases is in the types of data they manage. The majority of methodologies for data base design consider only data that represent objects of the real world (e.g. person's age and sex, the code of manufact piece, the address of an agency). Since these data represent elementary objects of the real world, they are considered as elementary

Compstat 1986

data: the conceptual schema of elementary data will be called in the following elementary schema. In statistical applications, data are described at different levels of aggregation, from elementary facts of the reality to complex aggregations, such as classifications, time series, indexes. All these types of data are usually called summary data.

In a methodology for conceptual design of statistical applications we have to produce both the elementary schema and the conceptual schema of summary data (in the following statistical schema). Considering together the two points of view provides the designer a global view of all types of data managed in the application; in such a way he can perceive:

1. which aggregations can be obtained from elementary data;
2. which elementary data are to be collected to perform the aggregations of interest.

The methodology makes use of two data models for the representation of data: for elementary data the Entity-Relationship model, widely used in data base applications (see Chen 1976, Ceri 1983); for summary data a new model is used, named CSM (Conceptual Statistical Model) designed to be an effective trade-off between expressive power and simplicity of use: for a description of the model see Barcaroli et al 1986.

The paper is organized as follows. In section 2 we present the methodology for conceptual design of statistical applications. In section 3 open research topics and further developments are discussed.

2. THE METHODOLOGY

The general approach of the methodology is to fragment the overall activity in two steps:

a. proceed first to the design of the elementary schema; this is done to get from the beginning of the design activity a stable and complete description of classes of objects involved in the statistical application.

b. design of the statistical schema through incremental production of the schemas of aggregations, that are subsequently merged into the global statistical schema.

In fig.1 a detailed description of the methodology is shown, we now examine the single steps.

1. Analysis of user requirements

   1.1. Distinguish types of requirements;
   1.2. Analyze requirement, remove ambiguities.

2. Design of the draft elementary schema

   Build an elementary data schema:

   2.1. Build the ER schema starting from requirements referring to elementary data;
   2.2. Enrich the ER schema with requirements referring to summary data;

3. Design of the skeleton statistical schema

   Choose the generalizations in the statistical schema, and build a first skeleton schema;

4. Design of the statistical schema

   For each aggregation defined in requirements:

   4.1. Check the completness of the elementary schema
   4.2. Build the statistical subschema of the aggregation.
   4.3. Integrate the statistical subschema into the skeleton schema;

Fig 1: General structure of the methodology

<u>Requirements analysis</u> - The goal of the initial step is to collect and analyze requirements. Since requirements are usually expressed in terms of heterogeneous descriptions (forms, natural language, record formats, approximate knowledge) and refer both

to elementary facts and to aggregations, during this phase they are to be analyzed in order to get an homogeneous description, put in evidence ambiguities, synonims, and homonyms, finding the most important concepts and relationships among them.

Design of the draft elementary schema - The design of the elementary schema is split in two phases; the first phase takes into account elementary requirements, building a first draft skeleton schema, while the second phase enriches this initial description with new concepts expressed in aggregations. The design of an elementary schema has been thoroughly investigated in the literature (see Ceri 1983, Olle Verrijn-Stuart Sol 1983, Tardieu Rochfeld Colletti 1983).
The design of the elementary schema is now influenced by the knowledge of the aggregations of interest, that represent an important starting point for discovering new concepts. The suggestion here is to enrich the elementary schema with new classes of objects mentioned in the list of aggregations. To avoid proliferation of classes in the elementary schema, only the most relevant classes should be inserted. E.g. at the end of the previous step, an entity PERSON could be defined in the schema, that represent the class of all persons. Meanigful aggregations concerning persons could involve "Persons with 18 years and over", "Persons within 6 and 21 years", "Foreigners": all these classes are potential candidates for new entities in the elementary schema. In the above example, we could decide to enrich the schema with an entity FOREIGNER, while other classes could be mentioned only in the statistical schema.

Design of the skeleton statistical schema - At this point of the design, we have a clear perception of the elementary concepts. The suggestion now is to start the design of the statistical schema extracting from the elementary schema a first group of classes of data. These classes inherit the generalizations defined in the elementary schema on the corresponding entities. Starting from this first kernel of classes, we may enrich the schema with new classes, obtained from them with suitable generalizations and/or specializations.

Design of the statistical schema - The incremental design of the statistical schema is guided by the analysis of the aggregations; for each aggregation expressed in requirements, the methodology performs three main steps:

1. Completness check
2. Design of the aggregation statistical subschema.
3. Integration

During the COMPLETNESS CHECK we verify the completness of the elementary schema with respect to the aggregation. Checking an aggregation against the elementary schema may leave the schema unchanged (when all the information needed by the aggregation is represented in the schema), or cause some enrichement or restructuring activity. Completeness check may be done for simple aggregations by an inspection of the elementary schema, for complex aggregations we may select the elementary subschema involved and modify it in order to obtain a subschema in which concepts are expressed as they are seen in the aggregation.

During the DESIGN OF THE AGGREGATION STATISTICAL SCHEMA a formal description of the aggregation is built, in terms of the CSM model and, in case, of a query language. This activity can be decomposed into three main phases:

1. choice of the class of object involved in the aggregation
2. specification of category attributes
3. choice of the statistical computations.

During INTEGRATION the statistical schemas are incrementally merged to the skeleton schema, obtaining at last the global statistical schema.

3. CONCLUSIONS AND FUTURE RESEARCH

In this paper an effort was made to face some of the relevant issues that may arise in the design of statistical databases. We have adapted methods and guidelines of general methodologies for conceptual database design to specific issues of statistical applications.

Future research directions concern: the adaption of the methodology to work in several application environments, the development of automatic tools to support the designer. A first approach in this direction appears in Di Battista Tamassia 1986.

REFERENCES

Barcaroli G. et al. - CSM: A Conceptual Model for Statistical databases, Technical report ISTAT - DIS 1986.

Ceri S. (ed.), Methodology and tools for data base design. North Holland, 1983.

Chen P.P. - The Entity-Relationship model: toward a unified view of data - ACM Transactions on Data Base System, Vol 1, N.1.

G. Di Battista, R. Tamassia - An interactive graphic system for designing and accessing statistical data bases. Compstat Conference, Roma 1986.

T.W. Olle, H.G. Sol and A.A. Verrijn-Stuart (eds.) Information Systems Design Methodologies: A Comparative Review - Proc. of the IFIP WG 8.1 Working Conference on Comparative Review of Information Systems Design Methodologies, Noordwijkerhout, The Netherlands, North Holland 1982.

H. Tardieu, A. Rochfeld, R. Colletti - Conception d'un systeme d' information: construction de la base de donnees - Edition d' Organization, 1979.

Towards Natural Integration of Database Management System and Statistical Software

By P. J. Jasiński, B. Butlewski, and S. Paradowski, Poznań

SUMMARY
The paper presents an integrated software architecture of the data processing system which consists of relational data management tools and selected statistical functions. A complete set of ReDS software facilities is available at the uniform level of end-user interface. Statistical component of the system can be tuned easily to meet the changing requirements of the user.
KEYWORDS: relational database, statistical data analysis, integration

1. INTRODUCTION

The user of a database, when faced with a truly large collection of data, usually is not satisfied with simple retrieval and counting; instead, he requires more sophisticated data analysis methods, i.e. some level of statistics (called in database fied "aggregate" functions). Aggregate functions offerred in todayss standard database management systems [DELO85] are rather poor and provide the user with such limited options as minimum, maximum, sum, average and very little more. In such a case the user is being forced to prepare appropriately formatted data files and after that to process them using specialized, separate statistical package. Similar difficulties may be observed as far as statistics is primarily concerned. In some simple cases data and file management facilities provided within statistical software package fulfill the user's needs. However, serious complications and/or extensions in logical data structures usually cannot be coverd using these facilities and require the use of database system. Independently of the view-points presented, such a way of solving problems of data management and statistical analysis is called "integration". In fact, this approach means no more than a cooperation between two rather independent software packages (dBASE and SPSS for PC-like machines, SICLA and PEPIN [JOMI84] and others) as illustrated in Fig 1a.

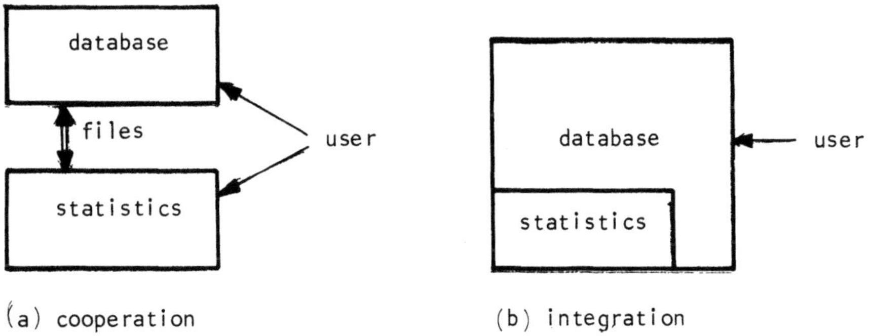

Figure 1. Architecture of database and statistics - two approaches

Compstat 1986

In this paper we present a software architecture in which the user is free from seeing two separate software packages and bi-directional transferring of data files (in ASCII or other form). In other words, natural integration means the total processing system that is available at the uniform end-user interface level (Fig. 1b).

2. CHOICE OF DATABASE METHODOLOGY

The well-know foundation of relational databases is that data are formatted into a set of two-dimensional, normalized tables which rows represent entities and columns properties of corresponding entities [DATE81, DELO85]. The table below shows the terms used in databases and in statistics.

general term	databases	statistics
table	relation (file)	matrix (of observations)
row	tuple (record)	object (observed)
column	attribute (field)	variable (observed)

The terms are different but the underlying structural concepts are undoubtfully compatible. Therefore, the authors firmly believe that relational methodology should remain the kernel of data management in a truly integrated processing system aimed at various (mostly scientific) data collections and statistical studies. An extensive, powerful set of operations (selection, projection, natural join, algebra of attributes, sum, difference, cross-section and so on) justify even stronger our opinion and choice.

3. STATISTICS AS A DATABASE FUNCTION

A classical assumption of relational database theory emphasizes the structural uniformity of base data and output results. The user deals with relations (base, derived, view and so on) and recives the answer to his query in form of a relation, too. This results from formal foundations of relational algebraic operations and, although theoretically elegant, does not satisfy important practical requirements. The simplest example is the class of queries involving cardinalities. This operation does not exist in the set of relational operations but owing to its great useability has became a standard function of relational databases (e.g. SELECT COUNT.... in SQL [DATE81]). Despite the fact that statistical computations are far more complex than simple counting there is no contraindication not to treat such computations as functions recalled from the level of the end--user language. Basic solutions and their implementation results are outlined in the next section of this paper.

4. DESIGN AND PROTOTYPE IMPLEMENTATION

Data management kernel of ReDS (Relational Data System) software consists of structure and semantics data definition tools (bases on RM/T approach [DATE83]) and a language DIAL (Data Inquiry and Administration Language) which provides the user with a fully relational set of operations and exhaustive updating facilities. Each DIAL query or command starts with a task-specific directive (SELECT, COUNT, CREATE, MODIFY, UPDATE, PERFORM,

DELETE) followed by logically connected specifications of monadic (selection, projection) and dyadic (usually natural join) operations.

The directive PERFORM remains a key-point of this paper subject. The lexical contents and syntax structure of a corresponding type of queries are as follows:

PERFORM → function → ON → projection_list → FROM → derived_relation_specification ;

where derived relation specification stands for a composition of dyadic (natural join or set-theoretic) and/or selection conditions applied for one or more relations involved. The set of statistical functions available is rigorously limited to those which are undoubtedly in frequent use and indeed suitable for a given application class.

Taking into account the end-users way of thinking, the authors of this prototype design clearly subdivide statistical functions according to their applicability to quantitative (numbers) versus qualitative (codes, char strings) variables. The line of functions for quantitative variables consists of:
- DESTAT (minimax, means, variation, standard deviation, median),
- HISTOGRAM,
- SCATTERGRAM (with Pearson r),
- ANOVA (with Duncan, Tukey and Scheffe tests),
- GROUPCOMP (t-Student, nonparametric Wilcoxon, Man-Whitney U and Kruskal-Wallis tests).

The corresponding line for qualitative variables is as follows:
- BARCHART,
- CROSSTAB (with Pearson Chi-square),
- GROUPCOMP (Chi-square).

Shown below are two simple examples (histogram of ages of male patients with acute myocardial infarction, groups comparison in a simple clinical trial of drug effectiveness) of queries, i.e. the way in which the user may recall statistical functions of ReDS system.

```
PERFORM histogram ON age FROM PATIENTS WHERE sex = "m" & DIAGNOSIS
                             WHERE icd_code = "410.1".

PERFORM groupcomp ON blood_pressure FROM NIFEDIPINE_TEST WHERE
                     drug = "nifedipine" VERSUS drug = "placebo".
```

After input of any query involving statistics the system performs all necessary relational operations and overtakes control under further clarification dialogue. In the case of histogram such a dialogue concerns the display of minimax boundaries founded for the attribute(s) in question, a possibility of further reduction (histogram zooming), a choice between the number of compartments and compartments widths and, in the latter case, the determination of compartments widths. The dialogue for the function of groups comparison (note a special operator VERSUS, which produces disjunctive subsets) covers a message about cardinalities of groups compared (with a warning in the case of significantly different cardinalities). Necessary tuning of computations, dependent on the cardinalities recognized, are introduced automatically. After that, a table look-up (e.g. t-Student distribution) provides the user with an alternative answer: no difference or the difference between groups analyzed is statistically significant at the (0.05 - 0.001) level.

An important feature of ReDS which is effectively utilized within COUNT directive and statistical functions concerns a schema of self-describing, so called systems' relations (TABLES, COLUMNS and RELATIONSHIPS) and, specifically, a systematic notion for default (not known or unavailable) values of selected attributes in a given database schema.

A prototype of ReDS software, written entirely in OMSI PASCAL, is available for LSI 11 based machines (03 and bigger) with RT-11 and RSX-11M operating systems. Fundamental implementation and programming works were carried out between June 1984 - December 1985 [JASI84, JASI85] . ReDS software has been applied in some research projects of clinical medicine (7-yaer retrospective study of acute myocardial infarction patients, evaluation of different rehabilitation strategies and others) and proved the attainment of primary conceptual and design goals.

5. CONCLUDING REMARKS

The idea to link relational database software and statistical software is neither new nor original. The software architecture presented results from our earlier works [JASI81] and exhibits natural similarities to other independent projects or concepts [HAUX82, HAUX83, HAUX84, JOMI84, KLON84].

In the work presented stress has been put on natural immersion of statistical functions in the framework of relational database management system and user-friendly interface.

Some aspects of this proposal may meet with criticism. For example, the set of statistical functions proposed is only a part of those available in packages such as SPSS, SAS and others. It is true, but ReDS software is not intended for professional statisticians. A typical ReDS user, primarily interested in the collection and rather simple data analysis can treat statistcal functions as an extra-powerful tool for data analysis. On the contrary, the user primarily concerned with statistical analysis can treat the database part of ReDS software as a very flexible tool for data management. Also, he may ask for application-oriented tuning of statistical functions set available and such a tuning remains relatively simple task. Both views are consistent with the preliminary design assumptions of ReDS software.

Current works are focused on the re-implementation of ReDS software on 16-bit IBM PC compatible machines. Also, the authors would like to study problems of the use of expert systems methodology, for example to advice the user to apply appropriate statistical methods

REFERENCES

[DATE81] Date, C.J.: An Introduction to Database Systems (3rd Edition). Addison-Wesley Publ.Co., Reading, MA, 1981.

[DATE83] Date, C.J.: An Introduction to Database Systems (Volume II). Addison-Wesley Publ.Co., Reading, MA, 1983.

[DELO85] Delobel, C. and Adiba, M.: Relational Database Systems. North-Holland, Amsterdam, 1985.

[HAUX82] Haux, R.: A Programming Technique for Software in Statistical Analysis. In: COMPSTAT 1981 (Physika Verlag, Vien) 266-271.

[HAUX83] Haux, R.: Statistical Analysis Systems. Construction and Aspects of Method Design (Part 1). Statistical Software Newsletter 9 (1983) 106-115.

[HAUX84] Haux, R.: Statistical Analysis Systems. Construction and Aspects of Method Design (Part 22). Statistical Software Newsletter 10 (1984) 14-27.

[JASI81] Jasiński, P.J., Krug, H., Paradowski, S., Zerbe, F. and Talkowska, B.: CARDIOCOD: a query data-base for clinical cardiology. Med.Inform. (1981) vol. 6, no. 4, 239-249.

[JASI84] Jasiński, P.J.: Project of self-contained relational database software: design and clinical applications outline. In: MIE-84, Lecture Notes in Medical Informatics, 24 (Springer-Verlag, Berlin) 1984, 194-199.

[JASI85] Jasiński, P.J.: Performance improvements of relational database software reconsidered for medical applications. In: MIE-85, Lecture Notes in Medical Informatics, 25, (Springer-Verlag, Berlin) 1985, 514-518.

[JOMI84] Jomier, G., Kezouit, O. and Ralambondrainy, H.: SICLA-PEPIN: A System Integrating Data Analysis and Realtional Data Base Management System. In: COMPSTAT 1984 (Physika Verlag, Vien) 263-267.

[KLON84] Klonk, J. and Rassman, B.: The architecture of a data processing system to support statistical studies. Med.Inform. (1984) vol. 9, no. 2, 125-134.

Easy-Link: An Expert Natural Language Interface to a Statistical Data Bank

By G. Lella, S. Pavan, and P. Bucci, Pomezia

SUMMARY

The paper describes the general framework of the EASY-LINK system: an expert interface in Italian natural language, which allows an intelligent interrogation of the economic, territorial, statistical data bank "SUPERSTAT" developed by SARIN.
After a discussion on the problems of misuse of statistical packages by naive-user and on the challenge for Artificial Intelligence techniques in statistics, attention is focused on the system's capabilities and architecture.
Our approach is to increase the user's gain by making the computer's response more informative, and hence to reduce his uncertainty about his problem.
The system consists of two modules: the natural language interface and the expert system on data bank enquiry integrated in an only system where processing takes place in a cooperative way.

KEYWORDS: Natural Language Interface, Cooperative Interface, Expert System.

1. INTRODUCTION

Recently researchers in Information Science have started to examine closely the developments in Artificial Intelligence (AI) for possible applications of its techniques to problems within statistical packages.
The "naive user" of these packages is someone who is an expert in his own field but does not have much knowledge of statistics and knows nearly nothing of how to use the several data analisys methods, the ground information, the different query and manipulation languages.
AI can give some potential solution to these difficulties using expert systems technology and natural language interfaces. Expert systems (ES) are knowledge-based systems which attempt to solve problems in a way similar to a human expert.

Compstat 1986

In a statistical domain a variety of tasks can be tackled by expert system technology:

- Apply in a proper way advanced statistical tools as regression analysis Gale (1982).
- Select an appropriate statistical method providing a framework for formalizing strategies to understand the user's needs Hakong (1985).

Another possible solution, in the frame of man-machine interfaces, is to front-end the online systems with a user-friendly interface that would assist the end user, while performing inquiries and computations.

The aim of this paper is to provide some details of a cooperative interface in Natural Language (NL): the EASY-LINK system for the access of non-technical users to the statistical data bank "SUPERSTAT".

EASY-LINK is an expert interface modelled according to the general criteria of the second approach outlined above which includes both linguistic coverage and expert competence, offering a cooperative interaction environment.

In the design phase a system has been devised which could both act as a consultant for the user while searching the data (that is to say an expert system in the use of the ground information) and understand the Italian natural language while communicating with the user.

The dialogue itself should increase user's performances by offering cooperation, understanding and explanation of both the NL and ES modules behaviour.

2.THE "SUPERSTAT" DATA BANK

The SUPERSTAT data bank Serra (1985) which has been designed, developed and marketed by SARIN is a service through which a numerical data base on the Italian socio-economic condition and a statistical software package are made available to process ground information which can eventually be integrated by the user's own information.

The information base is constituted by a series of data drawn from the best qualified services (Ministry of Labour, Transport, National Electrical Company, etc..), and by indicators on the Italian socio-demographic reality and its economical activities and production situation, geographically distributed according to province and city council.

3. THE NATURAL LANGUAGE INTERFACE

In the EASY-LINK system, the NL sentence is parsed by the linguistic analyzer and before being translated into the most appropriate SUPERSTAT commands (e.g. selection and printing of simple statistical information, building up of indexes and tables) is processed by the expert interface.
Many features of the linguistic analyzer have been taken from the PARNAX system, developed at CSELT. In Comino (1983) there is a description of it with particular emphasis on its characteristics and capabilities as a robust and effective natural language processor to a data base.
The following description simply summarizes the main features of our analyzer and focuses on the Internal Representation Language (IRL), output of this module, which is the fulcrum of the EASY-LINK system.
The analyzer combines conventional syntactic analysis based on the dependency grammar model, with a compositional semantic validation using dependency rules. The central structure (IRL) is a conceptual frame, a form of representation that achieves a considerable expressive power while maintaining structural simplicity.
The function of the analyzer is to select the appropriate sense of input sentence, in order to perform a "surface" comprehension that embraces every linguistic aspect of the sentence using a limited theory of the application domain.
The result is the filling of the IRL structure, where semantically meaningless components have been pruned and appropriate semantic variables have been added. Each filled slot of the frame is therefore associated with a crucial kind of information of the corresponding sentence fragment.
The IRL structure thus offers a large number of chances and leads to the subsequent "deep" comprehension processes, that will be discussed in the following section.

4. THE EXPERT CAPABILITIES

In many cases the user query will not map directly onto the SUPERSTAT data bank. EASY-LINK has at its disposal the capabilities to check whether this query is a meaningful request or not. If it is meaningful then the system translates the IRL into the queries program to access SUPERSTAT.
If not, EASY-LINK, instead of simply telling the user that his question

(even if properly formulated in a formal way) is not "processable" will try to transform the input sentence into something meaningful, semantically close to the original request. All these desirable features are supplied by the ES module which allows an intelligent use of the data bank and has been designed according to the needs and skills of users.

Many of the design decisions, implicitly or explicity make assumptions about the nature of users and about their approach to the SUPERSTAT data bank.

The SUPERSTAT data bank is an example of environment in which the user reaches the final result through a series of steps which provide intermediate results and allow him to verify his choices.

A typical situation may be that of the construction of the table whose rows correspond to the selected entities (regions, provinces or towns) and in whose columns appear the data corresponding to the attributes (statistical indicator) one is interested in.

During the construction the user accomplishes substantial changes to the structure of the table which reflects a continuous development of his plan of action: removal, insertion and substitution of the column with the values of a new indicator if he needs these new data.

Similar operations may also be executed on the rows of the table conditioning further the entities selection.

The problems, that the casual user encounters with this kind of approach, are not caused as much by the manipulation of the data structure as by the fact that he doesn't have sufficient knowledge of the contents of the data bank. Furthermore, he doesn't have the needed expertise to evaluate the estracted data and so he can't select the proper attribute to his own enquiry.

He could be in the dark about the presence of more useful statistical indicators or could not realize the chance of the defining of the new indicator (mean, variance, deviations) that can show the connection among data. Such a framework, made of intuitive guesses, "common sense theories" and experience based statements, is the justification for the ES module as such. This module constitutes the kernel of the system: it permits a good indicators selection, it proposes a suitable format of the output and evaluation of the used strategies and furthermore it allows the user to carry out an incremental interrogation.

This module starting from the analyser's output (IRL), selects out of it the main facts on which the inference engine formulates some hypothesis.

If these will be confirmed either through user dialogue or through the solution of possible conflicts and inconsistencies, the system will be able to infer the exact definition of the user's goals.

The design of the expert module follows the general organization of a rule-based system centered on the notion of task Brajnik (1985).

To perform its cooperative behaviours it uses a knowledge base that from a technical point of view is divided into a number of smaller ones:

- EXPERT KNOWLEDGE: it contains the reasoning schemata which allows EASY-LINK to assimilate the premises for understanding. Furthermore it considers all the aspects that refer to the communication process itself and to its preconditions (metacommunication). Expert knowledge takes the form of usual production rules.
- DOMAIN KNOWLEDGE: here is embedded all the ground knowledge of the domain of the SUPERSTAT data bank. It consists of two semantic networks: the first concerning geographical information, the second about information on the economic indicators and related terms.
- KNOWLEDGE ON THE USER (USER'S MODEL): It considers those aspects of input/output behaviour which are relevant for the user view of the problem. The user's model is a conceptual frame that describes how the users are modelled and a set of rules that determines how this structure is to be used.

5. AN EXAMPLE OF INTERACTION

An example of dialogue (translated in English) encompassing all the features of the system underlying the ES dimension, is given below.

The hypothesis is that of a firm wishing to select an indicator to spot the selling degree of an expensive consumer good (valuable furs).

U- WHAT IS THE INCOME OF DIFFERENT TOWNS IN THE PISA PROVINCE

S- Are you interested in the income:

1) available from families

2) produced

U- 1

S- families income is not shown either at provincial or city level

<u>Hp</u>. are you interested in knowing the consumption of goods or services at city level?

U- YES

S- I have the following groups of indicators of city consumption: which one best fits your needs?

1) electricity
2) car sales
3) TV subscribers

....

At the end of the dialogue the beginning sentence will have changed as follows:

U- SHOW A TABLE ON THE CITY PERCENTAGE DISTRIBUTION IN THE PISA PROVINCE OF THE CARS ABOVE 1.800 c.c. REGISTERED IN 1983.

6. CONCLUSIONS

We are now testing the complete linguistic module, which is presently running in a FRANZLISP version on VAX 11/730 at SARIN and we hope to complete the whole system by the end of 1986. The experimental work that we are carrying out has disclosed the possibility of extending EASY-LINK to future developments both by widening ground knowledge through the introduction of demographic and territorial data (up to now we have been considering economic data only) and by producing ex novo ES acting as consultants on the use of the advanced statistical functions (regression, cluster analysis) that SUPERSTAT can support.

REFERENCES

1. Brajnik G., Guida G., Tasso C. (1985) "An expert interface for effective man-machine interaction", in Bolc L. (ed.), "Cooperative interactive systems", Springer Verlag, Berlin
2. Comino R., Gemello R., Guida G., Rullent C., Sisto L., Somalvico M., (1983), "Understanding natural language through parallel processing of syntactic and semantic knowledge: an application to data base query", Proc. of the 8th Int. Joint Conf. on Artificial Intelligence, Karlsruhe, West Germany
3. Gale W.A., Pregibon D., (1982) "An expert System for regression analysis" in Heiner K.W. (Ed.) "Computer Science and statistics; Proc. of 14th symposium on the interface", Springer Verlag, New York
4. Hakong L.,Hickman F.R. (1985) "Expert systems techniques: an application in statistics" in Merry M. (Ed) "Expert System '85", Cambridge Press
5. Serra B. (1985) "Applicazioni SAS ad una banca dati territoriale", SAS User Group Italia, Bologna

Farandole: A RDBMS for Statistical Data Analysis

By M. Leonard, J. J. Snella, and A. Abdeljaoued, Geneva

Abstract

We propose a RDBMS (Relational Data Base Management System) for the management of statistical data.The contribution of FARANDOLE is mainly to free the statisticians from the problems related to the manipulation of the files' physical organization in one hand, and to check automatically the validity of the data in another hand.
An interface with well known data analysis programs such as SAS, SPSS, MODULAD will enable FARANDOLE to include the functions of these programs within its own RDBMS functions.
The kind of data used ,as well as the operations executed over these data,differ from "classical" data .
Therefore new functions of a RDBMS oriented towards data analysis are to be implemented.
The data model of FARANDOLE is a relational data model which has been enhanced in order to capture in a better way the needs of data analysis. It is possible with FARANDOLE to implement complex data structures and to store results of data analysis' methods.
Concerning the storage of the data, we make use of a particular file organization : the transposed files. This kind of file organization considerably simplifies the specific operations made on statistical data and makes the compression of data easier. The need for data compression comes from the important number of missing and redundant data in the field of statistical data analysis.
A specific language is developed in order to enable the statisticians to declare a data scheme using their usual vocabulary and concepts.

Keywords: RDBMS, Data Analysis, Relational Data Model, Transposed File Organization.

1. INTRODUCTION.

A <u>RDBMS</u> (Relational Data Base Management System) enables data management independently from the physical structure of the files containing the data, and makes data manipulation safer and easier. The DBMS already existing on the market mainly deal with data in the Management field. Statistical data however differ in their structure and in their size,and the usage which is made of these data is different too : most often, a large proportion of available data is required by some method, which computes a large number of results, which may be used in turn as data for another method.
This is why a specific DBMS is needed in order to manage efficiently this kind of data (Shoshani 83), (Bates 82), (Littlefield 83).
<u>FARANDOLE</u>,the RDBMS we present, is specific for statistical data. It includes the functions of data analysis packages and it enables the statistician to apply an analysis method to some set of observed variables without any knowledge of the files' physical structures. The SICLA-PEPIN system works in this way (Jomier 84). In this paper,we will only present the data model handled by FARANDOLE.

Compstat 1986

2. The Data Model of FARANDOLE.

The FARANDOLE RDBMS is based upon the <u>relational model</u> (Codd 70) caracterized by its simplicity compared to other data models such as the hierarchical and the network models.
The relational model is well suited for the domain of statistical data analysis. This model is based on the relation's concept : a relation may be represented by a table with a number of columns (<u>attributes</u>) and a number of lines (<u>tuples</u>). For data analysis, after data collecting, the basic element is a table which contains the observations on the variables. Thus, the concepts are the sames in the two fields : in a relational DBMS specific to statistical data analysis, relations should rather be called <u>tables</u>, attributes should be called <u>variables</u> and tuples should be called <u>observations</u>.

RELATION : Attributes	TABLE : Variables
<----- Tuple ----->	<- Observation --->

Several concepts have been added to the basic relational model to meet the needs of data analysis : the concepts of sub-relation, relationship relation, set relation, weak relation and array. In order to introduce these concepts , we will first present the different kinds of variables which FARANDOLE takes into account.

2.1 Variables.

A variable is defined upon a <u>domain</u>, which is of a given type. For some types of domains, multiple values and unkown values are allowed, and sub-variables may be derived.

2.1.2 The different types of variables.

- <u>quantitative variables</u> : the domain is either of the integer or real type , and upper and lower bounds may be set.
  Example : AGE is INTEGER in the range (0,150)
- <u>enumerated type</u> : the domain is a finite set of word values
  Example : the domain of SPORTS is (TENNIS, SOCCER, KARATE, BASKET-BALL)
- <u>temporal type</u> : the domain is a time interval, measured by a specified unit of time (year, week, day, ...)
- <u>string of characters</u> : values of variables of this type are strings of characters of fixed length. For instance, NAME is a string of 24 characters at most.
- <u>text type</u> : an unlimited chain of characters.
- <u>aggregate variables</u> : a unique variable may be defined as representing a whole group of variables. For example, if LEISURE-TIME, TRANSPORTATION-TIME, WORKING-TIME are variables, a variable NON-LEISURE-TIME may be defined as representing the group of variables TRANSPORTATION-TIME and WORKING-TIME .

- statistical variables : a statistical variable is defined over an other variable,called the reference variable;the values of the statistical variable are computed from the values of the reference variable through a statistical function. For example, if AGE is a variable, AVERAGE-AGE may be a statistical variable defined over AGE, whose values are computed as the sample mean of AGE values.

2.1.2 Multi-valued variables.

Variables whose domain is either of the integer, real, enumerated, temporal or string of characters , may be single or multi-valued. For example, the variable SPORTS-THAT-YOU-LIKE may be multi-valued : an individual may indicate one or several sports that he (or she) likes (e.g. SOCCER,TENNIS).

2.1.3 Unkown values.

Values of quantitative variables and variables of an enumerated type may be unknown. There are three types of unknown values : no-answer (coded "."), missing value (coded "?"), and impossible value (coded "!"). FARANDOLE supports these three types of unkown values, for queries as well as for insertions of data. For instance,the user may ask how many impossible values have been recorded in a given sample,or ask for the rate of no-answers for a particular variable.

2.1.4 Sub-variables.

A sub-variable is defined with respect to a quantitative variable or an enumerated type variable. Its set of values is included in the set of values of its reference variable.
Example: the sub-variable BALL-SPORTS ,whose domain is (SOCCER, BASKET-BALL), may be defined as a sub-variable of SPORTS, previously defined.

3. RELATIONS

FARANDOLE takes into account several kinds of relations : regular relations, sub-relations, weak relations,set relations and arrays.

3.1 Regular relation

As mentionned in section 2., a regular relation is simply a table of observations on a given set of variables (a line is associated to an observation, a column to a variable).

3.2 Sub-relation

This concept has been introduced in Smith (1977). A sub-relation is generated from a reference relation through a first order predicate. An entity e of the reference relation is an entity of the sub-relation if and only if the predicate is true when evaluated on e. Conversely, any entity of the sub-relation is always an entity of the reference relation. Usually, a sub-relation has its own set of variables.
Example: let the reference relation be the regular relation (or table) PERSON, whose variables (attributes) are NAME, JOB and AGE;let YOUNG-SECRETARY be a sub-relation generated from PERSON by the predicate : "AGE < 20 AND JOB = SECRETARY" . The following figure shows some tuples of the two relations :

PERSON (reference relation)

NAME	JOB	AGE
JOHN	ENGINEER	19
ROBERT	SECRETARY	18
RICHARD	CLERK	36
ANTHONY	SECRETARY	21
PEGGY	STUDENT	18
CHRIS	SECRETARY	17

Predicate : AGE < 20 AND
JOB = SECRETARY

YOUNG-SECRETARY (Sub-relation)

NAME	TYPING-SPEED
ROBERT	20
CHRIS	18

FARANDOLE guarantees that no tuple belonging to the table PERSON can be inserted in the relation YOUNG-SECRETARY if the predicate is false, providing in this way a powerful automatic check of data integrity.

3.3. **Weak relation .**

Weak relations are very useful whenever,in a survey, the answer to a question takes the form of a double entry table. For instance, if people are asked : indicate your appreciation (BAD,FAIR,GOOD,EXCELLENT) on this set of products (A,B and C) according to the following criteria (TASTE , COLOR , PACKING),a typical answer (e.g.,from a person named ROBERT) will be a table like the following.

PRODUCTS

	ROBERT	A	B	C
	TASTE	GOOD	FAIR	FAIR
CRITERIA	COLOR	EXCELLENT	GOOD	BAD
	PACKING	BAD	EXCELLENT	GOOD

FARANDOLE provides a way to suitably represent this answer by a weak relation whose attributes are CRITERIA, PRODUCT and APPRECIATION, along with the attributes inherited from the reference relation. The reference relation could be the regular relation PERSON above, in which case such a table of answers would be expected for every person in the survey. It could be the sub-relation YOUNG-SECRETARY, in wich case this table of answers is expected (and stored) for young secretaries only.

3.4 Set relation

A set relation is defined on a reference relation (RF) which may be a regular relation or a sub-relation. A set relation may be viewed as a classification of the entities of the associated RF : an entity of a set relation corresponds to a group of entities of the RF. The variables of the set relation are statistical variables defined over the variables of the RF, and some other single-valued variables.
For instance,let CLUSTER be a set relation, with PERSON as reference relation, AVERAGE-AGE as statistical variable defined over variable AGE of PERSON, and CLUSTER-NUMBER as property variable. The following figure illustrates some entities of the set relation CLUSTER.

PERSON (reference relation)

NAME	JOB	AGE
JOHN	ENGINEER	19
ROBERT	SECRETARY	18
RICHARD	CLERK	36
ANTHONY	SECRETARY	21
PEGGY	STUDENT	18
CHRIS	SECRETARY	17

CLUSTER (set relation)

CLUSTER-NUMBER	AVERAGE-AGE
1	18
2	25

3.6 Arrays.

Two types of arrays are considered :

-square arrays, whose rows and columns correspond to the same variables (a typical use is to store correlation matrix between variables),

-rectangular arrays, whose rows correspond to values of an enumerated type variable, and columns to the values of another variable (to store cross-tabulations,for instance).

A square array may be diagonal,upper or lower triangular. An array may be sparse.

Square array : Rectangular array :

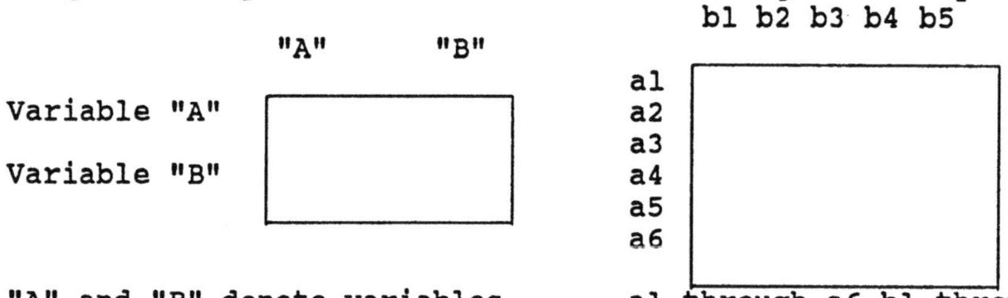

"A" and "B" denote variables.

a1 through a6,b1 through b5 denote domain values .

4.CONCLUSION.

FARANDOLE offers the following functions:

* a specific language for the definition of data in the usual data analysis terms (Abdel 84).
* a data manipulation language which automatically checks data integrity.
* a query language which allows the queries on aggregates.
* the transposed file organization. This organization is used in the statistical RDBMS RAPID (Hammond 83); it reduces the time of access to the set of observations on a variable, and is adapted to the compression techniques made necessary by the fact that statistical data are redundant and sparse. (Bates 82), (Eggers 81).

References

(Abdel 84)
Abdeljaoued Boujemaa Aicha (1984), Le Language de Déclaration de structures de FARANDOLE, Rapport interne du C.U.I, Universite de GENEVE.

(Bates 82)
Bates D.; Boral H.; Dewitt D.J. (1982), A Framework for Research in Database Management for Statistical Analysis, Proceedings ACM-SIGMOD Conference Orlando.

(Cood 70)
Codd E.F (1970) , A Relational Model of Data for Large Shared Data Banks, ACM, vol. 13, June.

(Eggers 81)
Eggers S.; (1981) Olken.F; Shoshani.A ,A Compression Technique for Large Statistical Database, Proceeding of the International Conference on Very Large Data Bases, Cannes.

(Hammond 83)
Hammond R.G. (1983), RAPID: A Statistical Database Management System ,Computer Science and Statistics, Proceedings on The Fifteenth Symposium on the Interface , James E.Gentle North-Holland.

(Jomier 84)
Jomier G. (1984), Kezouit O. , Ralambondrainy H., SICLA-PEPIN : A System Integrating Data Analysis and Relational Data Base Management System, in the 6e Proceedings in Computational Statistics: COMPSTAT, Prague , p 263-267, Edition Physica-Verlag,Vienne.

(Leonard 84)
Léonard M. (1984), Méta-Structure de FARANDOLE, Notes Techniques,C.U.I, Universite de GENEVE.

(Little 83)
Littlefield R. (1983), Cowley P.J., Some Statistical Data Base Requirements for the Analysis of Large Data Sets, Computer Science and Statistics: Proceedings on the Fifteenth Symposium on the Interface, James E.Gentle North-Holland .

(Shoshani 82)
Shoshani A. (1982), Statistical Databases: Characteristics, Problems and Some Solutions, Proceedings of the International Conference on Very Large Data Base , Mexico.

(Smith 77)
Smith,J.M;Smith,D.C.P, Database Abstractions : Aggregation and Generalization. ACM Tods, Vol.2 , Nb.2 , 1977.

Model- and Method Banks (MMB) as a Supplement to Statistical Data Banks

By K. Neumann, Berlin

SUMMARY

Using large data bases and computer aided technologies (CA Analysis, CA Interpretation, ADA, EDA) creates the need for more sophisticated and faster accessible statistical models and methods. One possible solution is to organize them in the same way as it proved to be succesful with the data. This can be done in different variants, e. g. as a special part of existing data banks or as seperate MMB either on the meta level or containing the real models and methods.

MMB are expected to ease the acces to models and methods in all phases of statistical analysis so helping the end user to analyze large data sets faster, more reliable and more complex. They are computerized tools using modern information processing technologies.

The paper describes the steps to organize MMB and the use of them which could lead to knowledge bases as part of expert systems.

Key words: Computer aided analysis / Data Banks / expert systems / methods / models.

INTRODUCTION

The long-term development of statistical computing was a steady change in the balance between statistical needs for computing capacity and their availibility (Sint, 1984). Incorporated in this often contradictionary relation was and is the interdependence of data and methods to transform them into reliable statistical information. For official statistics especially, even the extended use of electronic data processing for a long time above all improved more or less the availibility and reliabili-

Compstat 1986

ty of masses of data, but not the application of sophisticated methods. Just the recent development especially in the software field seems to bring the practical usability of statistical methods and their combination to a growing variety of models to a higher level (Nelder 1984). But there is still a gap between the handling of data on the one hand and of models and methods on the other. Whereas the main characteristics of modern processing of statistical data are large data banks the models and methods are available only in a steadily increasing diversity of software. It can be expected that the storing of such software representing statistical models and methods in a similar way as was proved to be successful in respect to the data will guarantee similar effects.

DEFINITION AND PREREQUISITE

A MODEL- AND METHOD BANK (MMB) is characterized by

- a unified organization of models and methods and their storage with a minimum on redundancy
- an automated management of the stored models and methods either with a given DBMS of an existing data bank or with a special management system.

A MMB can be organized

- on the meta-level, i. e. as a storage of information on models and methods
- on the level of programmed models and methods, i. e. analogue to the storage of real data in a data bank
- on both levels.

(A fourth solution could be to store not only metainformation and programs, but also the complete reflection of the models both in description and graphics.)

According to the practical experience in the use of data banks one can expect that the best way is to start with a Meta-Model- and Method Bank (MMMB). The main advantage could be that the user of such a MMMB will be guided immediately to the available programs or program packages. A prerequisite for such a solution would be a generally comprehensible survey on statistical software, not only listing the existing software but also comparing them at least in respect to their main characteristics. The pioneer work in this filed was done by Ivor Francis, who presented his results in an exhibition during the ISI-meeting in 1977 (New Delhi). He improved his work later (Francis, 1981). The "Taxonomy of Statistical Software, a two-way clustering of twenty variables (items) and one hundred and seventeen cases (programs)" (Francis, 1981, p. 5) can act as a model for the meta-information on statistical software mentioned. The same is true in regard of a less complicated classification, used in a "register of software for statistical, survey and general science computing" (Cable, McKay and Rowe, 1985). It is grouping the registered software product according to the availibility (on Micro-, Mini-, Mainframecomputers) as well as to the following facilities:

- S : survey administration
- C : CATI (Computer Assisted Telephone Interviewing)
- E : editing
- M : data management
- T : tabulation
- A : statistical analysis
- G : graphics

The final classification of software used in a MMMB possibly may combine the advantages of both of the examples, being not extended too much but delivering as much as possible of information to the statistical user.

REALIZATION

Which could be the practical steps in the formation of a MMB ?

First step:

Creation of a retrievable documentation of all reusable programs and programpackages with an evidence of their places and an exact information of their direct or indirect accessibility.
"Retrievable" means that at least the main characteristics of

the documented programs and program packages are stored as keywords. By expanding the evident characteristics and by logical combining the describing keywords a thesaurus should be aspired to enable an extensive retrieval.

Second step:

Fixing an unified interface (or a limited number of standardized interfaces) for the data exchange (as well for real data as for meta-data, too).
This step has to include the elaboration of service programs to realize the data communication with all programs and program packages stored in the MMB.

Third step:

Decomposition of all available programs and program packages in

- elements (mapping the elementary functions realized in the models and methods which are incorporated)
- links between the elements.

This menas that those interfaces (to the data) established in the second step are expanded by interfaces between the elements. That presupposes a set of rules for the "composition" of the elements to create "new" programs (i. e. combination of elements).

Fourth step:

Installation of a special MMB-management system to

- manage all "elements" and "links" (inclusive updating)
- record all activities and results.

By a permanent evaluation of these records we may get little by little a "storage of experiences" (in the sense of a "knowledge-base" following the theory of artificial intelligence).

Fifth step:

Expanding the documentation mentioned in the first step from a simple evidence into a "lexicon", in which there are stored for all models and methods short and long definitions, references, graphics, statistics on use etc. In such a way the MMB would become an instrument comparable with subject-matter oriented data banks, which are very important in all fields of scientific information.

For these steps two prerequisites are necessary:

- the classification of models and methods as basis for the formation of keywords and thesauri
- the implementation of a user language containing facilities for
  . different users
  . different models and methods, elements and links, existing programs and program packages.

CONCLUSION

The creation and the use of statistical data banks has proved as a permanent research-intensive process. The same will be true in respect to the establishment of a MMB as their supplement. But simultaneously the availibility of MMB will have an important impact on research and development in statistics in general and on statistical computing especially.

REFERENCES

Bibby, J. (1984) EDA and large-scale data sets - an example from the British new earning survey, Eurostat News, Special number 1984, p. 165 - 168

Cable, D., McKay, E. and Rowe, B. (1985)
Software for statistical and survey analysis, prepared for the Study Group on Computers in Survey Analysis

Daudin, J. J. Tomassone, R. and Trescourt, P. (1984)
Large-scale survey analysis, Eurostat News,
Special number 1985, p. 189 - 202

Dekker, A. L. (1984) Comparing StatisticaL Software packages on tabulation efficiency
COMPSTAT 1984, p. 317 - 322

Dörney, J. (1985) From dialogue systems to expert systems / Conception for application of AI in statistical information management
ISI 1985, Contributed Papers, Book 2, p. 573 - 574

Francis, I. (1981) Statistical Software / A comparative review, North Holland 1981

Gale, W. A. an Pregibon, D. (1984), Constructing an expert system for data analysis by working examples, COMPSTAT 1984, S. 227 - 236

Molenaar, I. W. and Broersma, H. (1984), Interpretation of statistical software output: some behavioral studies, COMPSTAT 1984, p. 341 - 346

Nelder, J. A. (1984), The role of models in official statistics, Eurostat News, Special number 1984, p. 15 - 22

Neumann, K. (1984), Metainformation as a tool for integration, COMPSTAT 1984, p. 273 - 278

Neumann, K. (1985), Computer aided interpretation of official statistics - Trends in the GDR, ISI 1985, Contributed Papers, Book 1, p. 327 - 328

Pregibon, D. and Gale, W. A. (1984), REX: an expert systems for regression analysis, COMPSTAT 1984, p. 242 - 248

Sint, P. (1984), Roots of computational statistics, COMPSTAT 1984, p. 9 - 20

Sundgren, B. (1973), An infological approach to data bases (Urval nr. 7), Stockholm 1973

Sundgren, B. (1984), Conceptual design of data bases and information systems, Stockholm 1984

Sundgren, B. (1985 a), How to satisfy a statistical agency's need for general survey processing programs, ISI 1985

Sundgren, B. (1985 b), Statistical information systems design, Statistical Journal of the United Nations ECE, 3(1985), p. 175 - 184.

A Security Mechanism Based on Uniform Rounding to Prevent Tracker Techniques

By. H. Sonnberger, Wien

In the last decade powerful snooping tools like the tracker have been developed, with which a questioner is able to deduce confidential information about individuals by processing statistical queries. In the first section of this paper we discuss a security mechanism based on Bernoulli experiments. Experimental results show that although this method has good statistical properties the relative errors of the estimators are high. The idea of exclusion resp. multiple inclusion of individuals (belonging to the specified query set) together with uniform rounding leads to a security mechanism with small relative errors. However, the statistics are no longer unbiased so small query set compromise and trackers are no longer possible.

Keywords: Statistical data bases; Security mechanisms; Uniform rounding; Tracker techniques.

Introduction

In statistical databases queries formulated in query languages must supply statistical summaries about a population without revealing particulars about any one individual (Denning et al. 1979). Compromise (or disclosure) occurs when a questioner deduces from the responses of one or more queries confidential information of which he was previously unaware (Dalenius 1977).

During the last decade an increasing number of researchers have studied methods to control compromise but have found that these methods can either easily be subverted by snooping tools like the tracker, see e.g. Schlörer (1975), Denning et al. (1979) or are impractical to use, see e.g. Dobkin et al. (1979) or Yu and Chin (1977). The first security mechanism that could partially overcome these deficiences was the Random Sample Queries (RSQ's) proposed by Denning (1980) which we shall use for comparison purposes throughout this paper.

A database contains information about the attributes of N individuals (organizations, companies etc.). There are M attributes (also called variables) where each attribute A_j ($1 \leq j \leq M$) has $|A_j|$ possible values. An example of an attribute is Sex with two possible values, Male and Female. We let x_{ij} denote the value of attribute j for individual i. Statistics are computed for subgroups of records having common attributes. A subgroup is specified by a characteristic formula C which informally is any logical formula over the values of attributes using the operators "or" (+), "and" (·) and "not" ($^-$). An example of a formula is (Sex=Male)·(Status=Divorced) that can be abbreviated to Male·Divorced, which specifies all divorced men of the database. The set of records whose values match a characteristic formula C is called the query set. In the discussions of last years special attention has been devoted to the following types of statistics.

Compstat 1986

$$COUNT(C) = n_C$$

where $n_C = |C|$ is the size of the query set C, and

$$SUM(C,A_j) = \sum_{i \in C} x_{ij}$$

The responses to COUNT and SUM queries are used to compute relative frequencies and averages,

$$RFREQ(C) = COUNT(C)/N = n_C/N$$

$$AVG(C,A_j) = SUM(C,A_j)/COUNT(C)$$

Obviously SUM and AVG queries apply only to numeric data.

If the reported frequencies are true frequencies then Schlörer (1975) and Denning et al. (1979) have shown that it is generally not possible to protect confidential information, even if a query set size restriction mechanism is implemented. The authors give the conditions under which it is possible to mask a restricted query - one which comprises either too few or too many individuals, i.e. $|C| \leq k$ or $|C| \geq N-k$ with k being a database parameter - in a way that one can get the individual value from the result of two answerable queries.

For example take a characteristic C with $|C|=1$. If one can find a characteristic M (used as mask) with $C \cap M = \Phi$ and $k \leq |M| \leq N-k$ one can always get the true answer from

$$COUNT(C) = COUNT(C+M) - COUNT(M)$$

Denning et al. (1979) have extended this idea to the concept of a general tracker and have shown that it is also applicable to a more general type of queries, including also RFREQ, SUM and AVG queries.

A Security Mechanism based on Bernoulli Experiments

Most of the snooping techniques work only if the intruder has control over the composition of the query set. Conway and Strip (1976) suggest the following method as an effective means to render the query set intersection techniques impossible: First determine the query set size $n_C = |C|$ and then draw n_C records from the query set with replacement. This however means, that some of the values would be repeated and some would be omitted hence introducing enough uncertainty into the query set composition to make disclosure impossible. The probability that a particular record will appear in the sampled query set is given by

$$1 - ((n_C - 1)/n_C)^{n_C}$$

If $n_C=100$ for example, this probability is only approximately 60 %. So it is no longer possible to isolate a single record because the intruder never knows whether the record in question is in the query set or not.

The method as it stands has one major drawback; it will work only for queries on the values of certain attributes, like SUM or AVG queries but not for COUNT queries as the true n_c is used as repetition factor in the sampling process. It should be obvious, however that the vulnerability of a database to COUNT queries is sufficient for a potential intruder as many interesting queries can be reformulated using only COUNT queries.

An obvious extension of this model which will work also for COUNT queries is not to report the true n_c value but to generate a random number n_c^* in the following way: First determine $n_c=|C|$. Then apply n_c independent Bernoulli experiments to each record of the query set with probability $1/n_c$ of success. The result are n_c random variables n_i that indicate how often record i is selected for the sampled query set. The questioner then receives the value $n_c^*= \sum_{i=1}^{n_c} n_i$. It can be shown that n_c^* is an unbiased estimator of n_c, and that the root-mean-squared relative error is $1/\sqrt{n_c}$ for sufficiently large values of n_c.

However, this security mechanism which will be called SM/B throughout this paper suffers from a deficiency in case of small query sets. From the set-up of our random variables $n_i \sim b(n_c, 1/n_c)$ we can see that in the special case of $n_c=1$ (which means identification of a single record) the intruder can be certain that a reported value of 1 reflects the true value. Therefore the following adaption of SM/B is necessary for small query sets: Let s be some threshold to be chosen by the database manager. If $0<n_c<s$ then apply s Bernoulli experiments with probability $1/n_c$ of success. With this setting one can expect that in case of small query sets, a record is selected more than once in the average, in the extreme case of $n_c=1$ even s times. This will produce a high relative error for COUNT queries making it impossible to isolate any single record. Our experimental results report relative errors of 2 - 34 % in case of COUNT queries and 0.1 - 30 % in case of AVG queries over a reasonably chosen salary field. Our results suggest that the relative errors of the estimators are larger than is necessary for a sufficient protection of confidential data.

To rule out the possibility of asking the same query many times and getting close approximations to the true values by averaging the responses, we let the seed for the random number generator depend on the query set. Therefore the intruder will always get the same response to any query comprising the same set of records. However, in this case we have to improve the basic idea of exclusion and multiple inclusions as with SM/B in such a way that the probability of a response with a large relative erro is minimized.

A Security Mechanism based on Uniform Rounding

Various authors, e.g. Nargundkar and Saveland (1972) and Achugbue and Chin (1978) studied rounding methods for preventing disclosure. We now combine rounding and the principle of exclusion and multiple inclusion of records and define the following ecurity mechanism SM/UR.

irst, determine $n_c=|C|$. Then using Uniform Rounding compute n_c^* as

$$n_c^* = \begin{cases} n_c & \text{if } r = \emptyset \\ n_c+b-r & \text{with probability } p=1/2 \\ n_c-r & \text{with probability } p=1/2 \end{cases}$$

here n_c is the integer to be rounded, b is the integer rounding base and $r=n_c$ mod b.

econd, for queries over attribute values like SUM or AVG, select b-r records to en-er the query set twice if $n_c^*>n_c$, or omit r randomly chosen records from the query et, if $n_c^*<n_c$. Compute the statistics from the modified query set. This yields a iased estimator for the query set size with bias b/2-r. Thus the bias depends on the nknown r but can be controlled via the rounding base b. For a b=5, as in our experi-ents, the bias is from the set {-1.5,-.5,.5,1.5}. If the r's are uniform distributed, he bias is

$$E(n_c^*-n_c) = 1/b \sum_{r=1}^{b-1} (b/2-r) = \emptyset$$

o our estimator for the query set size is unbiased irrespective of r. The conditional ariance is

$$var(n_c^*/r) = 1/2(b-r)^2+1/2(-r)^2 = b^2/4-br+r^2$$

hile the unconditional variance (assuming a uniform distribution for r) is

$$var(n_c^*) = 1/b \sum_{r=1}^{b-1} (b/2-r)^2 = (b^2-3b+2)/12$$

hus for e.g. b=5 we obtain a var(bias)=1. The root-mean-squared relative error can e shown to be $1/n_c$ for sufficiently large values of n_c.

xperimental Results

he security mechanisms SM/B and SM/UR were tested on a hypothetical database of N= 000 individuals and compared with the Random Sample Queries (RSQ's) suggested by enning (1980).

or each individual the following attributes were randomly and independently genera-ted according to the empirical distributions observed in the Austrian society: ex (2), Confession (4), Age Groups (8), Status (4), Number of Children (5), Education 5), Business Branch (10). Numbers in parentheses give the number of different values for the respective attribute e.g. there are four values for Status, single, married, divorced and widowed with the probabilities Pr(single)=0.19, Pr(married)=0.64, r(divorced)=0.12 and Pr(widowed)=0.05. Our sample should represent the Austrian so-ciety of age 20 - 60, age is reported in groups of 5 years. Table 1 and Table 2 list some results on subsets of SØ (male) that could be generated with one (·) operation

Table 1: COUNT Statistics for selected subsets of SØ using SM/B, SM/UR and RSQ's

	COUNT	SM/B	SM/UR	RSQ's(*)
SØ	506	462 (.087)	510 (.008)	501 (.010)
SØ·CØ	447	431 (.036)	450 (.007)	448 (.002)
SØ·C1	32	39 (.219)	30 (.063)	34 (.063)
SØ·C2	11	17 (.545)	15 (.364)	9 (.182)
SØ·C3	16	20 (.250)	20 (.250)	16 (.000)
SØ·FØ	100	120 (.20)	100 (.000)	99 (.010)
SØ·F1	324	331 (.022)	320 (.012)	327 (.009)
SØ·F2	57	51 (.105)	55 (.035)	55 (.035)
SØ·F3	25	21 (.160)	25 (.000)	24 (.040)

(*) This column is the result of two steps: first using p=0.9375 n_c^* is computed and then this value is divided by p, to give an unbiased estimate of COUNT(C).

Table 2: AVG Statistics for selected subsets of SØ using SM/B, SM/UR and RSQ's

	AVG	SM/B	SM/UR	RSQ's
SØ	7002	7577 (.082)	6961 (.006)	7092 (.013)
SØ·CØ	7009	7344 (.048)	7005 (.001)	7037 (.004)
SØ·C1	7406	5716 (.228)	7578 (.023)	7406 (.000)
SØ·C2	5916	5871 (.007)	4682 (.209)	6026 (.019)
SØ·C3	6728	6259 (.070)	6828 (.015)	7176 (.067)
SØ·FØ	6189	5241 (.153)	6189 (.000)	6178 (.002)
SØ·F1	7156	7135 (.003)	7196 (.006)	7134 (.003)
SØ·F2	7031	6949 (.012)	6879 (.022)	7448 (.059)
SØ·F3	8181	8823 (.078)	8181 (.000)	8421 (.029)

We report only a subset of our results to give an idea of the magnitude of relative errors and the difference between the three security mechanisms (full results are available upon request). As expected, SM/B behaves worst in allmost all tests, while SM/UR and RSQ's give in the average relative errors of the same order of magnitude for COUNT queries. The pattern for AVG queries is similar to COUNT queries: SM/B is worst in almost all tests. The full test results however support the conclusion that SM/UR is slightly better than RSQ's with respect to the accuracy of AVG queries

Compromising under SM/UR

Our security mechanism takes control of the composition of the query set away from the user. One question remains: Is it still possible to achieve disclosure?

We discuss two approaches: small query sets (of size Ø or 1) and trackers. Contrary to RSQ's, small query sets seem to create no severe problem. Even if the intruder has enough preknowledge to achieve a query set of size 1, he will only get a response of Ø or b with equal chance.

he following conditional probabilities are of special interest to the intruder

$$Pr\ [n_C=\emptyset \mid n_C^*=\emptyset]$$

$$Pr\ [n_C=1 \mid n_C^*=\emptyset]$$

$$Pr\ [n_C=1 \mid n_C^*=b]$$

he first two probabilities are 1/b, which is 20 % for our implementation (b=5), as a esponse of ∅ can result with equal probability for all cardinalities up to b-1. The hird case is even worse to judge as any of the 2b-1 values {1,...,2b-1} can be ounded to the value b.

similar situation arises for the intruder in case of the tracker technique. We ttempted several tracker compromises in our experimental database of N=1000 indiviuals. The target was a random individual uniquely identified by some formula C. e changed the tracker formula for each attempt as repeated attacks with the same racker yield identical answers due to the selection of the seed for the random numer generator.

e have attempted 70 different tracker attacks of the Schlörer type discussed above o compromise the true salary value and we got a mean absolute relative error of ore than 600 %.

eferences

chugbue J.O., Chin F.Y. (1978), Output Perturbation for Protection of Statistical Databases,Dept. Computing Science, Univ. Alberta, Canada.

onway R., Strip D. (1976), Selective Partial Access to a Database, Proc. ACM Am. Conf., 85-89.

alenius T. (1977), Towards a Methodology for Statistical Disclosure Control, Statistik tidskrift 15, 429-444.

enning D.E., Denning P.J., Schwartz, M.D. (1979), The Tracker: A Threat to Statistical Database Security, ACM Trans. on Database Syst. 4,1, 76-96.

enning D.E. (1980), Secure Statistical Databases with Random Sample Queries, ACM Trans. on Database Syst. 5,3 291-315.

enning D.E.(1982), Cryptography and Data Security, Addison-Wesley.

obkin D., Jones A.K., Lipton R.J. (1979), Secure Databases: Protection Against User Inference, ACM Trans. on Database Syst. 4,1 97-106.

argundkar M.S., Saveland W. (1972), Random Rounding to Prevent Statistical Disclosure, Proc. Am. Stat. Assoc., Soc. Stat. Sect., 382-385.

chlörer, J. (1975), Identification and Retrieval of Personal Records form a Statistical Data Bank, Meth. Inform. Med. 14, 7-13.

u C.T., Chin F.Y. (1977), A Study on the Protection of Statistical Data Bases, ACM SIGMOD Int. Conf. Management of Data, 169-181.

Teaching of Computational Statistics

A Data Analysis Course in a Computer Science Faculty

By T. Aluja-Banet and M. Marti-Recober, Barcelona

SUMMARY

We discuss here the "what", "how" and "why" of that course. First we examine the nature of Data Analysis, this concept being very broad and employed with different senses. We present a general scope of the different techniques covered by this concept and their relations and uses.

1. INTRODUCTION

Since their creation in 1976 the Informatics' Faculty of Barcelona offers a course on Data Analysis (DA). This course is given in the Operations Research and Statistics Department. The syllabus corresponds to an advanced course on applied statistics focused on the use of Multivariate Data Analysis (MDA). The curriculum of the Faculty is described and discussed in Martí et al. (1981).

Probably the main discussion about the usage of DA techniques is due to the distinction between descriptive (or data-analytic) methods and probabilistic (or model-oriented) ones. We shall stress the fact that the two approaches had been conceived for different purposes and they are complementary rather than competitive. As Tukey (1980) points out both approaches are necessary, both must be taught and the descriptive approach must precede the probabilistic one.

We present a taxonomy of MDA methods, as a preliminary step to decide what methods, what precise topics and in what sequence are to be included in the course.

Compstat 1986

Finally we discuss the teaching methodology employed in the course, their relationships to the research activities of the Department, and we include the course's contents.

2. MULTIVARIATE DATA ANALYSIS: EXPLORATORY AND CONFIRMATORY APPROACHES

Different definitions have been proposed to precise the scope of MDA techniques. Textbooks usually present MDA as a collection of techniques, being their common characteristic the fact that they study p-dimensional random variables; furthermore, we can add that they generally deal with a large number of individuals and they require the use of computers.

Apart from that, there are a great variety of techniques. Probably the main difference can be found between the probabilistic or model-based methods, which permit a confirmatory or explanatory approach, and the algebraic or geometric methods, intended for a descriptive or exploratory purpose. Especifically in the case of the MDA methods there is, at some extent, a controverse about both approaches, presenting as competitive what is really not. Sometimes one technique may be enough to perform a statistical analysis, but more often a statistical analysis is an iterative process of learning going from the exploratory approach to the confirmatory one.

As Cox (1981) argues, the main steps in a statistical study are: (i) initial data manipulation after data collection; (ii) preliminary analysis; (iii) identification or model especification; (iv) estimation; (v) model checking. Phase (ii) is of exploratory or descriptive-oriented nature, phase (iv) is clearly of confirmatory or model-oriented (probabilistic) nature; in phases (iii) and (v) usually descriptive techniques are used.

Tukey (1980), points out, about model specification, that in most cases "ideas come from previous exploration more often than from lightning strokes". This is the same idea when Box (1979) states "prior to using the methods that most text books emphasize, there is a very important and largely neglected phase of activity which Fisher called specification and which has also been called model identification. This involves informal techniques of analysis of data, many of them graphical, aimed at looking at the data in a preliminary and exploratory way in order to help understand what questions should be asked and what tentative models might be entertained".

In the MDA specific case this remarks are specially relevant due to the difficulty to explore a large data matrix. But also due to the fact that "the more the matrix to study contains a greater number of variables, the more the field of realistic probabilistic hypothesis will be reduced" (Lebart (1973)).

This leads us to consider the exploratory and confirmatory approaches as two phases of the same analysis, correponding to two different stages of the research; first we want to see which data we have, to represent it graphically, to detect abnormalities, to suggest patterns; second we want to confirm our hypotheses, to test them and to explain the pattern detected. Thus, we think that both approaches are complementary rather than competitive, both are necessary and have to be taught in a course as the one we present.

Another point of view to consider is the origin of the data under study: observational or experimental. The two phase analysis is especially useful in the former case, to analyse previously collected data, i.e. socioeconomic data, or surveys' information.

3. A SCOPE OF MULTIVARIATE DATA ANALYSIS

Many of the MDA techniques had been developed in the twenties and thirties, other were developed more recently. In both cases the main factor for the practical and increasingly use of these techniques is the growth of computing power and resources and their availability at very low cost.

There are many different MDA techniques that in fact are related in different ways and points of view, particularly from the mathematical structures they are based upon. Many of them had their origin in different application fields but they are supported by similar statistical methods.

Hanley (1983) divided the MDA techniques into those in which the variables involved (i) are all of "equal status" or (ii) can be divided naturally into two sets. In the first group he includes Principal Component Analysis (PCA), Factor Analysis (FA) and Cluster Analysis. For the second group he uses three criteria of classification:

a) response variables (outcomes or dependents) and stimulus variables (predictors or independents),
b) univariate and multivariate variables, and
c) type of coding used for the variables: categorical (discrete in Hanley's terminology) and continuous, and a mixture of both when necessary.

Those criteria allows him to present a taxonomy of parametric statistical methods with a high condensed view in a table in wich we can identify the next main families of methods:(i) multidimensional contingency tables for categorical variables; (ii) multiple and multivariate regression and canonical analysis, for continuous variables; (iii) logistic regression, discriminant analysis and MANOVA and MANCOVA, for mixed variables.

To go further we present here an alternative taxonomy of multivariate methods according to the following five criteria of classification:

- the data matrix's nature we are using: individual* variables, proximity between individuals or frequencies.
- the stage of study we are carrying on: with an exploratory or confimatory purpose.
- the homogenity of variables: homogeneous (one set) or forming several groups.
- the homogenity of individuals: one or several source populations
- the type of coding used for variables: categorical, continuous or mixed.

These criteria allow us to present the next clasification:

PROXIMITY MATRIX

- Metric :	Principal Co-ordinate Analysis Principal Components Analysis
- Non metric :	Multidimensional Scaling

FREQUENCIES MATRIX

	Simple matrix	Multiple matrix
- Exploratory:	Contingency table Corresp. Analysis	Mult. conting. table Mult. corresp. Anal.
- Confirmatory:	Log-linear model	Log-linear model

INDIVIDUALS*VARIABLES MATRIX

- Exploratory analysis

	Variables				
	One group		More than one group		
	Cont.	Categ.	Cont.	Categ.	Mixt.
One population:	PCA FA Cluster.	MCA Cluster.	Canon. Corr. Analysis Gener. Can. Corr. Anal.		
Several populations:	Discrim. Anal. Clustering				

- Confirmatory

	Cont.	Categ.	Cont.	Categ.	Mixt.
One population:	FA	Log-lin. model	Multiv. Regress. Causal Models	Multiv. Logist. Regress.	
Several populations:	Discrimi. Anal.				MANOVA MANCOVA

This presentation means the addition of one more group to Hanley's methods, for the treatment of categorical variables with exploratory purpose, formed by the Correspondence Analysis, Multiple Correspondence Analysis and Clustering methods. As Volle (1985) pointed out, Multiple Correspondence Analysis can be used advantageously with continuous variables recoded as ordinal.

4. THE DATA ANALYSIS COURSE

The main goal of the course is to provide students with enough knowledge to confront the problems they are likely to encounter in real world, without loosing a presentation of the theoretical support of each MDA technique.

The students have a good level of statistical theory, since they had before two statistical courses as prerequisite; they have also a good computer and data management practice and are able to complement the shortcomings of available statistical packages and also to interpret and report the results of the analyses.

DA is an optional fourth level course in the Informatics Faculty's curriculum, in Applied Multivariate Statistics.

We consider of a great importance exposing the students to real problems in such a way as Chattfield (1982) has pointed out, that is a "problem oriented" fashion rather than a "technique oriented" one. We achieve this by bringing to class selected aspects of the research projects conducted by the Department assuring a real involvement of students in applied problems. The software used is mainly the SPAD (1985) package for the exploratory phase and the BMDP package or similar ones for the probabilistic methods.

5. PROGRAM OF THE COURSE

Of course, it is difficult to present all the techniques above mentioned in one course, possibly it wouldn't be convenient indeed; a course is more than a collection of techniques; thus, the syllabus of the course is a logical consequence of our concept of MDA, presented in sections 2 and 3, and the teaching methodology.

The central focus of the course is the data (and consequently the statistical problem) under study, in fact, the program course tries to emphasize a statistical methodology; that is, given a data set, we start its analysis with an exploratory approach and later we use the confirmatory one. Apart from that, since there are different techniques suitable for continuous variables or categorical variables, we divide the course in two parts, one dealing with continuous variables on the other with categorical ones.

1. Introduction to Multivariate Data Analysis
2. Continuous data
   - Exploratory approach: Principal Components Analysis, Clustering, Factorial Discriminant Analysis
   - Confirmatory approach: Probabilistic model, Related sample distributions, Multivariate linear model, Discriminant Analysis
3. Categorical data
   - Exploratory approach: Correspondence Analysis, Multiple Correspondence Analysis, Clustering
   - Confirmatory approach: Contingency tables, Loglinear models, Generalized linear models (logistic regression)

This presentation is the result of our experience coming from a course presented as a collection of techniques that evolved gradually to a case oriented one, stressing a methodological approach such as presented previously.

REFERENCES

Box G.E.P. (1979) "Some problems of Statistics and every day life", J. of Am. Stat. Ass., v. 74, n. 365, p. 1-4.

Chatfield C., Collins A.J. (1980) "Introduction to Multivariate Analysis", Chapman & Hall, London.

Hanley A.J.(1983) "Appropriate uses of multivariate analysis", Ann. Rev. Public Health, v. 4, p. 155-180.

Lebart L. Tabard N. (1973) "Recherches sur la Description Automatique des Données Socio-économiques", CREDOC, Paris.

Martí M., Puigjaner R., Saltor F. (1981) "Barcelona Faculty of Informatics Curriculum: Its Planning, Implementation and Evaluation", WCCE, Ed. B. Lewis & D. Tagg, p. 767-774, North Holland.

SPAD (1985) "Système Portable pour l'Analyse des Données", Lebart, Morineau et coll. CESIA, Paris.

Tukey J. W. (1980) "We need both exploratory and confirmatory", The Am. Stat. v.34, n.1, p 23-25.

Volle M. (1985) "Analyse des Données", Economica, Paris.

Teaching Computational Statistics on Workstations

By B. Francis and J. Whittaker, Lancaster

SUMMARY

A statistical computing laboratory of ten workstations with 32-bit architecture has recently been installed at the University of Lancaster. This article discusses the potentiality of these machines for teaching statistics.

KEYWORDS: Statistical teaching, Computer statistical packages, Graphics workstations, GLIM.

1. The Computers in Teaching Initiative

In 1984, the Computer Board of Great Britain provided money to evaluate the benefits of the new technology for teaching in a wide variety of University subjects; more money was provided by the Computer Board and the UGC in the following year; providing total funding of over £4 million. An important aim of the initiative was to stimulate new teaching methods, and to encourage the development of software tools on the new computer workstations and microcomputers which are becoming available.

Statisticians at Lancaster University were successful in being funded to set up a statistics laboratory of ten Whitechapel MG-1 graphics workstations running 42-nix (a variant of UNIX™). The following sections describe the philosophy of teaching statistics at Lancaster, and how this philosophy will be incorporated in the development of new statistics courses using the workstations.

UNIX is a trademark of Bell Telephone Laboratories

Compstat 1986

2. The Lancaster Environment

The staff and their research interests: Statisticians are spread throughout the University at Lancaster with major groups in the Centre for Applied Statistics and the Department of Mathematics. The activities of these two groups differ in terms of the proportion of time devoted to teaching, research and consulting but unite through a common committment to applied statistical modelling. Expertise resides in the computer modelling of large scale, multivariate statistical systems, especially models for the analysis of data from sample surveys, generalised linear models for contingency tables, models for time series, stochastic processes and survival analysis.

Teaching and Computing Activities: The exciting part of the statistician's life is working with real data: as Sherlock Holmes is reported to say: *'Data, Data, Data!'* He cried impatiently. *'I can't make bricks without clay.'* It is strongly felt at Lancaster that is is important to integrate student teaching with currently active research: practical training requires computation with realistic data sets and fast computer response times.

Until the early eighties computing resources in the University were positively primaeval. Things then took a remarkable upturn with the acquisition of interactive facilities driven by two VAX processors, and from its initial installation the system has been fully exploited for its capacities in both teaching and research.

The change in computing resources has substantially modified teaching styles over the past few years, most notably in the direction of project oriented work. Standard undergraduate courses are given by the Mathematics Department. Approximately 50 students take the Statistics option in their first year which occupies one third of their time. They acquire a working knowledge of Basic, Minitab and GLIM. For the latter package, a successful teaching environment has been set up based on "learn" and information files (see Francis et al, 1985). The second year is more theoretical apart from one short course in Applied Linear Regression. A smaller, varying number select third year options, such as Mutivariate Statistics or Time Series, which both involve heavy usage of statistical software packages. Many specialist courses in statistical and mathematical modelling are put on by both groups; there have been recent intensive continuing education courses on GLIM, survival analysis, time series, variance components and graphical modelling.

Today, the situation has been reached where during the University term the Vax is frequently saturated resulting in very severe deterioration of the service. This prohibits the use of large data sets and graphical display in regular terminal sessions. Furthermore student word-processing is not encouraged due to inadequate terminal availability. The workstation laboratory is therefore an important addition to resources, freeing the statisticians to run and develop courses as they wish without the constraints of a central service.

3. The Lancaster Philosophy

The statisticians at Lancaster share a common belief in the methodology of applied probability modelling and the likelihood school of inference. In principle, the analysis of large scale data sets proceeds by writing down the likelihood function on the basis of the probability model and the sampling scheme. This philosophy solves problems, for example, in factor and latent structure analysis, in survival analysis, in the analysis of hierarchical variance component models, without any unwarranted ad hoc data reduction and simplification. Heavy computational demands in data analysis are occasionally made because of necessary numerical procedures such as repeated inversion of large matrices and numerical integration, and of computationally intensive graphics.

4. The Whitechapel MG-1 Workstation

Each workstation has a powerful 32-bit processor, with 2Mb of memory and high-resolution graphics, giving speed and power comparable to a small minicomputer. Each MG-1 is also provided with window management software, providing a friendly user interface which is capable of running many application programs simultaneously. A typical screen is shown in Figure 1. A mouse controls the manipulation of windows on the screen, and a keyboard is also available for traditional input. Each teaching machine has a 22Mb hard disk, and the ten are linked by ETHERNET, providing access by each machine to a printer and the campus network, as well as avoiding the duplication of software. The The MG-1 runs the 42-nix operating system, which is a variant of UNIX^{TM} with added windowing software. The advantages of UNIX are well known; Perlman (1982) provides an account of the usefulness of UNIX for data analysis. Software tools include GKS (Graphics Kernel System, level 2B) and DED, a mouse orientated screen editor, which is particularly easy to use for students new to computing. The MG-1 already supports the statistics package GLIM, and GENSTAT (particularly relevant for modelling time series) is being installed on the machine by the Lancaster team.

The workstations therefore allow analysis of realistic data sets while preserving the fast computer response times vital to both class demonstrations and individual student work. Above all, they allow high quality graphical output now essential for the interpretation of statistical models.

5. Software and Teaching Developments

Windowing software provides an excellent opportunity to provide the student with a more informative learning environment. Figure 1 shows a simple example of such a system, based around the statistics package GLIM. Four windows are displayed. The central window is running an interactive GLIM session; surrounding windows provide extra information on the package and advice and/or exercises for the student to carry out. The window in the bottom left-hand corner provides access to the operating system, allowing the user to examine directories, edit files, while still running the statistics software. A common complaint of students new to statistical computing is that they become 'lost' - not knowing if they are communicating with GLIM, the operating system or the campus network; the above approach solves these problems.

Figure 2 shows the typical contents of a window constructed during a session on the MG-1. Each curve is a plot of the relative likelihood function for the success probability in binomial samples of size 40. The

FIGURE 1: A screen dump made during an MG-1 session showing an improved teaching environment for GLIM

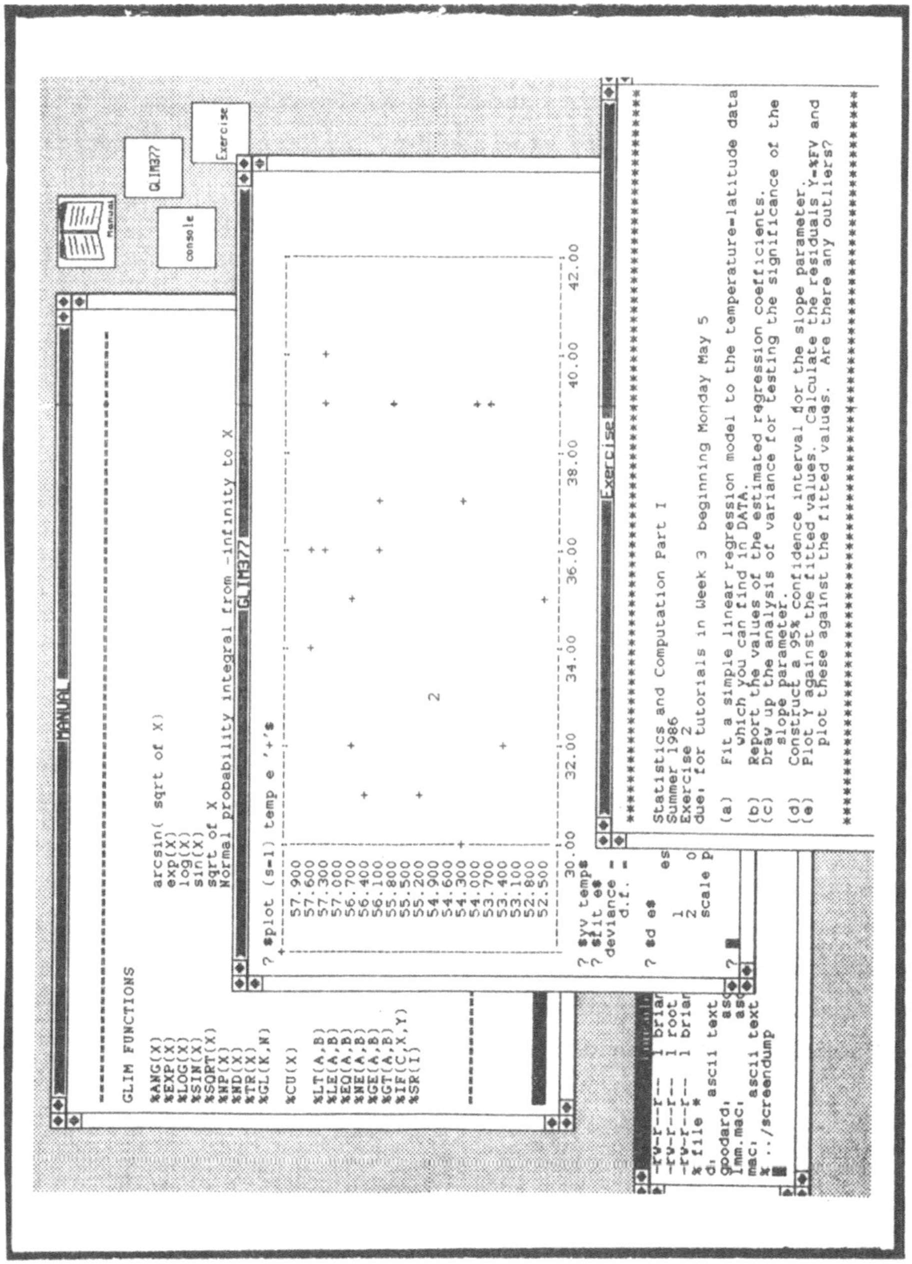

exercise was to plot the different curves corresponding to different observed numbers of successes in order to see how the inference might differ.

FIGURE 2: Relative Likelihood functions for Binomial samples of size 40 for different observed numbers of successes.

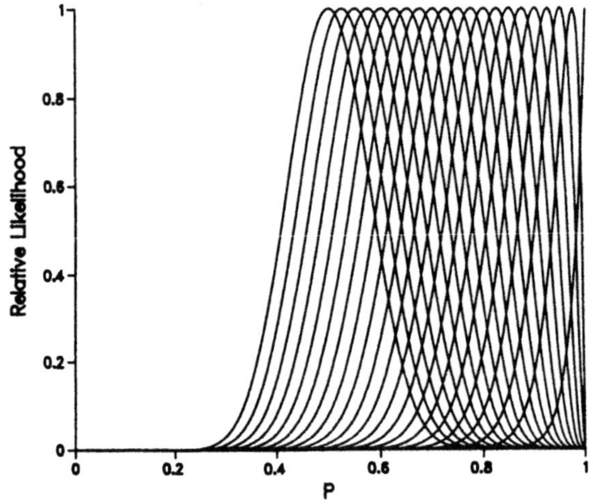

6. The future

An important priority in developing the workstations is to redesign existing instructional material to emphasise the simulation of sampling variation, likelihood function computation and plotting, and iterative maximum likelihood estimation. Full use will be made of the existing statistics packages GLIM and GENSTAT; a course text for GLIM is currently in preparation (Aitkin et al, 1987). Development of new courses will also involve the commissioning and writing of new software able to take advantage of the technology of the machine.

References

Aitkin, M.A., Anderson, D.A., Francis, B.J. and Hinde J.P. (1987). Statistical Modelling in GLIM3, Oxford University Press (to appear).

Francis, B., S.Thompson, J.Whittaker (1985). GLIM Newsletter 9. Teach-Yourself-GLIM in an interactive environment.

Perlman, G. (1982). Data Analysis in the UNIX environment. Computer Science and Statistics: Proceedings of the 14th Symposium on the Interface. Springer Verlag, New York.

Teaching Regression Analysis Using MIMOSA – an Interactive Computer System for Matrix Calculus

By S. Puntanen and H. Kankaanpää, Tampere

Summary: This paper describes the use of MIMOSA in the teaching of regression analysis. MIMOSA is an interactive computer system for matrix calculations developed at the Computer Centre of the University of Tampere.

Key words: Computer; Matrix; Regression; Teaching of statistics.

1. INTRODUCTION

This paper describes a software package, named MIMOSA, and its use in teaching a course in regression analysis. The MIMOSA system is a greatly extended and improved version of the MATRIX system described by Puntanen and Pukkila (1982). Both of these systems were developed at the Computer Centre of the University of Tampere; MATRIX was implemented on a DEC 2060 in 1981 and MIMOSA just recently in 1985. MIMOSA is written in Fortran 77 language.

The main point in the use of MIMOSA is that the matrix expressions for the usual statistics desired in linear model analyses can be written directly into the MIMOSA program using almost standard matrix algebraic notation. Thus students can become very much aware of what the "black box" is doing for them.

Briefly, the goal of the course in regression analysis for second- or third-year statistics students is that after the course they should be capable of carrying out a small empirical investigation where the main statistical technique is regression analysis. Thus we emphasize the practical skills: students should also be able to work with such a model matrix X which is not only an Arabic letter but also a matrix full of figures. Having this goal in mind, we have felt it very important that the following levels be taken into account, as regards numerical manipulation:

Working

- by hand with very small data,
- with statistical packages using both small data and large real data,
- by hand with very small data using matrices,
- with matrices using the computer as an aid.

The last two points are not necessary for a first-year student, but for the development of a deeper understanding of the concepts there still remains a

Compstat 1986

certain black box if the student always carries out his regression analysis by a standard statistical package.

The four points mentioned above are not so trivial as they may seem at first sight. And not all statistics teachers may agree on the necessity of these points. However, we would like to emphasize e.g. the role of real empirical data in the course of regression analysis: the presence of real data is a *must*. It belongs to the *real* nature of a *real* statistics course that *real* data are present. Accepting this principle it then follows that in the course of regression analysis an easy-to-use statistical package is needed by which all regression calculations can be carried out. On the other hand, students have to become familiar with certain mathematical formulae, and to express these mathematical formulae in civilized terms, the matrix notation is necessary. However, the numerical calculations themselves may not be best done using direct matrix calculus; even so, in this paper we do not pay much attention to the numerical optimality in carrying out calculations for regression. Namely, from the students point of view, that is not so central a matter in this context.

The fourth point above is the very core of this paper. We think that there is a certain gap between pure mathematical matrix manipulation by pen and paper and the practical numerical use of matrix formulae really to calculate something from data. The emphasis in the use of MIMOSA is on providing a more concrete feeling of the effects of different formulae and slight changes in them. Problems are now in essential statistical points: the user has only to know the right formula which can then be activated by MIMOSA.

2. CALCULATIONS

In this section we consider how calculations can be carried out in MIMOSA. The first phase is naturally to input the data and there are various ways to do this. For example one way of feeding the 3×2 matrix **A** with the columns (1, 2, 3)' and (4, 5, 6)' to MIMOSA, is the following:

```
A:= ROW(COL(1, 2, 3), COL(4, 5, 6));
```

The system contains standard matrices like `I` = an identity matrix, `E` = {1} and `O` = {0}. The notation `C:= B(2:10, 3:7);` indicates that **C** consists of the rows 2, 3,...; 10 and of the columns 3, 4,..., 7 of **B**. Furthermore, `F:= B(:7, :);` implies that **F** consists of the rows 1,..., 7 of **B** and of all the columns of **B**.

The system has a collection of standard functions - `TRANS`, `INV`, `CENT` (for centering) etc. - available to all users. So for example, in the case of the linear model $\mathbf{y} = \mathbf{X}\beta + \varepsilon$, the ordinary least squares (OLS) estimates for the regression coefficients can be calculated as: `INV(TRANS(X)*X)*TRANS(X)*Y;`

An extremely useful feature of MIMOSA is the user's possibility to define his *own functions*, like the function BHAT, as follows:

```
FUNC BHAT(X, Y) = INV(TRANS(X)*X)*TRANS(X)*Y;
```

Writing

```
A:= BHAT(X, Y);
```

the vector A will be the OLSE of β. The command OUTPUT BHAT > REGRAN; stores the function BHAT in the file REGRAN, whence it can be fed in whenever needed. Defining new functions the user can create a collection of functions; a kind of *function bank*. In the following some functions of a bank are shown.

```
FUNC MEAN(A) = TRANS(COLSUM(A))/ROWS(A);       VECTOR OF THE MEANS OF THE VARIABLES
FUNC SPM(A) = TRANS(A)*CENT(A);                SUMS OF (CORRECTED) PRODUCTS
FUNC COV(A) = SPM(A)/(ROWS(A)-1);              COVARIANCE MATRIX; ROWS(A)= # OF THE ROWS
FUNC COLSUM2(A)=TRANS(COLSUM(SQ(A))); COLUMN SUMS OF THE SQUARED ELEMENTS; B=SQ(A)⇒bij= aij^2
FUNC VAR(A) = COLSUM2(CENT(A))/(ROWS(A)-1);    VECTOR OF THE VARIANCES
FUNC STDEV(A) = SQRT(VAR(A));                                          EE(R,C) ={1}RxC
FUNC XUNIT(A) = ROW(EE(ROWS(A),1), A);         ADDS THE COLUMN OF 1'S AS THE FIRST COLUMN
FUNC INTRA(ROO,N)=ROO*EE(N,N)-(ROO-1)*II(N); INTRACLASS CORRELATION MATRIX; II(N) = INxN
FUNC COR(A) = INV(DIAGVM(STDEV(A)))*COV(A)*INV(DIAGVM(STDEV(A))); CORRELATION MATRIX
FUNC BHAT(X,Y) = INV(TRANS(X)*X)*TRANS(X)*Y;
FUNC YHAT(X,Y) = X*BHAT(X,Y);
FUNC RES(X,Y) = Y-YHAT(X,Y);
FUNC SSE(X,Y) = TRANS(Y)*RES(X,Y);
FUNC MSE(X,Y) = SSE(X,Y)/(ROWS(X)-COLS(X));
FUNC COVB(X,Y) = MSE(X,Y)*INV(TRANS(X)*X);     ESTIMATED COVARIANCE MATRIX OF THE OLSE
FUNC DEVB(X,Y) = SQRT(DIAGMV(COVB(X,Y)));
FUNC VIF(X) = DIAGMV(INV(COR(X(:,2:))));       VARIANCE INFLATION FACTORS
FUNC TTEST(X,Y) = DIV(BHAT(X,Y),DEVB(X,Y));    T-STATISTICS;     A:=DIV(B,C) ⇒ aij = bij/cij
FUNC FTEST(X,Y,K,D) = BEGIN                    F-TEST FOR THE HYPOTHESIS  H: K'b = d
   A:= TRANS(K)*BHAT(X,Y)-D;
   B:= INV(TRANS(K)*INV(TRANS(X)*X)*K);
   Q:= TRANS(A)*B*A;
   FTEST:= (Q/COLS(K))/MSE(X,Y);
END;
FUNC BLUE(X,Y,V)=INV(TRANS(X)*INV(V)*X)*TRANS(X)*INV(V)*Y;
FUNC BLUEC(X,V)=INV(TRANS(X)*INV(V)*X);        COVARIANCE MATRIX OF THE BLUE WHEN cov(ε)=V
FUNC OLSEC(X,V)=INV(TRANS(X)*X)*TRANS(X)*V*X*INV(TRANS(X)*X); COV MATRIX OF THE OLSE
FUNC H(X) = X*INV(TRANS(X)*X)*TRANS(X);        THE HAT MATRIX
FUNC M(X) = I-H(X);
```

Compared with the MATRIX system, MIMOSA offers e.g. the IF-THEN-ELSE - structure and also much wider and more flexible possibilities to define functions. As an example, let us consider the calculation of the externally Studentized residuals, whose formula is [Cook (1977), Hoaglin and Welsch (1978)]

$$STUDE_i = (y_i - \hat{y}_i)/[MSE_{(i)} m_{ii}]^{1/2},$$

where $MSE_{(i)} = [SSE - (y_i - \hat{y}_i)^2/m_{ii}]/f$, $\mathbf{M} = \mathbf{I} - \mathbf{H}$, $\mathbf{H}$ is the hat matrix, and f is the degrees of freedom. Using MIMOSA, the Studentized residuals can be calculated quite conveniently as follows:

```
FUNC STUDE(X,Y) = BEGIN
  DF:= ROWS(X)-COLS(X)-1;
  RESI:= RES(X,Y);     RESI2:= SQ(RESI);     RESI CONTAINS THE RESIDUALS y_i - ŷ_i
  A:= DIAG(M(X));      B:= INV(A)*RESI2;     B CONTAINS (y_i - ŷ_i)^2/m_ii
  SIG:=MINUS(SSE(X,Y), B))/DF;               U:= MINUS(T, V), T IS A SCALAR ⇒ u_ij = T - v_ij
  STUDE:=DIV(RESI, SQRT(A*SIG));             SIG CONTAINS THE MSE_(i)´S
END;
```

Note that the function STUDE could also have been written as a one-line statement, without the BEGIN-END structure; it is simply for clarity that we have divided STUDE into the parts above. The squared Cook´s distance can be calculated in MIMOSA in the following way:

```
FUNC COOKM(X,Y) = DIAG(H(X))*INV(DIAG(M(X)))*SQ(STUDE(X,Y))/COLS(X);
```

The k largest eigenvalues and the corresponding eigenvectors of a symmetric matrix $\mathbf{V}_{n\times n}$ can be calculated by writing `B:= EIGV(V, K);`, which implies that $\mathbf{B}$ will be an $(n+1)\times k$ matrix such that $\mathbf{B}' = (\mathbf{D}', \mathbf{C}')$, where $\mathbf{D}_{1\times k}$ comprises the k largest eigenvalues of $\mathbf{V}$ and $\mathbf{C}_{n\times k}$ consists of the corresponding eigenvectors. The following example shows how the user can take a nonnegative definite square root of a nonnegative definite matrix in MIMOSA:

```
FUNC ROOT(V) = BEGIN
  B:= EIGV(V,RANK(V));
  D:= TRANS(B(1,:));                     D CONTAINS THE POSITIVE EIGENVALUES OF V
  C:= B(2:,:);                           C CONTAINS THE CORRESPONDING EIGENVECTORS
  ROOT:= C*SQRT(DIAGVM(D))*TRANS(C);     ROOT(V)*ROOT(V) = V
END
```

Since the partial correlations are correlations between certain residuals, we observe that the function

```
FUNC PCOR(A,B) = COR(RES(XUNIT(B),A));
```

calculates the partial correlations between the variables in the data matrix $\mathbf{A}$ after the elimination of the variables in $\mathbf{B}$. Thus the function PCOR serves as an example of how simply the calculations in MIMOSA can be carried out. Note that the use of XUNIT is needed in PCOR.

3. SIMULATIONS

"What is meant by the variance of the estimator of a regression coefficient?" This is an example of the fundamental questions to which statistics students should necessarily be able to give a clear answer. Without understanding this concept the correct interpretation of the nature of test statistics is impossible. The usual way to try to concretize these kinds of concepts is to use phrases like "If this experiment were repeated 100 times, then we would observe..." However, without the computer it is very inconvenient to "repeat the experiment 100 times". Similarly, the standard assumptions regarding the error term ε are relatively difficult to concretize. In the course of regression analysis students should not only carry out their regression calculations from given data but also create data from a given model and then estimate the parameters, and of course, do this, say 100 times.

MIMOSA offers effective possibilities for random number generation which much facilitates illustration e.g. of the meaning of the sampling distribution of the OLS estimators. If we write `U:= NORMAL(A, B);`, then it means that each element u_{ij} of U is generated from $N_1(a_{ij}, b_{ij})$ and all u_{ij}'s are independent. Hence the function `FUNC NORMA(N,NRO) = NORMAL(OO(N,NRO), EE(N,NRO));` generates *NRO* independent observations from $N_n(0, I)$, where n = `N`. Furthermore, the function

```
FUNC LMYI(X, B, S2, NRO) = X*B*EE(1, NRO) + SQRT(S2)*NORMA(ROWS(X), NRO);
```

generates *NRO* observations for y from the linear model $y = X\beta + \varepsilon$, where β = `B` and $\varepsilon \sim N_n[0, (s2)I]$. If we put `YG:= LMYI(X, B, S2, NRO);`, then the *NRO* OLSEs for β can be calculated writing simply `C:= BHAT(X, YG);`. On the basis of the matrix C the student can check, for example, how well the unbiasedness of the estimator is fulfilled, how close is the covariance matrix of the simulated values to the theoretical value $s2(X'X)^{-1}$, what happens to t-statistics, etc.

It is also useful to define a function

```
FUNC NOR(U, V, NRO) = U*EE(1, NRO) + ROOT(V)*NORMA(ROWS(U), NRO);
```

which generates *NRO* observations from $N_n(\mu, V)$, μ = `U`. Thus the function

```
FUNC LMY(X, B, V, NRO) = NOR(X*B, V, NRO);
```

generates *NRO* observations for y from the linear model where $\varepsilon \sim N_n(0, V)$. When a student has generated y-values from $N_n(X\beta, V)$, then he can estimate the regression coefficients using both the OLS and the generalized least squares method. He can check whether the Gauss-Markov theorem indeed works: How bad is the OLSE compared with the BLUE? What happens to test statistics when standard assumptions do not hold?

Note that we can also generate the OLSEs of β directly, without first generating the y-values, using a function like

```
FUNC NORB(X, B, V, NRO) = NOR(B, INV(TRANS(X)*X)*TRANS(X)*V*X*INV(TRANS(X)*X), NRO);
```

4. CONCLUDING REMARKS

In this paper we have described a computer system which we have found very useful in teaching linear statistical models. It is clear that there also exist other computer systems with properties similar to those of MIMOSA. However, there are some features in MIMOSA which in particular make it attractive for teaching purposes. One important thing is that it is extremely simple to learn and use; no more is required than that a student must be able to write the formulae in matrix terms - no programming skill is needed. MIMOSA is not a programming language like APL, but for teaching purposes it is "close enough" to a programming language. As was mentioned in the introductory section, MIMOSA is a kind of *a formula activator*. When using it, the problems are in essential points, that is, what is e.g. the correct algebraic formula one should use. When the student finds the formula he merely activates it; he has not to write standard programming language code.

In our course every student makes his own function bank and that bank he will use for numerous data sets. In recent years indeed a great number of books on regression have been published offering many illustrative data sets. Using a system like MIMOSA the student can very easily test whether he is really able to get similar results. Also our students have to carry out an empirical project in which they collect their own data sets and apply regression analysis to them using both MIMOSA and standard statistical packages.

The simulation in MIMOSA is quite easy and thus it offers good possibilities to illustrate the central concepts associated with regression analysis. When generating observations from a model using MIMOSA, the student has merely to write down the model in matrix terms - and that is enough; thus there is no improper coding. Students can really "see" how the observations were born, and they can then calculate whether the results obtained are in balance with the theory.

No program should be a dead end, and in particular this is true in regression analysis. MIMOSA has flexible connections e.g. with statistical programs developed for plotting and other graphics.

REFERENCES

Cook, Dennis R. (1977): Detection of influential observations in linear regression. *Technometrics*, 19, 15-18.

Hoaglin, David C. and Welsch, Roy E. (1978): The hat matrix in regression and ANOVA. *The American Statistician*, 32, 17-22.

Puntanen, Simo and Pukkila, Tarmo (1982): The use of MATRIX, an interactive computer system for the analysis of matrices, in the teaching of statistics. *COMPSTAT 1982, Proceedings in Computational Statistics, Part II (Supplement): Short Communications, Summaries of Posters*, Physica-Verlag, Wien, 223-224.

Practical Use of Color Imaging in Automatic Classification

By N. Ohsumi, Tokyo

SUMMARY

Some proposals are presented for utilizing color information in data analysis in computer graphics environments. The possibilities presented by using color in data analysis and the problems involved are presented first, then the systematization of color models in data analysis is dealt with. A proposal is submitted regarding the user interface software in color graphics system design. Examples of data analysis using color are provided showing color imaging patterns of data matrices, color plots of principal component scores, and so on.

KEYWORDS: Color models (RGB model, HLS model), color monitor display, principal component analysis, colored pattern matrix, automatic classification, colored vector, areal clustering system

1. Introduction

Research and application of graphical representation techniques as a method of visual communication of statistical information is now being carried on actively. Notable improvements have been achieved in the hardware and software environments supporting such techniques and a large number of statistical graphing program packages have become easily available. These advances in hardware and software environment have given great impetus to the use of color in the preparation of statistical graphs and maps.

In the area of data analysis, too, a notable increase is observed in proposals and experiments on the use of color. However, the use of color is still open to question because color science is inexact and there is no firmly established, systematic color theory yet. Nevertheless, the use of color graphics in data analysis is quite tempting to many researchers because it is natural for us to see things in color.

This paper presents some proposals for exploratory methods that can be applied in using color to examine classification problems. All ideas advanced in this paper are practical, interactive ones that can be realized *with a microcomputer and a low-cost color monitor display,* and focus on the importance of color handling software engineering to data analysis.

2. Interface between color models and data analysis

The use of color must be examined and developed with a clear recognition of the differences between the perceptional model and the descriptive model. The proposals made in this paper are intended to open up a new approach to the use of color in data analysis, with due consideration to the points enumerated in past reports.

Compstat 1986

The color models used in computer graphics are descriptive models built on the principle of color harmony, advocated by Ostwald and Munsell. Color handling carefully based on these color models is a minimal condition for the use of color in data analysis. The representative color models cited in past reports dealing with color graphics are summarized in Fig. 1. The HSV and HLS color models descriptionally represent human color perception, using three aspects of color (H = hue, L = lightness or V = value, S = saturation). Thus there are best suited to data analysis methods because they are intuitive, yet algorithmic. The latest advanced color graphics workstations can display as many as 16 million colors, but in data analysis the freedom of color handling is more important than number of displayable colors. It is quite troublesome in the RGB model to specify the mixture rates of various colors to obtain the desired color. Specifying the color using the HSV or HLS models suits human perception better allowing the user to select a color on the basis of various intuitive parameters such as hue, intensity, brightness, lightness, tints, shades and tones.

This leads us to the conclusion that it is necessary to develop color transformation algorithms for linking the RGB model which is suited to color monitor control, and the HSV/HLS models. Figure 2 shows such an architecture, which can be fully realized with a microcomputer and a graphics workstation. Even more important is an easy-to-use program that provides a user-friendly interface. From this viewpoint, we have developed new programs for utilizing color models along the lines shown in Fig. 2. Each of these programs services as a basic tool for data analysis using color, and all ideas described in the following presuppose the use of these color systems and programs. These algorithmic color systems do not accurately represent the human color perception system, but are certainly more realistic and practical than the use of graphical representations or statistical graphs in which meaningless coloring is added.

3. Use of color information in automatic classification

Graphical representations are often used in automatic classification to display data, such as symbolic scatter plots, dendrograms and other tree representations, minimum spanning trees, cluster ellipses. Classification involves grouping the analytical data, which leads directly to coloring. However it presupposes that the process of coloring is free of subjective judgement. Thus main objective of this paper is to evaluate the effectiveness of using color in automatic classification by means of a system comprising color transformation algorithms linking predetermined color models and automatic classification programs.

3.1 Color imaging patterns of multivariate data

Classification of objects using a given multivariate data matrix for graphical representation of the results must answer to the following questions.

1) Which of the variables actually served for discrimination of objects?

2) Do we interpret the relationships among the variables?
3) Do we interpret the influence on the classification results caused by differences in clustering methods?
4) Do we avoid possible misinterpretation of the results based on the method used for drawing the dendrogram?
5) Do we know the influence of the selected slimialirty/dissimilarity or standardization on the clustering result.

To solve these problems, reexamination and idenfication of the clustering result against the original data set must be conducted. It is in this process of evaluation that the use of color will prove highly effective. The following basic procedure for color utilization in this process is quite simple.

Step 1: The objects of a given n objects by p variables multivariate data matrix X = (X_{ij}) (i = 1, 2 ... n; j = 1, 2 ... p) are grouped.

Step 2: The objects of the data matrix X are reordered according to the clustering results.

Step 3: Reordering of variables is conducted for representing the relationships (similarity or dissimilarity) among variables in color. Variables reordered can be conducted by the principle component analysis described below.

Step 4: The matrix obtained by reordering the objects and the variables is defined by a matrix X*.

Step 5: The matrix X* is converted to the *colored pattern matrix* Y represented by three aspects of color (H, L, S) using a transformation method computable with the color model. The colored pattern matrix Y thus obtained is transmitted through the RGB model to the color monitor display as *color patterns*.

The simultaneous two-way clustering method of the data matrix explained above offers the following advantages.

1) The characteristics of the multivariate data matrix resulting from its clustering can be grasped at a glance.
2) The characteristics of the original data matrix subjected to neither value transformation nor data modification can be directly observed in color.
3) Moreover, the method is easy and simple.

Some important points in connection with the above steps are explained below.

3.1.1 Conversion of reorganized data matrix X* to colored pattern matrix Y

Conversion of the reorganized matrix X* to the colored pattern matrix Y using the HLS (or HSV) color model is expressed as

$$X^* = (X_{ij}^*) \xrightarrow[\text{HLS color model}]{} Y = \{Y_{ij};\ H_j,\ L_{ij},\ S_j\} \quad (i = 1, 2, \dots, n;\ j = 1, 2, \dots, p) \tag{1}$$

where

H_j : hue, $0 \le H_j \le 360$; L_{ij} : lightness, $0 \le L_{ij} \le 1$

S_j : Saturation (see 3.1.3).

3.1.2 Coloring dispersion within a variable - Ulitization of lightness -

The dispersion of the observed values X_{ij}^* (i = 1, 2, ... , n) of a variable j is subjected to linear rescaling so that it will correspond to the lightness and come within the interval [0, 1].

$$L_{ij} := (X_{ij}^* - \min_i X_{ij}^*) / (\max_i X_{ij}^* - \min_i X_{ij}^*) \quad (2)$$

$$(i = 1, 2, \ldots, n;\ j = 1, 2, \ldots, p)$$

If necessary, a counter variable is generated according to the significance of th characteristics, so that the lightness and the observed values will correspond, i.e., the lightness will increase or decrease according to increasing values of t data.

$$X_{ij}^* := \max_i X_{ij}^* + \min_i X_{ij}^* - X_{ij}^* \text{ (for some variable } j) \quad (3)$$

3.1.3 Reordering variables - Utilization of saturation and hue -

A variety of cases can be conceived for coloring variables according to their intended use.

- Variables resembling (or expected to resemble) each other in characteristics ar alloted colors with similar hues in close-by positions.
- The complement of a hue is taken for variables differing from each other in characteristics.
- The hue is fixed and the characteristics of each variable are expressed only by a difference of lightness.

This freedom to select the interrelation among variables is important, but it is equally important to provide a procedure for objective determination of the relationships among variables.

Option 1: The saturation is fixed, and the hue only is varied for each variable shown below.

$$S_j := r \text{ (normally } r = 1)$$
$$H_j := \theta_j,\ \theta_j = 2\pi (j - 1)/P \quad (j = 1, 2 \ldots P) \quad (4)$$

Taking the saturation at $r = 1$ is identical to using only a certain specific pure color on the surface of the cone of HLS color model (see Fig. 4).

H_j of the variable j can be given so as to form equal angles to the number of variables p, but it can also be given according to subjective judgement of the similarity/dissimilarity among variables or on the basis of any information available in advance. In other words, the variable ordering need not be the same as that of the original data matrix. It may be also conceivable to use the variable ordering on the dendrogram obtained from hierarchical clustering of variables.

Option 2: Determinaing saturation and hue based on the similarity among variable

Free setting of the saturation and hue offers various advantages, but may also cause confusion when interpreting. For this reason, a method of automatic conversion of interrelationships to saturation and hue based on the principal component analysis is proposed. It satisfies the following conditions.

- Variables are reordered with consideration given to their relationships, and are made to correspond to the changes in hue.

- The contribution of each variable is expressed as changes in tint, for example, variables with a high contribution are expressed more brightly.

The process operates as follows:

Step 1: A given data matrix X is subjected to principal components analysis and converted to a component loadings matrix.

Step 2: The configuration of component loadings obtained by choosing two components ℓ, ℓ' is made to correspond a holizontal slice of the HLS color model, as shown in Figs. 3 and 4. Next, the fundamental colors are arranged consecutively on the periphery of the unit circle (Fig. 3), with red on the right side followed counterclockwise by yellow, green, cyan, blue and magenta.

Step 3: H, L and S are given at the following transformed values for the observed values x_{ij} of a variable j.

$$H_j := \theta_j,\ 0 \leq \theta \leq 360\ ;\ S_j := d_j,\ 0 \leq d_j \leq 1 \qquad (5)$$

L_{ij}: = Same way as that of Eq. (2)

The distance d_j from the origin (Fig. 3) is defined as the saturation for a variable j, and θ_j as the horizontal angle in the counterclockwise direction. Ordering of variables selected consecutively in the counterclockwise direction from an arbitrary baseline is adopted. H, L and S thus determined are distributed along the dotted line in Fig. 4 for the variable j.

With this method, the color effectively brings out the significance of the multivariate data. Moreover, the data can be represented by a 3-dimensional color space. Colors acquire the following significance.

1) Variables with similar hues have similarity relationships while those with complementary hues are dissimilar.
2) Variables approaching grey have small component loadings and those with highly saturated hues have a high contribution. In other words, the degree of contribution of variables can be visually observed as changes in tints.
3) The shades of a color represent the dispersion and spread of observed values. Outliers especially appear conspicuously.

3.2 Example of a simple experiment

Below is the experimental result of colored pattern matrix based on Roger de Piles' data used for clustering by Davenport et al ([2]). The original data was a collection of aesthetic judgements of 56 painters, less two who did not give some scores.

3.2.1 Determination of hue and saturation by principal component analysis

The results of principal component analysis can all be represented on the graphics display. Figure 5 indicates the first and second principal components, and shows the configuration of component loadings, order of variables, and values of hue and saturation. The order of variables was obtained by specifying a baseline, with

X_2 (drawing) coming first, followed by X_4 (expression), X_1 (composition) and X_3 (coloring). The differences in hue show that $\{X_2\}$ and $\{X_3\}$ are separated from $\{X_4, X_1\}$. It is possible to see approximately which hue represents each variable, e.g., X_3 is close to green. The saturation shows a large value along both the first two axes, so that all four variables present a bright hue.

3.2.2 Sorting of objects by automatic classification

Conventional techniques such as Ward's method and the k-means method can be applied for classifying. After the objects are classified and the variables reordered, the original data matrix is converted to a reorganized matrix X*, which is then transformed to a colored pattern matrix Y. Figure 6 is a monochrome copy of the colored pattern matrix obtained by Ward's method. The colored pattern thus makes it possible to obtain a firmer grasp of the features of each cluster than is possible from dendrograms or plots of principal component scores alone.

3.2.3 Coloring principal component scores

Each multivariate vector of the original data matrix is represented by a colored belt, which have called a *colored vector*. Each colored vector is then located at the coordinates of the corresponding principal component score plotted along the specified principal components ℓ, ℓ'. Thus, the values for each multivariate vector, in this experiment four, can be seen at glance at the location of the composite score for the multivariate vector and the original four dimensions are reduced to two. When compared with the dendrogram obtained by Ward's method, the colored pattern matrix makes it possible to obtain an immediate characteristics of each painter.

The important features of the color imaging method are:

- The order of variables can be fixed by a single calculation worked out for principal component analysis. However, the order of objects varies from one clustering method to another, so that the degree of clustering requires visual inspection.
- The color assigned to each element of the matrix X* remains the same even if the clustering method is changed, eliminating confusion in color judgement and evaluation.
- When other principal components are chosen, the variable relationships in the configuration of such new components change in relation to the hue and saturation, so that these changes can be observed as significant color changes.
- In addition to the relationships among variables, the interaction between variables and the variables that are contributed in clustering the objects can be observed as the variation in lightness.

4. More sophisticated applications

The development of color graphics systems and supporting software can open up greater possibilities of data analysis using color. Using our color handling system, we have conducted a variety of practical analyses, including, area information analysis using the *areal clustering system* ([5] , [6]). The system can represent multivariate data having coodinates on a map as changes in color and makes it possible to grasp such changes intuitively. It is expected that the concept of the system will be applied in multicolor display of choropleth maps, trend surfaces and contour mapping in the area of quantitative geography as well as in coloring of various output data in the research of computational geometry.

5. Conclusion

Although color science still involves many aspects awaiting further study, the use of color is progressing rapidly because it promises to enrich visual information both quantitatively and qualitatively. It is incumbent on us to develop a new graphics system suited to data analysis by introducing the mechanism of color perception and the knowledge of color models so far made clear. We earnestly hope that this paper will prove helpful in paving the way in this direction.

References

[1] Becker, R.A. (1983), Integrating Color into Statistical Software, Computer Science and Statistics: Proc. of the 14th Symposium of the Interface, 220-223.

[2] Davenport, M. et al. (1972), The Statistical Analysis of Aesthetic Judgment: An Exploration, Applied Statistics, 21, 324-333.

[3] Foley, J.D., Van Dam, A. (1982), Fundamentals of Interactive Computer Graphics, Addison-Wesley.

[4] Nicholson, W.L., Littlefield, R.J. (1983), Interactive Color Graphics for Multivariate Data, Computer Science and Statistics, Proc. of the 14th Symposium of The Interface, 211-219.

[5] Ohsumi, N. (1983), Practical Techniques for Areal Clustering, Data Analysis and Informatics III (eds., E. Diday and others), North-Holland.

[6] Ohsumi, N., Mizuno, K. (1984), Microcomputer Graphics Clustering Techniques for Urban Experimental Evaluation, Proc. of the XIIth International Biometric Conference, 148-156.

Color models
- Hardware-oriented models
  - RGB model: TV monitor display, etc.
  - YIQ model: TV broadcasting, etc.
  - CMY model: Color printing, etc.
- Software-oriented models
  - HSV model
  - HLS model

Fig. 1 Color Models for Graphics

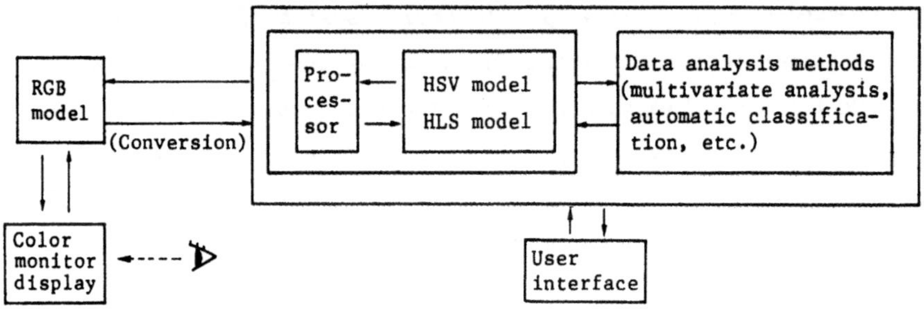

Fig. 2 Schematic of Color Handling System for Data Analysis

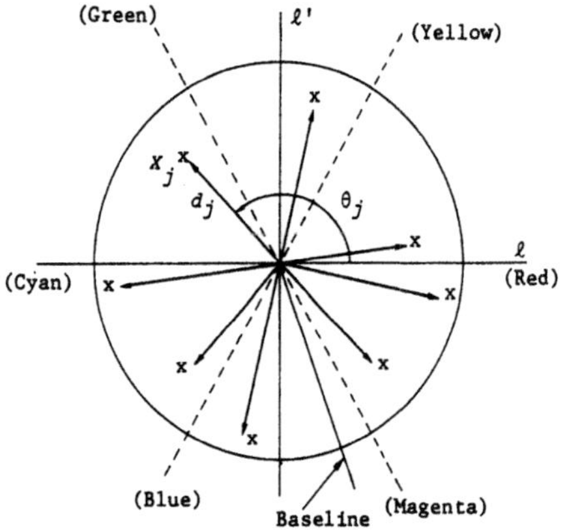

Fig. 3 Configuration of Component Loadings

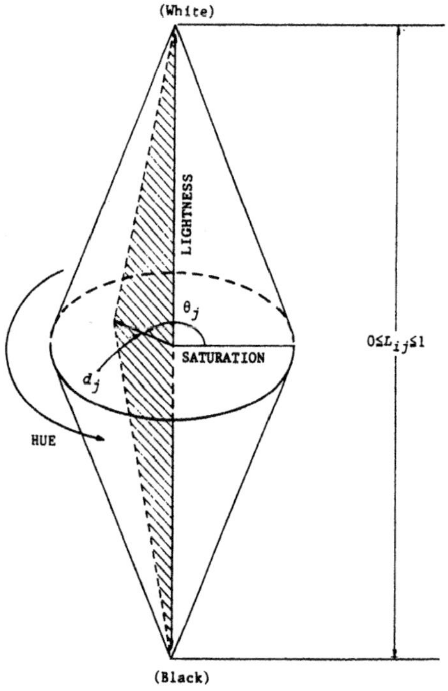

Fig. 4 Relationship between HLS Color Model and Data

```
X & Y ? : 1 2 ← Indicate the axes
                of the principal
                components

Order : Var.  :    Hue     :  Sat.

  1       2       341.69     0.89
  2       4        20.78     0.91
  3       1        33.60     0.90
  4       3       124.24     0.94
Do you want to print ?(y/n) : Y
```

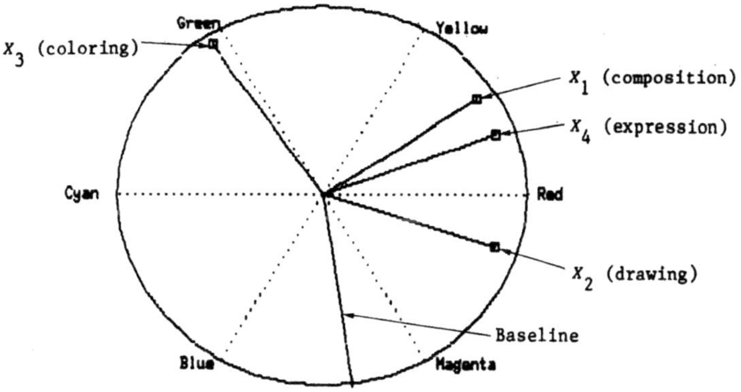

Fig. 5 Output Example of Piles' Data

By giving the baseline shown in the figure, the order of variables is determined: X_2 (drawing), X_4 (expression, X_1 (composition), and X_3 (coloring).

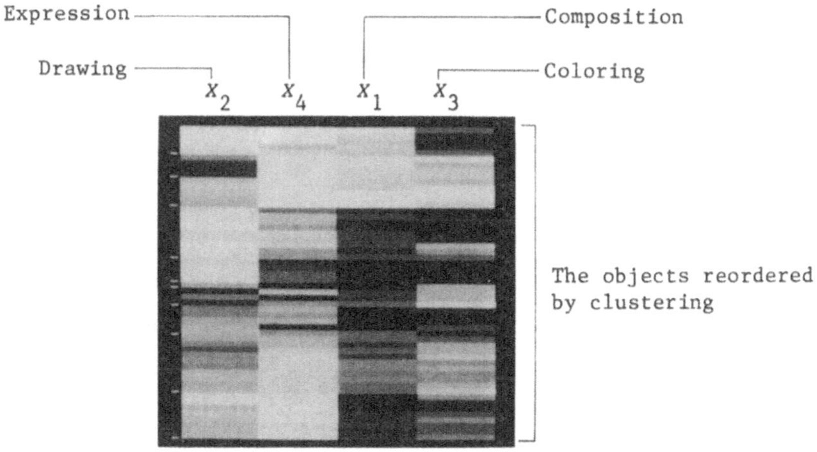

Fig. 6 Example of Colored Pattern Matrix

Algorithm and Software of Multivariate Statistical Analysis of Heterogeneous Data

By G. S. Lbov, Novosibirsk

Abstract

A new approach to the analysis of statistical data is proposed. The essence of the approach is in the use of the class of logical decision functions of heterogeneous variables (Boolean, nominal, rank and quantitative ones) for data analysis. On the basis of the present approach the methods of solving various problems of statistical analysis have been developed: discriminant and regression analysis, cluster-analysis, analysis of multivariate time series. The above problems for shortness will further be called data analysis problems.

Key mords: Data analysis, logical functions, miltivariate features.

1. Introduction

The experience of solving problems in medicine, sociology, geology, etc. shows that rather often when solving the problems of data analysis it is necessary to take into account the following peculiarities of their solution: the problems have to be solved under the conditions of high a priori indefiniteness about the kind of probability distribution in a space of features; a great number of features chosen to describe objects (up to 100 and more);a small size of sample (the number of objects under consideration is comparable with the number of

Compstat 1986

features); multivariate features; there might be gaps in data tables.

The methods destined to solve the problems of this class should satisfy certain requirements. They should admit sufficiently weak restrictions upon the form of probability distributions (the universality of methods), give stable solutions under the conditions of high demension and small sampling (statistical stability). In fact, all the known methods of statistical analysis use one or another form of proximity measure between the objects in the space of features. However, introduction of various proximity measures in a space of multivariate features is connected with methodological difficulties when calculating the distance between the discriptions of two objects one has to deal with the results of measuring a fortiori non-comparable magnitudes. Methods of analyzing multivariate data should take this fact into account. Finally, it should be noted that the results of data analysis may be used not just to predict the values of features of new objects, but they may serve as the main source of information about the structure of objects under consideration, about their intrinsic cause-and-effect relations (this is especially important at the initial stage of studying complicated objects). Therefore, the statistical conclusions (decisions) should be represented in the form most suitable for them to be interpreted (in a language close to the natural language of logical statements). In fact the creaters of methods have not paid attention to this aspect of data analysis.

These reasons created the necessity to search for new approach to solving various problems of data analysis possessing the above peculiarities. In the author's opinion, the most acceptable approach is the one using the class of logical decision functions. First this class was used to solve the problems of discriminant analysis for Boolean features only [1,2]. Then this approach was extended to solve the same problem for the case of multivariate features [3,4]. The class of logical decision functions of multivariate features was later used to solve various problems of data analysis [5,6,7] .

2. The class of logical decision functions

Let Γ be a general population of objects under consideration. To describe an object $a \in \Gamma$ we use a system of features $X = \{X_1,\dots,X_j,\dots,X_n\}$. Denote by $\mathcal{D}_j$ a set of possible values of the feature X_j : for Boolean features $\mathcal{D}_j = \{$no, yes$\}$; for nominal features $\mathcal{D}_j$ is a set of some names (symbols); for rank features $\mathcal{D}_j$ is a set of ordered marks; for quantitative features $\mathcal{D}_j$ is some interval of values on the real axis. A multidimensional space of features is assigned by the Cartesian product of the above sets, i.e. $\mathcal{D} = \prod_{j=1}^{n} \mathcal{D}_j$. Denote by $x_j = X_j(a)$ the value of the feature X_j for an object a . To the object $a \in \Gamma$ there corresponds a point $x = (x_1,\dots,x_j,\dots,x_n)$ in an n-dimensional space of features. Consider the partition of the set $\mathcal{D}$ into M subsets $\alpha = \{E^1,\dots,E^t,\dots,E^M\}$, i.e. $\mathcal{D} = \bigcup_{t=1}^{M} E^t$, $E^t \cap E^\ell = \emptyset$ for $t \neq \ell$; $M > 1$. It should be noted that in fact in all the known methods of data analysis to describe the decision function one used either the whole initial system of n features or some most informative subsystem of m features ($m < n$). The peculiarity of the given approach is the fact that to describe every subset E^t one uses its special subsystem of features $\{X_{j_1},\dots,X_{j_{m_t}}\}$, consisting of m_t features, i.e. $E^t = \tilde{E}^t \times \prod_{j \notin I^t} \mathcal{D}_j$, where $I^t = \{j_1,\dots,j_{m_t}\}$.
In the simplest case the subset $\tilde{E}^t$ is the Cartesian product $\tilde{E}^t = E_{j_1} \times \dots \times E_{j_{m_t}}$, where $E_{j_1} \subset \mathcal{D}_{j_1},\dots, E_{j_{m_t}} \subset \mathcal{D}_{j_{m_t}}$. Such a representation of the subset $\tilde{E}^t$ for t=1,...,M is a peculiar constraint upon the class of partitions of the set $\mathcal{D}$. Denote this class by Ψ_M^1 .
Denote by $\mathcal{J}(a, E_j) = [X_j(a) \in E_j]$ the predicate which degenerates into a statement for a fixed object $a \in \Gamma$ and for fixed $E_j \in W_j$. This statement may be true or false. The set W_j is a set of all subsets of the set $\mathcal{D}_j$, including the subset itself and excluding the empty one. In such a way the set W_j is formed for the Boolean and nominal feature; for the rank feature W_j a set of all subsets $\mathcal{D}_j$ is formed, the subsets E_j containing only the neighbouring elements; for the quantitative feature W_j a set of all possible subintervals of some interval $\mathcal{D}_j = [\alpha_j, \beta_j]$. is formed. Such predicates and the corresponding statements are said to be simple. Consider the predicate S as a conjunction of simple predicates, i.e. $S = S(a,E) = \mathcal{J}(a, E_1) \wedge \overline{\mathcal{J}(a, E_2)} \wedge \dots \wedge \mathcal{J}(a, E_n)$.

Note the equivalence of the predicates $S(a, E^t) \sim S(a, \tilde{E}^t)$.
The partition $\alpha \in \Psi_M^1$ is most conveniently described as a tree. By a tree we mean a root dichotomous tree for which to every inner node there is corresponds some predicate; to the branches coming from the inner node there corresponds the truth or falseness of the statement obtained as a result of replacement of features by their values. Thus, to a set of partitions Ψ_M^1 there will correspond some set of trees, each of them is characterized by M terminal nodes. The terminal nodes of a tree B are denoted by $b_1, \dots, b_t, \dots, b_M$.
The complexity of partitions (the complexity of a tree) seems to depend on the number of terminal nodes M and on the type of the predicate used in the nodes of the tree. For data analysis two types of predicates were mainly used: either simple predicates (in this case we obtain a class of partitions Ψ_M^1), or conjunctions of simple predicates (in this case we obtain a class of partitions Ψ_M^2). Note that, generally speaking in the nodes of a tree one can use even more complicated predicates and obtain sufficiently complicated partitions of an n-dimensional space of features.
If to every terminal node b_t of a tree B we ascribe a decision, then such a tree is called a decision tree r.

3. Data analysis problems

In the discriminant analysis to a node b_t we ascribe a class number $\omega_t \in \{1, \dots, k\}$, where k is a number of classes ($K \geq 2$); in the regression analysis - the regression function $f_t(x)$, in the cluster-analysis - the cluster number. For a decision tree r one may calculate the value of the criterion F(r) whose form is assigned in dependence of the type of the problem. For instance, for the discriminant analysis F(r) is the probability estimate of erroneous classification, for the regression analysis F(r) is a mean square of deviations predicted values to true values of the function. The best decision tree is said to be r* for which $F(r^*) = \min_{r \in R_M} F(r)$, where R_M is a set of all possible decision trees with M terminal nodes. The set R_M is called a class of logical decision functions. Methods of seach for the best decision r* were developed to solve the problems of discriminant, regression and cluster analysis [6].

The given class was also used to solve three dynamic problems.

The first problem. Let there be considered some set of objects $A \subseteq \Gamma$ in the time interval T. Each object may exist in one of two states: $\{\omega, \bar{\omega}\}$. For the object $a_i \in A$, $i = 1, \ldots, N$, we measure the features $X_1, \ldots, X_j, \ldots, X_n$ at different moments $t_1^i, \ldots, t_d^i, \ldots, t_{v_i}^i$. Besides, for any a_i there holds the condition: at the moments $t_1^i, \ldots, t_{v_i-1}^i$ the object a_i is in the state $\bar{\omega}$, and at the moment $t_{v_i}^i$ it passes to the state ω.

On the basis of the results of the given statistical material it is necessary for a new object $a \subseteq \Gamma$ in the state $\bar{\omega}$ to define, by time values of its features, the period of time in which it passes to the state ω.

The second problem. Considered is some set of objects $A \subseteq \Gamma$ in the time interval T. It is supposed that at the end of the interval T one may define the state of the object $\omega \in \{1, \ldots, k\}$, where k is a number of states. The statistical material is presented by a set of tables $\vartheta^1, \ldots, \vartheta^r, \ldots, \vartheta^k$. The table $\vartheta^k = \{x_{ij}^d\}$ where x_{ij}^d is the value of the feature X_j at the object a_i at the moment t_d. It is necessary, on the basis the results of analysis of the given statistical material for a new object $a \in \Gamma$ at the moment $t \in \{t_1, \ldots, t_d, \ldots, t_{V-1}$, to define, the estimate of probability of its transition to one of K states.

The third problem. Considered are n multivariate time series $\{X_j(t_d)\}$, j=1,...,n; d=1,...,V. The measurements of features are supposed to be made in equal time intervals. It is necessary, on the basis of the results of analysis of the given statistical material, to construct the decision rule to predict values of any feature X_j at the moment t_{V+1}.

At present a set of programs was developed to solve problems of discriminant and regression analysis, cluster-analysis, and to solve problems of dynamic forecasting [8].

References

1. Hunt E., Marin J., Stone P. (1966), Experiments in induction, New York.

2. Bongard M.M. (1967), Problem of recognition, Nauka, Moscow.

3. Lbov G.S., Kotyukov V.I., Manokhin A.N. (1973), On one algorithm of recognition in heterogeneous feature space, Computer Systems, 55, 98-107.

4. Muchalski R.S. AQAL/1 - Computer implementation of variable - valued logic sistem VL_1 and examples of its application to pattern recognition. Proc. First Intenat. Joint Conference Pattern Recognition. Washington. P.C.

5. Lbov G.S. (1978), Logical functions in problems of empirical prediction, Computer Systems, 76, 34-64.

6. Lbov G.S. (1981), Methods of heterogeneous experimental data processing, Nauka, Novosibirsk.

7. Lbov G.S. Logical Functions in the Problem of Empirical Prediction. Handbook of Statistics, vol.2; Classification, Patern Recognition and Reduction of Pimen sionality, North-Holland Publ. Comp., Amsterdam, New York, 1983.

8. Lbov G.S., Frolova T.I., (1983), Set of programs using a logical decision functions, Proc. Conference "Algorithmic and programming support of multivariate statistical analysis, Erevan.

Address-List of Authors

ABDELJAOUED, A., Universitè de Genève, Centre Universitaire D'informatique, rue du Lac. 12, 1207 Geneve, Suisse

ALUJA-BANET, T., Dept. Inv. Op. Estad., Fac. D'infomàtica, Barcelona-Univ., Politècnica de Catalunya, Barcellona, Spagna

ANDERSEN, E.B., Statistical Institute, Studiestraede 6, 1455 Copenhagen, Denmark

ANTOCH, J., Dept. of Statistics, Charles University, Sokolovsà 83, 186 00 Praha 8, Czechoslovakia

ASANO, C., Research Institute of Fundamental Information Science, Kyushu University, Fukuoka 812, Japan

BAINES, A., University of Leeds, Dept. of Statistics, University of Leeds, Leeds LS2 9JT, UK

BALDESSARI, B., Dip. di Statistica, Probabilità, e Statistiche Applicate, Università "La Sapienza", P. le A. Moro, 5, 00185 Roma, Italy

BATINI, C., Dipartimento di Informatica e Sistemistica, Università degli Studi di Roma "La Sapienza", Via Buonarroti 12, 00185 Roma, Italy

BELLACICCO, A., Dip. di Statistica, Probabilità, e Statistiche Applicate, Università "La Sapienza", P. le A. Moro, 5, 00185 Roma, Italy

BERZUINI, C., Dip. di Informatica e Sistemistica, Università di Pavia, Via Abbiategrasso 209, 27100 Pavia, Italy

BESSE, P., Laboratoire de statistique et Probabilités U.A., C.N.R.S. 745 U.P.S 118, rte de Narbonne, 31062 Toulouse Cedex, Francia

BOOMSMA, A., University of Groningen, Oude Boteringestr, 9712 GC, The Netherlands

BUCCI, P., Sarin, U.O. Ricerca, SS. 148 Pontina Km. 29.100, 00040 Pomezia (RM), Italy

BUTLEWSKI, B., Technical University of Poznan, Inst. of Cardiology, Academy of Medicine, 60965 Poznan, Polonia

CARLIER, A., Universitê Paul Sabatier Uer Mig, 118 rte de Narbonne, Toulouse 31062 Cedex, France

CARLSEN, F., Dept. of Mathematics, University of Bergen, 5000 Bergen, Norway

Caussinus, H., Laboratoire de Statistique et Probabilités U.A., C.N.R.S. 745 U.P.S. 118, rte de Narbonne, 31062 Toulouse Cedex, France

CELEUX, G., Inria, Domaine de Voluceau Rocquencourt, Le Chesnay 78153, France

CLITHERO, D.T., University of Leeds, Dept. of Statistics, University of Leeds, Leeds LS2 9JT, UK

Conley, W.C., Mathematics and Business Administration Departments University of Wisconsin - Green Bay, Breen Bay, Wisconsin 54302, USA

COOK, J.R., North Carolina State University, Raleigh, NC 27695-8203, USA

COPPI, R., Dipartimento di Statistica, Probabilità, e Statistiche Applicate, P. le A. Moro, 5, 00186 Roma, Italy

COSTA, P., Sperry Italia, V. Cristoforo Colombo 112, 00142 Roma, Italy

CRIBB, R.L., Dept. of Management and Dept. of Computer Science University of Western Australia and Westat Associates Pty Ltk., Nedlands, Western Australia

DAMBROISE, E., INRIA, Unité de Biométrie, 9, Place Viala, 34060 Montpellier, France

DARIUS, P.L., Laboratory of Statistics and Experimental Design, Fac. of Agronomy, K. Univ. Leùven, Kard. Mercierlaan 92, 3030 Heverkee, Belgium

DE JONG, V.J., University of Groningen Econometric Institute, P.O. Box 800, Groningen 9700 AV, The Netherlands

DENTENEER, D., Netherlands Central Bureau of Statistics, Department for Statistical Methods, P.O. Box 959, 2270 AZ Voorburg, The Netherlands

DI BATTISTA, G., Dipartimento di Informatica e Sistemistica, Università degli Studi di Roma "La Sapienza", Via Buonarroti 12 00185 Roma, Italy

DICKSON, J.M., Agricultural and Food Research Council Unit of Statistics, JCMB, The Kings Buildings, Mayfield Road, Edinburgh, UK

DIDAY, E., University Paris 9 and INRIA, Rocquencourt B.P. 105, 78153 Le Chesnay, France

DOBRIC V., University Computing Centre, Engelsova BB, Zagreb 41000, Jugoslavia

FERRANTI, G., Dipartimento di Informatica e Sistemistica, Università degli Studi "La Sapienza", Vai Buonarroti 12, 00185 Roma, Italy

FERRE, L., Laboratoire de Statistique et Probabilités U.A., C.N.R.S. 745 U.P.S. 118, rte de Narbonne, 31062 Toulouse Cedex, France

FICHET, B., Université Aix-Marseille II, Laboratoire de Biomathematiques, 27, Boulevard Jean-Moulin, 13385 Marseille Cedex 5, France

FINE, J., Laboratoire de Statistique et Probabilités U.A., C.N.R.S. 745 U.P.S. 118, rte de Narbonne, 31062 Toulouse Cedex, France

FRANCIS, B., University of Lancaster, Lancaster UK LA1 4YL, UK

FROESCHL, K.A., Institut für Statistik und Informatik, Universität Wien, 1010 Wien, Austria

GENG, Z., Research Institute of Fundamental Information Science, Kyushu University, Fukuoka 812, Japan

GEORGIEV, A.A., Technical University of Wroclaw, Wyspianskiego 27, 1-6, 50-370 Wroclaw, Poland

GERIG, T.M., North Carolina State University, Raleigh, NC 27695-8203, USA

GIRONE, G., Dipartimento di Statistica, Facoltà di Economia e Commercio, Via Camillo Rosalba, 70124 Bari, Italy

GORDESCH, J., Freie Univ. Berlin, Spechtstr. 15, 1000 Berlin 33, I

GORDON, A.D., Dept. of Statistics, University of St. Andrews, North Haugh, St. Andrews KY16 9SS, Scotland

GREEN, M., Centre for Applied Statistics, University of Lancaster, Lancaster, UK

GROSSMANN, W., Institut für Statistik und Informatik, Universität Wien, 1010 Wien, Austria

HAUX, R., University of Technology Aachen, FRG and Inst. for Prevention, Research and Social Medicine, Bremen, FRG

HENDERSON, J., Clustan Limited, 16 Kingsburgh Road, Edinburgh, Scotland

HEUCH, I., Dept. of Mathematics, University of Bergen, 5000 Bergen, Norway

HIETALA, P., University of Tampere, Dept. of Math. Sciences, P.O. Box 607, SF 33101 Tampere, Finland

HOLLE, R., Inst. für Med. Dokumentation, Stat. V. Datenverarbeitung, Im Neuenheimer Feld 325, 6900 Heidelberg, FRG

HUET, S., INRA, CRJ-INRA, Laboratoire de Biometrie, Jony en Josas 78350, France

JASINSKI, P.J., Technical University of Poznan, Inst. of Cardiology, Academy of Medicine, 60965 Poznan, Polonia

JIDA, J., Laboratoire d'Informatique, Université de Nice, Département informatique de l'IUT, 41, bd Napoléon III, 06041 Nice Cedex, France

JÖCKEL, K.H., Bremer Institut für Präven, St. Jürgen-Straße 1 2800 Bremen, FRG

KANKAANPÄÄ, H., Tampere University Central Hospital, Teiskontie 35, SF-33520 Tampere, Finland

KELLER, J.W., Netherlands Central Bureau of Statistics, P.O. Box 959, 2270 AR Voorburg, The Netherlands

KOWARSCHICK, W., Technische Universität München, Arcisstraße 21, 8000 München 2, FRG

KREDLER, C., Technische Universität München, Arcisstraße 21, 8000 München 2, FRG

KREINER, S., The Danish Inst. for Ed. Res., Hermodsgade 28, 2200 Copenhagen, Denmark

KRUSKAL, K., AT&T Bell Laboratories, Murray Hill, NY 07974, USA

KUH, E., Ccrems, Massachusetts Institute of Technology, Cambridge, Massachusetts 02139, USA

Kurihara, K., Research Institute of Fundamental Information Science, Kyushu University, Kukuoka 812, Japan

LANE, P.W., Rothansted Experimental Station, Harpenden - Herts - AL5 2JQ, Hertfordstire, UK

LARIZZA, C., Dip. di Informatica e Sistemistica, Università di Pavia, Via Abbiategrasso 209, 27100 Pavia, Italy

LBOV, G., Inst. of Mathematics, Sibirian Division of ANUSSSR, Novosibirsk, USSR

LELLA, G., Sarin, U.O. Ricerca, ss. 148 Pontina Km. 29,100, 00040 Pomezia (RM), Italy

LEMAIRE, J., Laboratoire d'Informatique, Université de Nice, Département informatique de l'IUT, 41, bd Napoléon III, 06041 Nice Cedex, France

LEONARD, M., Universitè de Genève, Centre Universitaire D'informatique, rue du Lac, 12, 1207 Geneve, Suisse

LISKI, E., University of Tampere, P.O. Box 607, 33500 Tampere, Finland

LONGFORD, N., Centre for Applied Statistics, Lancaster University, Lancaster LA1 4LY, UK

MARAZZI, A., Dept. de l'Intérieur et de la Santé Publique, Institut Universitaire de Médicine sociale et préventive, Rue du Bugnon 17, Lausanne, Svizzera

MARBACH, G., Dipartimento di Statistica Probabilità e Statistiche Applicate, University of Rome "La Sapienza", P. le A. Moro, 5, 00185 Roma, Italy

MARCOTORCHINO, F., Centre Scientifique, IBM-France, 36 Avenue Raymond Poincaré, 75016 Paris, France

MAROZZA, F., ISTAT, Via Cesare Balbo 16, 00184 Roma, Italy

MARTI-RECOBER, M., Dep. inv. Op. i. Estad., Fac. d'informâtica, Barcellona-Univ., Politècnica de Catalunya, Bardellona, Spagna

MARUBINI, E., Istituto di Biometria e Stat. medica, Fac. di medicina e chirurgia, Università degli studi, Via G. Venezian 1, 20133 Milano, Italy

MASSOTTE, P., IBM France, Usine de Montpellier, La Pompignane, Service 1542, BP 1021, 34006 Montpellier, France

MESSEAN, A., INRA, CRJ-INRA, Laboratoire de Biometrie, Jony en Josas 78350, France

Molenaar, I.W., Vakgroep Statistiek en Meettheorie, Oude Boteringestraat 23, 9712 GC Groningen

MORABITO, A., School of Medicine, Univ. of Milan, Istituto di Statistica medica e Biometria, Via Venezian, 1, 20133 Milano (Italy)

MURPHY, B.P., Dept. of Mang. and Dept. of Comp. Sc., University of Western Australia and Westat Associates Pty Ltd. Nedlands, Western Australia

NEUMANN, K., State Central Statistical Office, Hans-Beimler-Straße 70-72, 1026 Berlin, DDR

NUMMI, T., University of Tampere, P.O. Box 607, 33500 Tampere, Finland

OHUSUMI, N., Inst. of Stat. Mathem., 4-6-7 Minami - Azabu Minato-ku, Tokyo 106, Japan

OLDFORD, R.W., Massachusetts Institute of Technology, Bldg. E40 Rm 139, 1 Amherst Street, Cambridge Massachusetts 02174, USA

PARADOWSKI, S., Technical University of Poznan, Institute of Cardiology, Academy of Medicine 60-965 Poznan, Polonia

PAVAN, S., Sarin, U.O. Ricerca, SS. 148 Pontina Km. 29.100, 00040 Roma, Italy

PAYNE, R., Rothansted Experimental Station, Harpenden - Herts - AL5 2JQ, Hertfordstire, UK

PETERS, S.C., University of California, Berkeley, CA, USA

POSTEN, H.O., Dept. of Statistics, U-120, Univ. of Connecticut, Storrs, CT 06268, USA

PUNTANEN, S., Dept. of Mathematical Sciences, University of Tampere, P.O. Box 607, 33101 Tampere, Finland

REY, W.J.J., Philips Research Laboratories, P.O. Box 80.000, 5600 JA Eindhoven, The Netherlands

ROHL, J.S., Dept. of Management and Dept. of Computer Science, University of Western Australia, and Westat Associates Pty Ltk., Nedlands, Western Australia

ROSS, G., Rothamsted Experimental Station, Harpenden - Herts, UK

RUDAS, T., Inst. of Sociology, Eötvös University, Pesti B. u. 1., 1052 Budapest, Hungary

SAMAROV, A., Ccrems, Massachusetts, Institute of Technology, Cambridge, Massachusetts 02139, USA

SCARANI, C., Dipartimento Scienze Economiche, Strada Maggiore 45, 40126 Bologna, Italy

SCHADER, M., Institut für Statistik und Quantitative Ökonomik, Universität der Bundeswehr, Holstenhofweg 85, 2000 Hamburg 70, FRG

SCHMID, F., Institut für Statistik und Ökonometrie, Universität Hamburg, Von-Melle-Park 5, 2000 Hamburg 13, FRG

SEEBER, G.U.H., Institut für Statistik, Universität Innsbruck, Innrain 52, 6020 Innsbruck, Austria

SNELLA, J.J., Université de Genève, Centre Universitaire D'Informatique, rue du Lac. 12, 1207 Geneve, Suisse

Sonnberger, H., Institute for Advanced Studies, Stumpergasse 56, 1060 Wien, Austria

TALBOT, M., Agricultural and Food Research Council Unit of Statistics, JCMB, The Kings Buildings, Mayfield Road, Edinburgh, UK

TAMASSIA, R., Dipartimento di Informatica e Sistemistica, Università degli Studi di Roma "La Sapienza", Via Buonarroti 12, 00185 Roma, Italy

TANAKA, Y., Okayama University, 2-1-1 Tsushima-naka, Okayama 700, Japan

TARUMI, T., Okayama University, 2-1-1 Tsushima-naka, Okayama 700, Japan

TITT, E.M., Tartu State University, Vanemuise 46-232, Tartu 202400, Estonia, USSR

VICTOR, N., Inst. für Med. Dokumentation, Stat. V. Datenverarbeitung, Im Neuenheimer Feld 325, 6900 Heidelberg, FRG

VINCIGUERRA, F., CRAI, Via Tevere 15, 00198, Roma, Italy

WHITTAKER, J., University of Lancaster, Lancaster LA1 4YL, UK

WINSBERG, S., AT eT Bell Labs, 600 Mountain Av., Murray Hill, New Jersey, USA

WISHART, D., Clustan Limited, 16 Kingsburgh Road, Edinburgh, Scotland

YONCHEV, H.A., Department of Automation, Higher Institute of Chemical Technology, Sofia 1156, Bulgaria

GPSR Compliance

The European Union's (EU) General Product Safety Regulation (GPSR) is a set of rules that requires consumer products to be safe and our obligations to ensure this.

If you have any concerns about our products, you can contact us on ProductSafety@springernature.com

In case Publisher is established outside the EU, the EU authorized representative is:

Springer Nature Customer Service Center GmbH
Europaplatz 3
69115 Heidelberg, Germany

Batch number: 10726298

Printed by Printforce, the Netherlands